ENVIRONMENTAL CHANGE AND RESPONSE IN EAST AFRICAN LAKES

MONOGRAPHIAE BIOLOGICAE

VOLUME 79

Series Editors

H.J. Dumont and M.J.A. Werger

Environmental Change and Response in East African Lakes

Edited by

JOHN T. LEHMAN

Department of Biology,
University of Michigan,
Ann Arbor, Michigan, U.S.A.

KLUWER ACADEMIC PUBLISHERS

DORDRECHT / BOSTON / LONDON

A C.I.P. Catalogue record for this book is available from the Library of Congress

ISBN 0-7923-5118-5

Published by Kluwer Academic Publishers,
P.O. Box 17, 3300 AA Dordrecht, The Netherlands.

Sold and distributed in North, Central and South America
by Kluwer Academic Publishers,
101 Philip Drive, Norwell, MA 02061, U.S.A.

In all other countries, sold and distributed
by Kluwer Academic Publishers,
P.O. Box 322, 3300 AH Dordrecht, The Netherlands.

Printed on acid-free paper

Printed in the Netherlands.

This book is dedicated to
Yvette Edmondson and to Raelyn Cole,
with gratitude

Contents

IDEAL investigations of Rift Valley lakes

Preface

The idea for this book was born at the June 1996 meeting of the IDEAL Steering Committee in Milwaukee, Wisconsin. We had just completed a successful and stimulating special symposium during the annual meeting of the American Society for Limnology and Oceanography, and enthusiasm was running high for the production of a volume that could assemble in one place the scientific findings that were starting to emerge from East Africa. IDEAL, an International Decade for the East African Lakes, had ended one round of field investigations, many of which had been centered on Lake Victoria. As the climatologists, geologists, paleolimnologists, and biologists displayed their results and debated interpretations, it appeared that some paradigms were shifting, and that new explanations of climate history and modern processes were taking shape.

The Steering Committee endorsed the production of a volume that would draw together the different research results that were emerging and which would be representative of the scope of science issues that exist within IDEAL. This book follows in the spirit of *The Limnology, Climatology, and Paleoclimatology of the East African Lakes*, published in 1996, but has a somewhat different purpose. The previous publication also included original science results, but it was conceived to review the state of knowledge, identify critical problems, and point to new paths of inquiry. It accompanied the development of our first Science and Implementation Plan for the East African Lakes. This volume is a collection of new findings and a cross-reference to other new developments in the East African lakes. It retains the flavor of its predecessor by continuing to fuse together the findings of physical scientists and biologists, about both ancient and modern environmental conditions.

IDEAL is a consortium of scientists from diverse disciplinary backgrounds who are united by their intellectual passion for studying the great lakes of East Africa. As the list of contributing authors makes clear, this is a group that hails from institutions in the Northern Hemisphere as well as from inside East Africa itself. There is a universal recognition that the challenges presented by the lakes demand different academic perspectives and Interdisciplinary Scientific Talents. The resulting IDEALISTs have promoted inquiries into many interesting questions or problems presented by the East African lakes. Efforts are guided by a conviction that the East African high elevation tropics are a sensitive barometer of modern environmental change, and that improved understanding of the lakes will help decipher the history of climate and environments in the part of the world where homonids, bovids, sudids, and other major groups of animals and plants evolved and radiated.

Acknowledgments

Many colleagues generously donated their time and talent to provide review comments which improved the papers in this volume. The authors owe thanks especially to Paul Collinveaux, Dan Conley, Jack Dymond, Brian Eadie, Ned Grossnickle, John Halfman, Rob Hart, Alex Horne, Matt Julius, Daniel Livingstone, David Rea, and Craig Sandgren.

Laura Lee Gance has provided a thorough account of recent IDEAL activities through the *IDEAL Bulletin* and from records of Steering Committee meetings. The work presented in this volume was sponsored in part through the encouragement and support of program managers at the National Science Foundation, NOAA National Undersea Research Program, the National Geographic Society, the International Rotary Club, and internal funds of numerous universities and research centers.

The governments and people of East Africa have been gracious in their accommodation of this research and the scientists who conduct it. Our thanks extend beyond our colleagues and research collaborators to the numerous support staff, technicians, drivers, and vessel crews whose help and cooperation makes scientific study possible under otherwise challenging conditions.

INTRODUCTION

What is IDEAL learning about the great lakes of East Africa? As the chapters in this volume reveal, it is becoming clear that climate variations can be traced like a red thread through both ancient and modern environmental conditions of the basins. African lakes are sensitive to environmental factors, and climate often seems to operate as the master variable. The tropical lakes of East Africa do not experience the annual low light and cold temperatures that reset chemical and biological processes in Temperate Zone lakes each winter. The hypolimnia of stratified tropical lakes are perennially warm, and warm temperatures speed the rates of respiration and other biogeochemical processes. So, the chemistry and biology respond quickly and in harmony with the signals of climate and environmental change.

1. About This Volume

This book is a collection of original research reports from scientists who have been working in Africa under the consortium banner of IDEAL. The research problems range across such apparently disparate topics as lake level variation, production biology, rock magnetism, sediment and water chemistry, and isotope geochemistry. The topics are linked by common efforts to understand modern environmental processes, and to use that understanding to interpret past conditions.

Paleolimnology and its supporting disciplines offer all the challenges of a good detective story. There are no eyewitness accounts telling what the lakes and landscape were like when mountains of ice carpeted the northern continents 20,000 years ago, or telling us when the Nile first started to flow north from the equator. The circumstances under which species evolved and modern conditions started to emerge consequently have to be teased by sleuth and forensic science from the circumstantial evidence and proxy measurements that can be uncovered. Our desire to understand the past and to decode its signals has led to a partnership between neolimnologists, who study modern processes and mechanisms, and paleolimnologists, who search for telltale traces of past mechanisms in the sediments and rocks. Together, the partners can piece together accounts of past environmental variations through their resulting effects on lakes and surrounding ecosystems.

Many of the IDEAL investigations reported in this volume have focused on the modern and ancient condition of Lake Victoria. The lake attracted this attention because of ready logistics, simple structure of its basin, and intense interest sparked by its changed condition. The modern lake has diverged substantially from the condition that was recorded 35 years ago. Lake Victoria presently exhibits increased algal biomass and reduced water transparency from its historical condition. The species composition of the planktonic algae has changed, as well. There are now more cyanobacteria and fewer large diatoms and chlorophyte algae. The heavily silicified diatom *Aulacoseira* (formerly called *Melosira*) is now absent from the flora, and concentrations of dissolved silicate in surface lake waters have declined. Whereas rates of primary production have accelerated near the lake surface, rates of decomposition have depressed dissolved oxygen concentrations over as much as the deepest one-third of the water column. The lake is approximately 0.5 °C warmer than in 1960, and it exhibits reduced water transparency, stronger stability of thermal stratification, and decreased photic zone depth. Concentrations of phosphorus have increased in offshore waters, and concentrations of sulfate have been reduced. The species richness of its endemic cichlid fish fauna has declined, and the abundance of an introduced predatory fish, Nile perch, has increased to dominate the fishery.

These are the changes we know about securely, because documentation of earlier lake conditions were recorded. As such, the modern changes represent a calibration test for the paleo records that precede human observations.

IDEAL discoveries about Lake Victoria have included the realization that the lake was completely dry during the previous continental glaciation (Johnson et al. 1996), raising speculation about rates and mechanisms of its famous species endemism, and about the temporal origin of the White Nile, which is discussed in this volume. Here Curt Stager reports a detailed study of diatom stratigraphy which indicates that the modern Lake Victoria, in which the heavily silicified diatom *Aulacoseira* is rare, may not be as unique as was previously believed. The modern condition of the lake is explored with respect to climate-related variations and their effects. Analyses reveal that climate variations with respect to air temperature, dew point, and wind speeds during the past 35 years are consistent with changes in lake temperature, mixing, and biological condition.

2. Climate and Lake Victoria

Recent results show that phytoplankton of Lake Victoria are nutritionally limited by inorganic nitrogen during all times except deep mixing. Most of the time, algal biomass is high and mixing depths are shallow. When deep mixing events occur, however, they preclude sustainable high biomass levels because of feedback between light attenuation and biomass. The problem is effectively that of self-shading in a deep water column. With deep mixing, algal concentrations decrease, owing to significant dark respiration and loss rates, and the algae lose their previous control over dissolved nutrient levels. There is a defined relationship between algal biomass and mixing depth, with resulting effects on nutrients in the lake water.

Mixing depth is a key variable for understanding the changes that have occurred to Lake Victoria, but the master variables for the changes seem to lie in the historical and recent records of meteorological conditions. Weather conditions are linked with lake response through heat balance and momentum transfer effects on mixed layer depth. During the last 35 years there have been significant changes in mean air temperature, dew point, relative humidity, barometric pressure, average wind speed, and atmospheric visibility around Lake Victoria. The meteorological changes are consistent with other changes that have occurred in the lake. They are also consistent with the fact that the lake is elevated by more that one meter from the surface height recorded 35 years ago. According to Nicholson's analyses, the level of Lake Victoria has been rising and falling in response to climate variations for millenia, recorded in detail for the last 200 years, and in good approximation by the low stands of the Nile in Egypt even before that.

An emerging paradigm from IDEAL studies is that the physical and ecological conditions of Lake Victoria appear responsive to lake-atmosphere interactions. Changes in heat transfer functions have led to elevated water temperature. Slackened winds have caused less intense mixing, and the generally humid conditions and elevated lake levels have encouraged chemical weathering of the apatite-rich soils. Many residual soils of the region are derived from carbonatite, a basic volcanic rock that is easily weathered, and which is rich in many minerals, including phosphate. Deforestation and destruction of wetlands have played an important role, too.

Climate variations intensify and accelerate the effects of nutrient income. Biological and chemical changes in Lake Victoria are consistent with diminished vertical mixing and thus with increased physical stability of stratification. Mixing depths are influenced by heat and momentum transfers at the lake surface. Shallow mixing has permitted higher algal biomass to develop in nutrient rich surface waters, promoted the loss of heavy diatoms, and reduced the ventilation of deeper waters with oxygen.

Phosphate concentrations in the offshore region of Victoria are higher now, and the mineral appears to have a quarter century residence time in the lake, almost the same as for silica. Essentially all inorganic P is converted to organic form in the nearshore before the water exits the lake at the Source of the Nile (Lehman and Branstrator 1994). Consequently, phosphate levels set the constraint on maximum algal biomass levels of nearshore regions. The changes in the condition of the lake can be seen in the composition of its recent sediments. Landscape alteration during this century have affected the export of both biogenic P and Si to the sediments of the lake. Ngobi, Kelts, Johnson, and Solheid demonstrate that much of the recent sediment has come from the watershed. And Holtzman and Lehman show that the land-derived material is rich in apatite phosphates. It seems most accurate to regard climate change and eutrophication not as alternative explanations for lake condition, but as elements of a single explanatory theory about this lake.

The evidence accumulated in this volume supports an interpretation about Lake Victoria as follows. Modern climate conditions have contributed to a lake that is warmer and which mixes less deeply and less thoroughly than it did earlier in the century. Wind stress and surface cooling by radiative and evaporation losses, which are sources for the total kinetic energy of the mixed layer, have declined. Favorable water balance and elevated humidity have contributed to the flooding of previously riparian and terrestrial soils. The submerged soils, as well as the runoff from deforested land, are rich in phosphates because of the indigenous mineral content of easily eroded basic volcanic rocks (carbonatite). Weathering rates may have been accelerated further by strong acids now present in global precipitation. The net result is a phosphate-rich water column with a shallow mixed layer and weakened rates of both vertical and horizontal mixing.

Virtually every recent change in Lake Victoria from conditions recorded early in the century, except the acts of introducing Nile perch and water hyacinth, can be ascribed by cause and effect to these environmental variations. Heavy-walled diatoms were lost with their silicon to sediments under the relaxed mixing regime. In the warm, stably stratified surface water, N-fixing cyanobacteria had a heyday, making their own amino-N to accompany the rich P that was already reported to be in the water earlier in the century; and offshore P has increased even more since then. The biomass production that sank to deep water and decomposed there contributed to O_2 depletion which outpaced the sluggish rate of oxygen renewal by mixing. And the diversity of cichlid fish, driven from their deep, dark hiding places by lack of oxygen, fell easy prey to the new predator. Maybe in times past the species flocks would have diversified into new habitats, but not during this climate cycle.

Even reduced levels of sulfate in modern Lake Victoria are a consequence of the climatic change, because sulfate is used by microbes as a terminal electron acceptor in the absence of hypolimnetic oxygen. Rates of denitrification intensify also in the anaerobic hypolimnion, a fact which in turn guarantees that non-diazotrophic algae lack for N during long periods of stratification.

Given adequate nutrients, the shallow mixed layer can and does support elevated algal biomass. The increased quantities of surface algae in Lake Victoria in turn suppress dissolved nutrients. Phosphate is present for geological reasons at great stoichiometric excess over N in proportion to protoplasmic composition. As a result, nitrate is depleted differentially with respect to P in surface water. N-fixation by diazotrophic cyanobacteria cannot restore stoichiometric equilibrium for N because the amount of new organic N they produce is offset by denitrification in anoxic deep water. Consequently, the algae remain strongly N-limited.

Rates of vertical mixing restrain the resupply of Si to the surface layer. The modern diatoms are small, weakly silicified *Nitzschia* species, which maintain their buoyancy in the weakly mixed surface layer by commensal association with colonial cyanobacteria, *Microcystis*. The deep waters remain hypoxic or anoxic for much of the year because at warm water temperatures the richly organic sediments consume oxygen faster than it is mixed downward.

One corollary to this explanation is that the modern state of Lake Victoria is at least partially reversible if climate conditions revert to their pre-1960s levels. Deeper mixing depths could predictably suppress algal concentrations offshore through light limitation. More vigorous vertical mixing would promote the success of heavily silicified diatoms which export the products of their primary production from mixed layer to sediments more rapidly than do cyanobacteria or small diatoms. Increased ventilation of deep water resulting from mixing would replenish oxygen and restore some of the deep water fish habitat that has been lost.

For nearshore regions, however, the key to reversal of modern trends lies in reduction of the phosphate-rich material entering the lake from its catchment.

3. Food Webs and Trophic Dynamics

The prevailing view of the Lake Victoria food web is of a lake that supports a tremendously productive fishery built mainly on an introduced species, the Nile perch (*Lates niloticus*). Nearshore and in shallow water there is also another non-native fishery component, Nile tilapia (*Oreochromis niloticus*). Apart from these nonindigenous fish, there is only one species of the original fish fauna that attracts a significant fishery. The remaining native is the cyprinid *Rastrineobola argentae*, a minnow.

The remarkable fact about Nile perch is that the forage base has become so slim that piscivorous size classes now dine mainly on juveniles of their own species. The base of the cannibalistic pyramid is anchored on a freshwater atyid shrimp, *Caridina nilotica*. *Caridina* is regarded to be an herbivore or a detritivore, and the stable isotope analyses conducted so far have endorsed this trophic position for the animal. Nearshore, the shrimp are found among rooted and floating macrophytes, particularly the water hyacinth that now blankets much of the shore. Offshore, and in nearshore regions away from the macrophytes, the animals engage in diel vertical migration. During the day, the shrimp stay close to the bottom, but at night they swim up to the surface waters. This diel migratory behavior was originally recorded by Worthington (1931) during the 1920s, and it persists to this day, despite considerable reduction in the oxygen of deep water.

What is curious about the migrations is that the shrimp leave their detritus-rich habitat of the bottom to swim upwards to a region where they are more vulnerable to predation. As Branstrator and Ndawula point out in this volume, the Nile perch avoid regions of low oxygen, and their exclusion gives the shrimp a refuge deep in the lake. One hypothesis had been that the shrimp may be facultative planktivores (Lehman et al. 1996), which rise to the surface at night to feed when it is dark and risk of predation is reduced. Another idea had been that the shrimp could not withstand long periods of hypoxic conditions. In other words, the shrimp might rise to the surface each night to recover from an oxygen deficit. However, Branstrator and Ndawula now make it clear that the shrimp can live and grow at low oxygen levels for upwards of a week at a time.

We still don't know why *Caridina* takes even a reduced risk of mortality by migrating upward at night, but one thing is clear. The low oxygen refuge for the shrimp in deep water guarantees that Nile perch cannot deplete the shrimp population. Existence of a refuge habitat thus provides a stabilizing influence for a trophic relationship between the fish and their forage food. Otherwise the interaction might be subject to wild fluctuations in population levels owing to predator-prey cycles. The balance between refuge and exploitation is delicate, however. The larger the hypoxic refuge habitat for *Caridina* becomes, the less habitat remains for fish, and potential fishery yield becomes more threatened.

Given their emerging importance as the stable forage base for the Nile perch fishery, the biomass and secondary production of *Caridina* takes on a practical dimension of interest. Potential yield of the fishery is constrained energetically by the biological production of the shrimp, and by the efficiencies of trophic transfer to harvestable size classes of fish. Accordingly, Mbahinzireki, Lehman, and Ochieng in this volume extend and apply a production model for the shrimp that was developed in earlier IDEAL studies.

Besides *Caridina*, there is another invertebrate that plays a significant role in benthic-pelagic coupling within Lake Victoria. The vast swarms of lake flies that periodically cloud the horizon offshore are mainly composed of *Chaoborus*, the phantom midge, a true fly of the Order Diptera. *Chaoborus* is a predator during its aquatic larval stages, but it fasts during its short adult life of aerial existence, when it is thoroughly devoted to sexual reproduction. Larval *Chaoborus* also engage in diel vertical migration, but in their case it is known that they migrate from their refuge in or near the sediments to feed near surface on the zooplankton. Their targets in Lake Victoria are typically Crustacea, mainly juvenile copepods and copepod nauplii.

Chaoborus has a two month generation time in Lake Victoria. Cohorts are synchronized, and two overlapping cohorts that differ in age by one month are present simultaneously. As with *Caridina*, a production model was developed and applied to field data in order to assess trophic energetics. The chapter by Lehman, Halat, Betz, Ndawula, and Kiggundu in this volume reports the construction of the model and its application to lake conditions. What results is the first time series of production measurements for the invertebrate predator from offshore waters of the lake.

4. The Rift Lakes

Signals of climate and environmental variability are evident in Rift Valley lakes, as well. Sharon Nicholson reconstructs a 200 year record of lake level variations for Lake Malawi from a combination of historical accounts, oral traditions, and archeological evidence. The basin has evidently undergone major changes in water balance dating back to the late Pleistocene, as Ricketts shows from analysis of lake carbonate deposits.

Ng'ang'a, Muchane, Johnson, and Sturgeon provide insight to the mechanisms behind inferences of paleotemperatures and water balance from the stable isotope composition of carbonates by examining the differences between biogenic and abiogenic carbonates in Lake Turkana. The study has particular relevance for interpretation of climate signals that are often inferred from changes in the proportions of ^{13}C and ^{18}O in the mineral fractions. It appears that the frequency and intensity of algal blooms can have significant impacts on gas exchange rates and on the bulk isotope composition of dissolved inorganic carbon.

Beuning, Kelts, and Stager report pollen evidence from Lake Albert that indicates a period of tropical aridity preceded the initiation of the cool European Younger Dryas Interval at the end of the last Ice Age. Lake Albert seems to show abrupt signals of regional climate change. Lake Albert also seems to provide an analogy to the predicted effects of deeper and stronger mixing in Lake Victoria. Albert has modern phosphate concentrations even higher than those of Victoria, but it maintains a low algal biomass in its offshore region because of strong wind stress and sustained deep mixing. Even so, there may have been climate-induced reductions in the mixing strength of both Lakes Albert and Edward in recent years which contribute to reduced modern nutrient concentrations compared with historical records.

The African Rift Lakes, and Lake Edward in particular, have been identified as the next target for coordinated IDEAL research. It remains to be seen whether the patterns of climate variability and ecological response documented at Victoria are repeated and reinforced at other lakes. Edward and the Rift Lakes in general offer a chance to extend inquiries back even further in time because it is unlikely that the lakes dried up completely as Victoria did during the Pleistocene. Instead, there exists the possibility to apply the forensic skills of lake detective work across time scales that span the rise of the modern human species and its co-habitants among the environments of East Africa.

5. References

Johnson, T. C., C. A. Scholz, M. R. Talbot, K. Kelts, R. Ricketts, G. Ngobi, K. Beuning, I. Ssemmanda, and J. A. McGill. 1996. Late Pleistocene desiccation of Lake Victoria and rapid evolution of cichlid fishes. Science **273**: 1091-1093.

Lehman, J. T., and D. K. Branstrator. 1994. Nutrient dynamics and turnover rates of phosphate and sulfate in Lake Victoria, East Africa. Limnology and Oceanogrophy **39**: 227-233.

Lehman, J. T., G. B. Mbahinzireki, and L. M. Ndawula. 1996. *Caridina nilotica* in Lake Victoria: abundance, biomass, and diel vertical migration. Hydrobiologia **317**: 177-182.

Worthington, E. B. 1931. Vertical movements of freshwater zooplankton. International Revue der gesamten Hydrobiologie und Hydrographie **25**: 394-436.

HISTORY AND ONTOGENY OF IDEAL

DONNA A. LEHMAN
Department of Biology, University of Michigan
Ann Arbor, MI 48109-1048, USA

The International Decade for the East African Lakes was conceived as a scientific collaboration among limnologists world-wide to study the large lakes of Africa as archives of environmental and climate dynamics (IDEAL 1990). The scientists who formulated IDEAL share values that result in a program which is science-driven. Those who work under the IDEAL umbrella are a far cry from the "businessmen in white lab coats" who Frank Rigler and Rob Peters used to characterize as conspicuous consumers and managers of science funding and research enterprise. These are hands-on scientists who derive satisfaction from recognizing important problems, posing relevant questions, answering their questions in the field, and sharing their knowledge with fellow professionals and the public. IDEAL is an example of how compelling science can be accomplished without inflated budgets, grandiose administrative schemes, or competing egos. As such, it presents an instructive case study about the ontogeny of a scientific initiative.

1. Initial Efforts

The documented grass roots efforts of this initiative trace to Fall 1986, when the U. S. National Science Foundation and the University of Michigan sponsored a workshop entitled "Basic Issues in Great Lakes Research". More than fifty scientists, representing many nations and many subdisciplines of the aquatic sciences, attended the three day gathering at the Kellogg Biological Station in Hickory Corners, Michigan. The meeting was intended to identify research topics that were compelling and timely. The goal was to encourage a "unified and coherent" set of priorities and to develop an effective strategy for research (GLRD 1987). Workshop findings were published several months later.

Study of the African Great Lakes emerged as a dominant focus of meeting discussions that ranged across land-water interactions and evolutionary processes as well as plankton ecology, fish ecology, and food web structure. The workshop report proposed five major initiatives. First among these was raising and studying long sediment cores from ancient lakes, with the target lakes of greatest priority being those in Africa (GLRD 1987). Workshop participants explained in their report that the

J.T. Lehman (ed.), Environmental Change and Response in East African Lakes, 1–6.

technical feasibility of the project was beyond the capability of single investigators and that progress required effort by a coordinated community with institutional support.

Basic Issues in Great Lakes Research generated interest and enthusiasm for studying the African Great Lakes. Dr. Thomas C. Johnson, first at Duke University and later at the Large Lake Observatory of the University of Minnesota-Duluth, stepped forward with the energy and vision to direct this enthusiasm and helped to focus a number of separate initiatives into a cohesive plan. Johnson "took his show on the road" and presented seminars articulating his vision at colleges and universities, research centers, and government agencies. Aided by the additional energy and ideas of Dr. Kerry Kelts, then with the Swiss National Climate Program, Johnson approached the U.S. NSF for funds to host an international meeting. He was joined in his proposal by Duke University's dean of African limnology, Dr. Daniel Livingstone. The meeting was intended to test the appeal of the research initiative and to enunciate the grass roots ideas for a research program Johnson and Kelts were calling IDEAL.

Sponsored by the U. S. National Science Foundation and the National Climate Program of the Swiss Academy of Sciences (PROCLIM), a workshop was convened at the University of Bern, Switzerland from 29 to 31 March 1990. Eighteen participants and eleven observers met to share visions, articulate goals, and declare mutual cooperation for interdisciplinary research on the large lakes of the East African Rift Valley.

Participants of this embryonic IDEAL workshop listed their primary purposes as:

1. obtaining long, high-resolution records of climatic change and ecosystem responses in tropical East Africa.

2. providing a comprehensive training program that would result in collaborative efforts between African and northern hemispheric limnologists and paleoclimatologists (IDEAL 1993).

The initial vision called for work to collect and decipher the records of climatic variability from long sediment cores (IDEAL 1990). By the end of group discussions in Bern, however, the scope had grown to include neolimnology, evolution, and modern biogeochemical processes (Johnson and Odata 1996). Meeting participants declared a mission to create infrastructure and support systems for environmental research in the African Great Lakes.

Dr. Herman Zimmerman, Director of the Climate Dynamics Program at the National Science Foundation, offered unflagging moral support during the early stages of IDEAL planning. Zimmerman was nonetheless unsparing in his assessment of odds for success in the face of shrinking budgets, and Johnson wrote members of the steering committee that "Zimmie" told him that a large multidisciplinary program would be difficult to fund at NSF (T. C. Johnson, *vide* J. T. Lehman 14 February 1992). In consultation with Zimmerman, Johnson opted to submit a proposal to the NSF Climate Dynamics program. If funded, this would be the beginning of a field program in Africa by early 1993. Johnson's plan was to begin IDEAL in Lake Tanganyika or Lake Malawi where he had already undertaken initial research projects.

2. A Start at Lake Victoria

In April, 1992, however, Robert Hecky, at the Freshwater Institute in Winnipeg and John Lehman, at the University of Michigan, suggested that IDEAL begin its efforts at Lake Victoria where they had laid groundwork for the program (correspondence to T. C. Johnson, 27 April 1992).

Most of the rationale for Lake Victoria was scientific. Lake Victoria and its surroundings had undergone significant ecological changes over the last thirty years. These changes included: 1) a loss of forests, 2) lake eutrophication, 3) deep water deoxygenation, 4) changing water levels, 5) changing algal species, and 6) changes in mixing dynamics and productivity. All of these changes had pre-human analogs in the stratigraphic record and thus provided a calibration for the ancient conditions.

The rationale was also practical. An infrastructure for limnological research was already in place at Lake Victoria. The Uganda Freshwater Fisheries Research Organisation (UFFRO), headed by Dr. William Kudhungania, was located in Jinja, Uganda and offered ready access to the lake as well as proximity to an international airport. UFFRO had laboratory space, housing, and several boats, including the newly overhauled *R/V IBIS*. Dr. Kudhungania and his competent scientific colleagues and staff at UFFRO were welcoming and enthusiastic about the possibilities for training and collaboration with the IDEAL scientists.

With an infrastructure in place, with local colleagues willing to facilitate difficult arrangements, and building on significant scientific knowledge of the lake ecosystem, Hecky and Lehman proposed that IDEAL could demonstrate its ability to work in Africa and solve important problems in a timely and efficient way. For the paleolimnologists, the shallow water of Lake Victoria would make coring relatively easy and the results relatively straightforward to interpret. After cutting its teeth with Lake Victoria, IDEAL could move on the deep rift lakes which were the ultimate targets (communication to T. C. Johnson, 27 April 1992).

The first grants for preliminary science and continued planning commenced in 1991 and 1992 sponsored by both the U.S. National Science Foundation and the National Undersea Research Program of the U.S. National Oceanic and Atmospheric Administration. Funding was received by Johnson to organize and convene a second, more comprehensive, planning workshop and symposium in 1993. IDEAL members spent 1992 planning that meeting, conducting initial research, and preparing results and proposals for future evaluation.

3. NSF and Multidisciplinary Aquatic Science

Proposal development was often difficult and frustrating. The NSF, the U.S. agency primarily responsible for basic research in non-medical science, was the logical place to seek funding for much of the IDEAL project. The administrative structure of the NSF, however, was not suited to accommodating a scientific study with the multidisciplinary scope of IDEAL that was also regional in focus. The NSF was organized according to traditional scientific disciplines of the early 20th century. Biologists sent their applications to the Directorate for Biological Sciences. Geologists and Climatologists applied to the Directorate for Geosciences. Physical limnologists and freshwater environmental chemists had no home directorate for their research domain. Because it posed questions that spanned multiple traditional disciplines, there was no natural home for IDEAL within the existing NSF organization.

In a letter dated 22 September 1992, Tom Johnson vented his frustrations to John Lehman: "I am quite disappointed...and spent the weekend wondering whether to try to fight the NSF system and force cross directorate sponsorship...". Believing in IDEAL as a coordinated program he worried that submitting independent and uncoordinated proposals "gets us no further along than when we were individual, non-communicating investigators" and admitted "In fact, my writing a paleoclimate proposal for Lake Victoria is far more difficult to sell than for continuation of my studies at Malawi or Turkana". But Johnson remained committed and strong of heart. In that same communication, he told Lehman "Despite these shortcomings, I believe that the long-term benefit to the science of the African lakes and training of African collaborators will be furthered more if we do work and move en masse under the umbrella of IDEAL". He suggested a meeting between IDEAL scientists and NSF personnel to present research ideas and discuss cross-directorate funding. Others agreed and Johnson and Kelts, with the help of Zimmerman at NSF, organized the one-day event.

Entitled *The International Decade of East African Lakes (IDEAL): Need and Justification for Cross Directorate Support from NSF*, the meeting was intended to illustrate that IDEAL scientists were serious about pursuing research in a coordinated program. Objectives were 1) to present compelling scientific reasons for breaking the old disciplinary boundaries, and 2) to seek financial support cutting across directorates at NSF.

Both objectives were argued on 17 November 1992 at NSF offices in Washington, D. C. Invitations were issued to directors from programs including Biological Oceanography, Climate Dynamics, Ecology, Ecosystem Studies, Systematics, and International Programs. IDEAL was represented by Johnson, Kelts, Lehman, Hecky, Livingstone and several African students. Also invited were Dr. Les Kaufman, of the New England Aquarium and Dr. George Kling, of the University of Michigan. Both had previously conducted research on African lakes and had expressed interest in learning more about IDEAL. Kaufman had declared in a memo to Johnson on 28 October 1992 that to the "... community of North American biologists working in the East African great lakes...IDEAL is something of a mystery, if not a direct threat" (personal communication by L. Kaufman, 28 October 1992). Johnson's invitation to them stemmed from a desire to debunk the putative "mystery".

The meeting was well attended by NSF personnel and the presentations by IDEAL scientists included a historical perspective of research on the African lakes, a vision for IDEAL, the role of NSF, and various aspects of current research (memo from T. C. Johnson to meeting participants, 10 November 1992). Kaufman and Kling talked about their independent program ambitions.

At the end of the day's effort, NSF administrators had been offered a comprehensive vision of IDEAL and were given the opportunity to recognize the research themes of separate proposals within the context of the whole program. Within months, Johnson and Kelts received NSF funding for IDEAL research on Lake Victoria.

4. Science and Implementation Plan

Still needed was a concise written statement of the complete IDEAL science plan and its proposed implementation. This object was accomplished at the Second IDEAL Symposium held 18-23 February 1993 at the Sunset Hotel near the UFFRO laboratories in Jinja, Uganda. Graciously hosted by the late Dr. William Kudhongania, the science and planning meeting was convened by Tom Johnson, Dr. Eric Odada from the University of Nairobi, and Dr. George Kitaka of the International Hydrology Program of UNESCO. Funding provided by NSF and UNESCO allowed participation by about 125 scientists from around the world.

The goals of this IDEAL symposium were two-fold. First, participants would prepare research papers that would constitute a compendium of current knowledge about African lakes. Second, through discussion and synthesis, participants would identify current intellectual challenges and propose mechanisms to resolve them. In short, the participants developed a science and implementation plan.

The international scientific community that gathered at Jinja succeeded in defining IDEAL and the roles of its component scientific disciplines. The scientists also articulated the research priorities of IDEAL and addressed the technical and logistical difficulties presented by East Africa. Reports produced during the meeting were edited by Johnson and the final document, *IDEAL: The Science and Implementation Plan*, was issued in July, 1993 (IDEAL 1993). The research papers and literature reviews presented at the meeting were assembled in a volume titled *The Limnology, Climatology, and Paleoclimatology of the East African Lakes.* (Johnson and Odata 1996).

The IDEAL symposium of 1993 marked the inception of the International Decade, and science planning and research began in earnest. Since then, research programs have been funded in the U.S. and abroad allowing scientists to pursue inquiries about many aspects of the African great lakes. IDEAL enjoys the talents and enthusiasm of a Program Manager, Ms. Laura Lee Gance who, among many other things edits a biannual newsletter. The North American Steering Committee meets regularly to assess program goals and to revise them if necessary. It facilitates planning and collaboration on projects and helps organize technical coordination. Students from Africa, the United States and Europe are being guided by IDEAL scientists and a research training program at Lake Tanganyika will commence in 1998 for U.S. undergraduate college students. In June 1996, IDEAL presented a third multidisciplinary symposium at the annual meeting of the American Society of Limnology and Oceanography.

Publication of the first IDEAL book in 1996 has been called the birth announcement of this program. As with any birth, much of the fun is in the conception. Assuredly, the labor and delivery are a mixture of pain and joy. So with IDEAL. But also like parenthood, that pain is outweighed by joy that comes with successful results of vision and planning. With IDEAL there has been satisfaction in following a grassroots effort through its first steps. Even more satisfaction should lie ahead as the program reaches for new challenges and new discoveries in the Great Rift Valleys of East Africa.

5. References

GLRD 1987. Basic Issues in Great Lakes Research. Special Report No. 123 of the Great Lakes Research Division, University of Michigan.

IDEAL 1990. An International Decade for the East African Lakes. Workshop Report No. 1 on the Paleoclimatology of African Rift Lakes. Duke University Marine Lab Technical Report.

IDEAL 1993. IDEAL Science and Implementation Plan. PAGES Workshop Report Series 93-2.

Johnson, T. C. and E. Odata [eds]. 1996. Limnology, Climatology, and Paleoclimatology of the East African Lakes. Gordon and Breach.

HISTORICAL FLUCTUATIONS OF LAKE VICTORIA AND OTHER LAKES IN THE NORTHERN RIFT VALLEY OF EAST AFRICA

SHARON E. NICHOLSON
Department of Meteorology, Florida State University
Tallahassee, FL 32306, USA

1. Abstract

This paper describes the fluctuations of Lakes Victoria, Stefanie, Turkana and Naivasha over the last two centuries. A chronology of Lake Victoria back to 700 A.D. is also developed. These chronologies are based mainly on oral traditions of the local peoples, as described in various historical sources, and on reports of European visitors, settlers and explorers. In some cases actual historical levels have been reported. The historical fluctuations are meshed with the modern record to provide a picture of the fluctuations in lake levels until the late twentieth century.

The chronologies for Victoria and Stefanie contain much new material, permitting higher temporal resolution and better quantitative assessments, as well as extension of chronologies to the beginning of the 19th century. For Lakes Turkana and Naivasha, chronologies published by other authors are expanded and compared with those for Victoria, Stefanie and other African lakes. A long term chronology for Lake Victoria is developed using the record of the summer Nile flow.

These lakes show remarkably similar trends. The most important of these trends are low levels during the first half of the 19th century, very high stands in the last decades of the 19th century, and around the turn of the century a rapid fall to 20th century levels. The lakes returned to relatively high stands in the 1960s, but these generally ended in the 1970s.

2. Introduction

Fluctuations of East African lakes dramatically register the pulse of rainfall variability in the equatorial tropics. Lake Victoria, losing 90% of its water input via evaporation, is a particularly sensitive indicator of climatic change, having gone from complete desiccation some 12,000 years ago (Johnson et al. 1996) to a stand some 12 to 18 m above the present lake surface about 7,000 years ago (Stager et al. 1997).

Even recent fluctuations have been remarkable. An abrupt 2.5 m increase commenced in 1961 and continued until 1964, when lake level approached the high stands of the last century (Lamb 1966). This transgression of the lake, a result of

J.T. Lehman (ed.), Environmental Change and Response in East African Lakes, 7–35.

extremely high rainfall in its catchment (Flohn 1987; Nicholson 1996), occurred just prior to a major shift in the lake ecosystem. Fish populations were decimated, thermal stratification grew more persistent, diatom populations declined and significant changes in species occurred (Hecky 1993). The high stands of the lake may have contributed to these changes (Lehman 1997).

Historically, Lake Victoria had been of interest to European explorers, in part because of the hitherto elusive quest for the source of the Nile. The lake was "discovered" by Speke in 1850; its tremendous size became known when Stanley circumnavigated it in 1874. Much recent scientific research on Lake Victoria has been in the context of study of the River Nile. Meteorological interest arose when it was noted that precipitation in the catchment could not account for the dramatic rise in the early 1960s. Extensive water balance studies (WMO 1974, 1981) revealed that the lake creates a local circulation system that enhances rainfall, particularly nocturnal rains (Flohn and Fraedrich 1966; Fraedrich 1968, 1971). Other studies (e.g, Moerth 1967) demonstrated strong correlation between lake levels and sunspot activity, suggesting a link between solar activity and climate.

Lakes Turkana, Stefanie and Naivasha have also undergone changes over geologic and recent times (Lamb 1966; Nicholson 1978, 1981; Butzer 1971; Butzer et al 1972; Grove and Goudie 1971; Grove et al 1975; Gasse 1975; Nicholson and Flohn 1980). Like Victoria, they experienced low-levels or complete desiccation during the period of late Pleistocene aridity, but reached high stands in the early Holocene some 8,000 to 10,000 years ago. Their historical levels were also noted by early European explorers in East Africa.

Historical chronologies of these lakes have been established in numerous sources (e.g., Lamb 1966; Butzer 1971; Grove et al. 1975; Nicholson 1978, 1981, 1995; Vincent et al. 1979). In this article, additional documentary evidence expands these chronologies for Lakes Victoria and Stefanie and provides better quantitative estimates of fluctuations. Combining the historical record with modern data, this article reconstructs lake fluctuations back to the beginning of the 19th century. Comparison is made with the historical records of Lakes Naivasha and Turkana, using lake-level curves published by Vincent et al. (1979) and by Butzer (1971), and augmented with new documentary sources. For Lake Victoria, early fluctuations are suggested by comparison with the Nile record, going back many centuries, and with major famines and droughts in the lake district of East Africa.

2. Geography of the Lakes

The locations of the major East African lakes and rivers are depicted in Figure 1. The largest lake, Victoria, has a surface area of 68,800 km^2 and a mean depth of 40 m, the maximum depth being only 92 m. Lake Turkana, with a surface area of 8,860 km^2 and a mean depth of abou 28 m, is shallower and considerably smaller (Spigel and Coulter 1996). Lake Stefanie, just to the northeast of Turkana, is a shallow, now-dry, ephemeral lake. Lake Naivasha, in the eastern Rift Valley, has a surface area of about 150 km^2 and a mean depth of 6 m (Aloo 1996).

Figure 1. Location of various lakes mentioned in the text.

Mean annual rainfall in the region is shown in Figure 2. Most of the region would be characterized as arid or semi-arid. Around Lake Victoria, mean annual rainfall is mainly 1200 to 1600 mm. The authors deBaulny and Baker (1970) report mean rainfall over the lake as 1630 mm yr^{-1} but various estimates range from 1145 to 1810 mm (Datta 1981; Howell et al. 1988; Xin and Nicholson 1998). Estimates of mean evaporation over the lake vary from about 1370 to 1600 mm (Xin and Nicholson 1998). Lakes Turkana and Stefanie lie in a more arid region, with mean annual rainfall varying from about 200 to 800 mm. Spigel and Coulter (1996) give a value of 360 mm for mean annual rainfall over Lake Turkana itself and mean annual evaporation of 2340 mm. Over Lake Stefanie, precipitation is probably roughly the same as over Lake Turkana. Hastenrath and Kutzbach (1983) estimate annual precipitation over Lake Naivasha to be 600 mm, with annual evaporation on the order of 1500 mm. They estimate precipitation over the land within its catchment to be 900 mm; this figure is questionable, however, since there are few measurements in the Aberdare Mountains, where most inflow to the lake arises (Vincent et al. 1979).

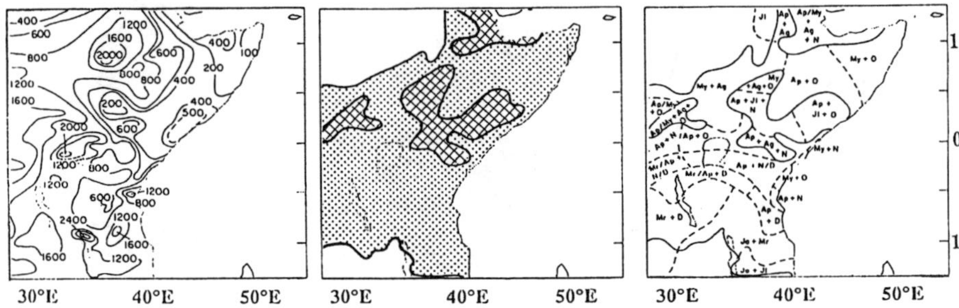

Figure 2. Rainfall characteristics over East Africa. Left: mean annual rainfall; Center: distribution of bimodal (dots) and trimodal (hatching) rainy seasons. Right: months of peak rainfall (from Nicholson 1996).

Although rainfall regimes within this region are diverse in both amount and seasonality (Fig. 2), the region is coherent with respect to interannual variability (Nicholson 1996). This implies both that the causes of rainfall variability are uniform throughout the region and that the lakes might exhibit similar fluctuations over time. However, the hydrologic regimes of the lakes and their responses to rainfall variability are not uniform (Vincent et al. 1979).

Lake Victoria receives inflow from 17 tributaries, the largest of which is the Kagera (Howell et al. 1988). Tributaries contribute only about 13% of the water entering the lake, the rest is by rainfall. The only outflow is via the White Nile. Lake Turkana (Rudolf) receives inflow mainly from the River Omo, its only permanent inlet. Next in importance is the Turkwel River, arising from rains near Mt. Elgon (Beadle 1974). Lake Naivasha is fed by three rivers arising in the Aberdare Mountains: the Malewa, which contributes about 90% of the inflow, the Gilgil, and the Karati. The lake also receives water through underground seepage (Vincent et al. 1979; Aloo 1996). Neither Naivasha nor Turkana have any known outlets, but Naivasha appears to have underground seepage (Åse 1987; Ojiambo and Lyons 1996). Lake Stefanie (also known as Ch'ew Bahir) receives inflow from the Sagan River (Butzer 1971; Grove et al. 1975), which drains the Ethiopian Rift south of Lake Chamo, and from the Dulei River. It currently has no outlet.

Lake Victoria is the source of the White Nile. Exiting the lake near Jinja, Uganda, it flows through Lakes Kyoga then Mobutu Sese Seko (Albert) before crossing into the Sudan. Near Khartoum the White Nile links with the Blue Nile, which drains the Ethiopian highlands, to form the main Nile flow through Egypt.

3. Reconstruction of Levels

Of the lakes considered, only Victoria and Naivasha have long series of gage measurements. Lake levels of Victoria have been recorded continuously at Jinja since 1899 and records are available from Entebbe starting in 1896 (Lamb 1966). Levels of Lake Naivasha have been recorded since 1910 (Vincent et al. 1979). Sporadic measurements are available for Turkana and Stefanie, although those for Turkana are nearly continuous between 1949 and 1961. Indirect indicators, usually historical references, are used to establish prior levels.

Historical witness to lake fluctuations is provided mainly by European explorers and settlers and by oral or written histories of indigenous peoples. There are direct references to the lakes, as well as complementary references to famine and drought in the vicinity of the lakes. These reports cannot fix actual lake levels, but can help to establish short-term rises and regressions. Much of the original material is difficult to access, appearing in century-old and often local archives. To the extent possible, original sources have been consulted. Additionally, summaries by Sieger (1887), Brueckner (1890), Lamb (1966), Butzer (1971), Grove et al. (1975) and Vincent et al. (1979) have been useful in reconstructing the lake levels.

Few direct measurements were made during historical times, but they serve to gauge fluctuations described relative to those years. Catholic missionaries at Buganda measured the levels of Lake Victoria for several years in the mid-1870s. The measurements indicate a stand some 8 feet higher than the 1898 levels. In 1872, Fischer found that the surface of Lake Naivasha stood about 1880 m above sea level (Fischer 1885). For Stefanie, a stand 6 to 8 m higher than modern levels is indicated both by the lake's geography and by lacustrine materials on a plain at this elevation. Other references frequently indicate a quantitative rise or fall of the lake during some time interval.

To establish the curves in Figures 3 to 7, material in chronicles is used to establish year-to-year trends. Fluctuations depicted in these curves are described in Results. The few quantitative entries are used to establish rough magnitudes of the fluctuations, as indicated in the 19th century lake level curves. For Victoria, some quantitative values are derived by the known correlation between its levels and the flow of the Nile River (e.g., Flohn and Burkhardt 1985). The correlation between the levels of Lake Victoria and the summer Nile discharge of the following year is 0.66 for the period 1898 to 1928.

The levels of the Nile in Egypt have been measured for several millenia. A nearly continuous record began on Rodah Island near Cairo in the year 641 A.D. and continued until the 20th century, but with considerable gaps, particularly in the period ca. 1470 to 1835. This record, published by Toussoun (1925), has been analyzed by numerous researchers (e.g., Hassan 1981; Riehl and Meitin 1979; Riehl et al. 1979; Herring 1979a; Cohen 1974). The maximum flood height is an indicator of the flow of the Blue Nile, and hence indicates rainfall over the Ethiopian highlands. Summer minimum levels are a rough indicator of equatorial rainfall near Lake Victoria, but they are also influenced by high floods of the previous year (Herring 1979a).

The Rodah measurements are not without problem (Popper 1951; Herring 1979a) A constant standard is lacking, since the gage was repaired and reconstructed at various times. Also, levels may not always have been recorded or reported accurately. The government may have inflated the figures, since taxes were levied according to the height of the flood. Accurate records required periodic cleaning of the nilometer. Finally, the river bed itself silted up over time. Despite this, it has been shown that the record can be used with a fair degree of confidence.

4. Results

4.1. LAKE VICTORIA

4.1.1. *The Last Two Centuries*

Lake Victoria maintained a low level during the early 19th century, peaked sharply in the late 1870s, then declined to 20th century levels. The decline was briefly interrupted by high stands in the early 1890s (Fig. 3). During the 20th century Lake Victoria was low until its dramatic rise in 1961 to 1962. Lamb (1966) and Flohn (1987) discuss this event. The lake has maintained higher stands since then.

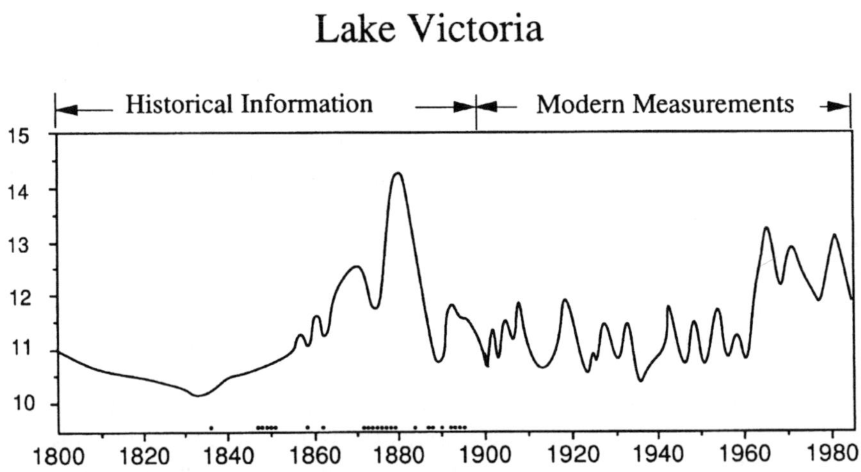

Figure 3. Fluctuations of Lake Victoria since 1800. Years of specific references are indicated at bottom with dots. Levels since 1896 are based on modern measurements.

Evidence for low lake levels early in the century includes tribal traditions of dry conditions in the 1820s and 1830s in many areas of Kenya, Uganda, Tanzania and Ethiopia (Appendix A). The 1830s drought was long and intense enough to impact local peoples seriously, forcing migrations of many groups. The height of the lake indicated at this time in Figure 3 is only a rough estimate based on the assumption that the lake was only marginally lower than during the driest years of the twentieth century. The continuation of arid conditions into the 1840s is suggested by Nile levels and by tribal traditions of extremely low levels of Lakes Turkana and Naivasha. A strong regression of both lakes probably means a significant regression of Victoria as well. Severe drought in Uganda around 1847 to 1851 suggests that the arid period, and low lake levels, would have continued until this time.

Lake Victoria began rising in the 1850s in response to wetter conditions that led to the expansion of forests and woodlands in areas of East Africa (Dale 1954). The lake supposedly rose rather steadily for two decades, although brief recessions are cited in 1858 and 1862. By about 1870, Victoria reached a relatively high stand, but this was interrupted by a more arid period around 1872 to 1875, as indicated by droughts and low Nile levels. Lyons (1906) indicates that Victoria was reasonably low in 1875. It rose rapidly again in the late 1870s, reaching a maximum in 1878. The average level during ca. 1876 to 1880, as reported by missionaries at Buganda, was about eight feet (0.5 to 0.7 m) higher than its 20th century peak in 1964. Thus, the lake stood at about 14 m then.

Lake Victoria fell rapidly in the 1880s, a response to numerous droughts, particularly late in the decade. It continued falling in the early 1890s, as drought continued, but a brief recovery of higher stands occurred ca. 1892 to 1895. Between 1895 and 1902 it fell about two-and-one-half feet (0.76 m). The fall was probably accelerated during the period 1897 to 1899, when unprecedented drought conditions struck. This fall is noted in historical documentary sources, as well as in the modern record of the lake.

4.1.2. *Long-term Chronology*

Prior to about 1800 there are no direct references to the stands of Lake Victoria. General trends can be established, however, using the Nile flow and references to oral histories of the local peoples. In the inter-lacustrine region of East Africa, rich oral traditions have allowed researchers (e.g., Herring 1979a; Cohen 1974; Webster 1979) to derive detailed, multi-century historical chronologies of the peoples. These chronologies include references to famines and droughts, as well as other phenomena such as migrations and cattle disease that are often linked to them. The chronologies provide indications of likely fluctuations of Lake Victoria in earlier centuries.

Webster (1979) outlines a chronology for the Central Lwo people, who inhabited mainly Uganda. He includes numerous references to famines and droughts. Information is plentiful since the early 16th century, but local chronologies for some Lwo groups are well established back to the early fifteenth century, at least. Major famines and droughts probably indicate relatively long and intense dry periods, because an individual dry year would likely not warrant mention in the oral traditions. Occasionally the traditions include references to periods of low Nile flow.

The chronology is genealogically dated, assuming a 27-year length for each dynastic generation and fixing dates, where possible, by references to solar eclipses. Dates are therefore imprecise. Webster (1979) has, however, reduced the margin of error on the dates by cross-referencing droughts noted in the reigns of kings of several groups and ascribing a date that is common to all reigns. Webster uses Herring's (1979a) analysis of the Nile record to further improve the dating of the famines and droughts. There is remarkable agreement between the Lwo famine and drought chronologies and the Nile record in recent centuries.

The early Lwo chronology suggests major droughts ca. 1587-1589 (the Nyarubanga Famine) and ca. 1621-1622 (the Great Famine) and a hiatus until the 18th century. The Nile levels (Herring 1979a) suggest that the period 1587 to 1623 was an almost continuous period of low rainfall, with the most extreme conditions occurring in 1588, 1601-1602, 1613, and 1617-1621.

Droughts and famines of the 18th century occurred ca. 1725-1729 (the Nyandere I famine), ca. 1749-1755 (the Nyandere II famine), ca. 1761-1769 (the Laparanat I famine), and during the 1780s (the Laparanat II famine) (Webster 1979). Famine in the 1750s was particularly severe in the south, where heat and dryness led to forest fires. Cohen (1974) and Herring (1979a) give further evidence of mid-eighteenth century droughts in numerous regions of East Africa, but these are less precisely dated. One example is from Koro tradition, indicating that the worst drought that ever occurred was within the period ca. 1740-1767 (Herring, 1979a). The 1720s famine was important because it provoked migrations and invasions. The lowest summer Nile levels of the 18th century occurred between 1726 and 1729. Nevertheless, famine and drought in the 1720s were probably less severe and of shorter duration than the Nyarubanga famine of the 1580s. The 1780s famine was also quite severe. There is additional evidence of very dry conditions in the late-18th century further east in what is today modern Kenya.

Further south, to the southwest of Lake Victoria, there are traditions from the Kibiza, Rwanda, and Karagwe groups of a very serious famine and drought early in the 19th century (within the time frame 1795 to 1830). This famine is referenced by only one of the northern Lwo groups, but it parallels the regressions of Lake Victoria, as described earlier, and Lakes Stefanie, Naivasha and Turkana (see sections 4.2 to 4.4), Malawi and Chilwa (Nicholson 1998) The few available Nile levels from this period also suggest extremely dry conditions. Moreover, the 1820s and 1830s was a period of continental aridity (Nicholson 1981, 1994).

Figure 4 sketches long-term fluctuations of summer Nile flow since 600 A.D., as published by Lamb (1977). These Nile fluctuations can be considered a reasonably direct indicator of the fluctuations of Lake Victoria. Flohn and Burkhardt (1985) developed a regression between the two and utilized Nile flow to derive recent historical levels of the lake. The considerable gaps in the Nile record, particularly in the 16th and 17th centuries, are quite evident. Relative variations depicted on the curve are probably accurate, but the magnitudes are approximate since there are no actual measurements of the level of the lake prior to the 19th century. The magnitude is based on the range of levels shown during the 19th century and on a comparison of this with the range in discharge over the same period.

Lake Victoria probably reached high stands around 800 to 940 and 1100 A.D. It was probably relatively low during most of the 13th and 14th centuries. It was again relatively high during the 15th and early 16th centuries, with a period of relatively low levels following until the late 19th century. The relatively arid conditions of the 17th and 18th centuries are probably those indicated in the geological record of Lake

Victoria for the period ca. 200-600 B.P. (Stager 1997). Low lake levels also occurred in Ethiopia about this time (Bonnefille and Mohammed 1994).

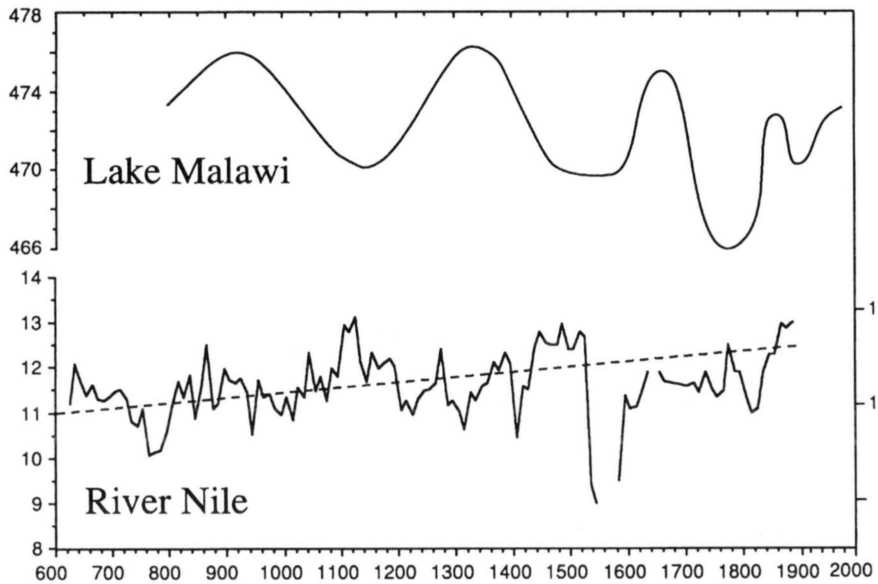

Figure 4. Fluctuations (m) of summer Nile minimum levels since 700 A.D. based on decadal averages and representative of long-term fluctuations of Lake Victoria. Roughly corresponding lake levels (m) are indicated on the right. Silting of the river bed has created a gradual rise of levels (dashed line) of 0.1 to 0.13 m per century. Lake fluctuations should be judged relative to the dashed line. Above: generalized curve of fluctuations of Lake Malawi (msl, from Nicholson 1998).

A comparison with a more generalized long-term chronology for Lake Malawi in southern Africa shows generally opposite trends in the two lakes (Nicholson 1998a). Exceptions are around 900 and 1700 A.D., when prolonged periods of high stands are indicated for both lakes, as well as the overall trends since about 1780 A.D. Then both lakes regressed substantially and maintained very low stands in the early 20th century. Both transgressed in the late nineteenth century and declined abruptly toward 1900, maintaining relatively low stands during most of the 20th century. The earlier pattern of opposition of these two lakes is consistent with the most typical patterns of modern rainfall anomalies (Nicholson 1986a, 1986b), which show strong opposition between equatorial and southern Africa during most years.

4.2. LAKE STEFANIE AND OTHER ETHIOPIAN LAKES

Grove et al. (1975) compiled extensive information on the historical fluctuations of Lake Stefanie. The lake curve in Figure 5 is based primarily on their material, as summarized in Appendix B. The only quantitative information derives from soundings made by Davidson Smith in 1895, none of which were more than 8 m deep. Grove and Goudie (1971) suggest that the lake was then generally about 6 m deep. Lacustrine materials on a gravelly flat along the eastern shore suggest the lake had maintained this depth for a long period of time.

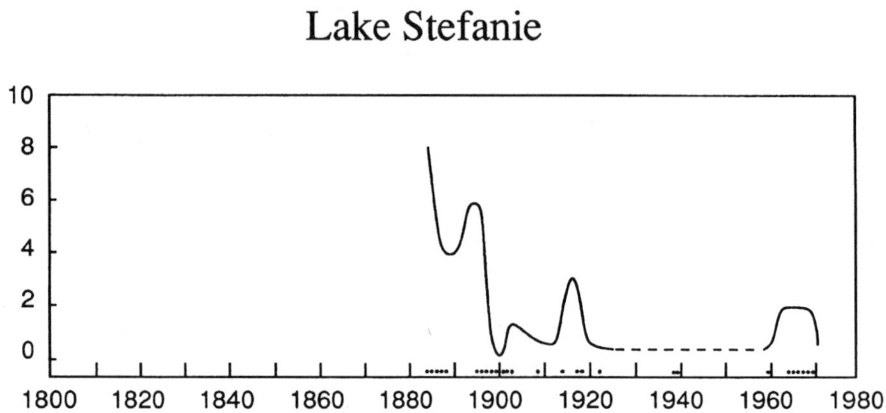

Figure 5. Fluctuations of Lake Stefanie since 1800. Years of specific references are indicated at bottom.

The information presented by Grove et al. (1975) starts with the year 1884; a report from Hobley (1914) indicates a rapid fall of Lake Stefanie between 1884 and 1888. Nevertheless, it was still a large lake when discovered by Count Teleki in 1888. A transgression occurred ca. 1895-1897, but drought resulted in extremely low lake levels by 1898 and 1899. By the early 1900s it was almost entirely desiccated. It remained low for several decades. Butzer (1971) suggests that it was essentially dry between 1898 and 1961. Visitors to the lake in the years 1901,1903, 1908, 1922, 1938 and 1939 variously described it as dry mud, dry marsh, or a small playa lake no more than 2 m deep.

This general trend is further substantiated from lakes in other parts of Ethiopia. Lakes Shallo and Awasa, north of Stefanie, and Lake Abhé, straddling the border between Ethiopia and Dbijouti, all had much higher stands in the latter half of the 19th century than during most of the twentieth century (Grove et al. 1975; Rognon and Gasse 1974). Numerous witnesses described rapid regression of Lake Abhé commencing toward the very end of the 19th century (Rognon and Gasse 1974). Around 1870 Lake Abhé had reached Heymour and therefore must have stood at a level of 35 m above the modern lake (Gasse 1975). Lake Shallo, now essentially dry, was an expansive body of water in the 19th century. Even as late as 1901 it was united with Lake Awasa during the rains to form a single lake. At that time Lake Awasa was some 3 m higher than during a recent wetter period in the early 1970s; it probably fell at least 5 m between 1900 and the 1940s, when Nilsson visited it. Nilsson (1940) noted that desiccation of the lake had proceeded for some time and speculated that the lake was in danger of becoming a papyrus swamp if this trend continued. By the 1950s the lake occupied only a 12 km^2 area within such a swamp (Grove et al. 1975).

Chronologies of famine and drought in Ethiopia compiled by Pankhurst (1966) and Cohen (1974) provide additional information. Pankhurst's list covers the 16th through

19th centuries and identifies causes of indicated famines. During the 19th century, famines shown to be associated with drought or crop failure occurred in 1828, 1829, 1835-1837, then not again until the late 1880s and 1890s (1888, 1889, 1898 and 1899). This chronology suggests that, as around Lake Victoria, relatively dry conditions occurred during at least part of the early 19th century and relatively good rains occurred in the mid- and late 19th century, with rainfall declining toward the end of the century.

There was apparently some recovery of higher stands in the 1960s. Lake Awasa showed a rising trend from the mid-1950s to 1972, a consequence of several factors including high rainfall in the 1950s and in later decades, a reduction of evaporation and increase in the number of rainy days (Makin et al. 1974). A subsequent regression began in 1972. Lake Chamo rose dramatically then, over 2 m between 1964 and 1968. Lake water flooded shoreline vegetation, including mature trees, thereby indicating a prolonged prior period with no flooding. Lake Abaya continuously overflowed into Lake Chamo from 1963 to at least the early 1970s, and Lake Chamo overflowed into the Sagan River (and hence into Lake Stefanie) beginning in 1965. Further north, Lakes Ziway and Langano attained high levels in the late 1960s and subsequently dropped rapidly. Lake Ziway continually overflowed into Lake Abiyata from 1966 to 1972, when Lake Abiyata started to fall. The flooding of Lake Abiyata in the 1960s drowned vegetation along its shores, including mature acacias and fig trees. As in the case of Lake Chamo, this indicates a prolonged prior period of low levels.

It appears that the Ethiopian lakes maintained relatively low stands throughout most of the early 20th century and began rising in the late 1950s or 1960s, reaching high stands in the late 1960s. More arid conditions, with regressions of the lakes, occurred in the mid- to late 1970s. This coincided with the onset of a remarkable arid episode in the West African Sahel, beginning with the well-known drought of the early 1970s.

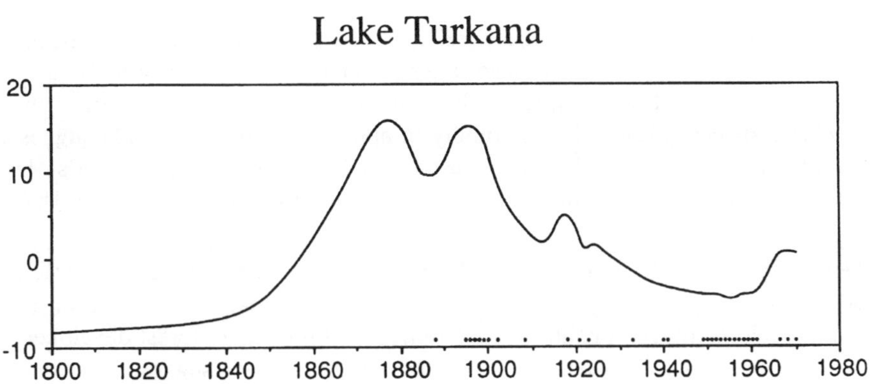

Figure 6. Fluctuations of Lake Turkana since 1800, modified from Butzer (1971). Years of specific references are indicated at bottom.

4.3. LAKE TURKANA

Butzer (1971) produced an excellent curve of the levels of Lake Turkana since 1880 by examining travelers' reports and maps in meticulous detail and comparing them with geographical indicators in the vicinity of the lake. That curve is reproduced in Figure 6, with detail added prior to 1885.

The first reports of the lake come from Hoehnel (1894), who described its state when he and Count Teleki discovered it in 1888. However, inferences about earlier times can be made from observations they recorded. From about the 1880s, there is enough information from travelers to give rough quantitative estimates of the lake level and to describe individual years. Lake Turkana rose from ca. 1888 to 1896 by some 5 to 6 m, but fell again abruptly after 1898. By 1921 it had fallen 14 m, compared with its maximum in 1896 and 1897. The lake remained low during subsequent decades, but began to rise in the 1960s, like other lakes of eastern Africa (Butzer 1971).

The year that von Hoehnel and Teleki visited Lake Turkana was exceptionally dry; only three small showers fell in the area during spring of 1888. There was a famine that year near Mt. Kulall on the southeastern shore of the lake, and everywhere from Lake Baringo to Ngaboto further south.

Nevertheless, evidence along the shore indicated that the lake level had been rising over the previous few years. In Alia Bay, large native villages were found on two low sand banks some 200 m offshore, i.e., on a partly-submerged beach ridge. Along the southeastern shore there were numerous groups of dead trees (including dum palms) standing in the shallow water, or at and just above the watermark. The trees presumably were killed by the rise of alkaline lake and ground waters. Similarly, von Hoehnel (1894) described and illustrated a string of widely spaced, dead tree trunks that once formed an extension of the Omo River and extended, half-submerged for almost 35 km into the lake. In his 1938 paper, von Hoehnel reports the distance instead as 24 km and states that "the further this tree line went from the shore the more decayed were the trees and the more sparse the row." This growth had once been a fringing woodland that extended along drowned levees.

Butzer (1971) concludes that the 1888 level of Lake Turkana was 8 to 10 m above its 1970 level. These observations also suggest that at some time prior to 1888 the lake had been appreciably lower long enough for mature trees to establish. Since the trees were widely spaced, bleached, and spindly in appearance, Butzer (1971) suggests that their submergence must have begun at least a decade earlier, during the 1870s. He thus concludes that Turkana had a higher stand of perhaps +15 m dated to the late 1870s and early 1880s.

Development of a fringing woodland probably requires a minimum of three to five decades, suggesting a previous period of low lake levels that lasted 50 years or more. This time scale seems reasonable, as mature trees but no mature forest developed along the lower Omo River between 1933 and the 1960s. Butzer suggests that these low levels occurred ca. 1765 to 1830, with subsequent higher levels ca. 1830 to 1860. This rise of the lake probably started somewhat later. There is direct reference in oral traditions to very low levels some time around the 1840s (Dale 1954; Appendix A) and other indicators of relatively dry conditions in East Africa in the 1840s. There were intense droughts and famine in Kenya in the 1830s and in Ethiopia in the 1820s and 1830s. About 1840 and earlier the flow of the Nile was particularly low; flow generally increased from the 1850s to the 1870s (Sieger 1887, citing Fritz; Appendix A). In any case, in the mid-19th century the lake could have been no higher than in 1970 (Butzer 1971).

von Hoehnel also noted sporadic indications of former higher lake levels. Near the southern end of Alia Bay there were numerous catfish in rain pools at some distance from the shore, presumably indicating a seasonal or short-term fall of Lake Turkana. Somewhat older are the *Etheria* encrustations seen on rocks about +20 or 30 m, near the southeastern corner of the lake, and the dry salt pans at +10 m, separated from the shore by a low sand ridge, found a little further north. Allowing for a +8 to 10 m lake level in 1888, as noted above, Butzer (1971) suggests these beach features may record older shorelines of Lake Turkana at +20 and +30 m.

As of the 1890s, information about Lake Turkana became much more plentiful. A lake level rise during the mid-1890s was documented by Donaldson Smith during the course of several successive visits between July 1895 and December 1896. In July 1895 Teleki's camp from 1888 had been inundated and the nearby village abandoned in favor of higher ground. Butzer concludes that water was 15 m higher than its 1970 level and 5 m higher than in 1888. Numerous other supporting indications exist, including river depth and inundation of forests.

Between 1897 and 1903 individuals or groups visited at least once in each year. Butzer (1971) interpreted lake levels from their maps and comments, which provided clear evidence of falling lake levels. Harrison said Lake Turkana fell 12 feet (4 m) during the year prior to March 1900. There were definite signs of a recession of the lake by April 1901, and an even greater recession by June 1902, when fish skeletons littered former lake beds. Visitors in 1903 and 1905 also described the area, but provided insufficient information to be certain of the level of the lake. Clearly, though, there were no signs of extreme change with respect to its 1902 level.

Other visitors and maps enable estimates for several years between 1908 and 1941 (Table 1). Then commencing in 1949 irregular gage readings became available. Butzer (1971) bases his curve (Fig. 6) on these measurements. Particularly noteworthy is the rapid rise of Lake Turkana at the end of 1961 and into 1962. The lake was presumably higher in 1962 than it had been for some thirty years, since Fort Wilkinson, built around 1930, was partially submerged. The lake gage at Ferguson's Gulf was also submerged in 1962, so the lake was at least 65 cm higher than the seasonal high of late 1950.

Merging this historical information with geological evidence, Butzer (1971) also produces a long-term chronology of Lake Turkana. He suggests two possible chronologies, based on "long" and "short" estimates of the time required to produce various changes. Although he suggests the long response is more likely, his short response chronology is remarkably similar to that for Lake Victoria and is in excellent agreement with the Nile record. Hence, it appears to be more appropriate. In addition to fluctuations already described, the short response chronology indicates a major transgression ca. 1680 to 1765 (with the lake at +8 to 15 m) and moderate levels (0 to +8 or +10) during the period 1530 to 1680 and 1500 to 1530.

TABLE 1. Estimated levels of Lake Turkana with reference to level in 1968 (Butzer 1971).

Year	Level (m)
1888	9.5
1895	14.5
1896	15.0
1897	15.0
1898	14.0
1899-1900	11.5
1902	8.0
1908	2.5
1918	5.0
1921	1.0
ca.1923	1.0 to 2.5
1933	-1.5
1940-41	-3.5
1949-53	-4.0
1954-56	-4.5
1957-61	-4.0
1966	-0.5
1968	0.0
1970	0.5

4.4. LAKE NAIVASHA

Figure 7 shows the long-term fluctuations of Lake Naivasha. The curve from 1888 onward is taken from Vincent et al. (1979), who used the work of Sikes (1935) to estimate levels prior to 1910. Similar reconstructions, with minor modifications, were also published by Åsa et al. 1986, Åse 1987, Ojiambo and Lyons 1996 and Verschuren 1996. Information in their reconstructions is cited in Appendix A. We have added a tentative trend at the beginning of the century and modified the curve in the 1880s.

The earliest part of the curve is imprecise and is based on only a few direct references to Lake Naivasha. The first reference is from Dale (1954), who reports that according to tribal tradition Lake Naivasha was reduced to a puddle around 1840 to 1850 (Appendix A). Another Masai tradition describes a very dry period immediately before the arrival of white men; during this time cattle grazed on Crescent Island, indicating that it was then a peninsula (Sikes 1936; Åse 1986). There is also an indication that around the middle of the 19th century, the Masai were herding cattle on dry land in what is now the middle of Lake Naivasha (Edmonson 1877). Droughts and famines in Kenya and northern Tanzania in the 1830s (Appendix A), as well as low Nile levels (Cohen 1974), likewise suggest low levels then.

Increasingly wetter conditions are suggested by expansion of forests and woodlands in the 1850s. Presumably a trend of rising lake levels commenced at that time. The lake was rising at the time of Fischer's visit in 1883 (Fischer 1885). According to the reconstructions of Sikes (1935) and others, the lake was reasonably low then but continued to rise until the mid-1890s, reaching a peak around 1895. There was a relatively steady decline until the 1940s, with a rapid fall between about 1897 and 1900 and moderately low levels around 1908 to 1915. A period of particularly low levels persisted throughout the 1940s and 1950s, then the lake rose in the early 1960s, like others in the region. Its level has varied by about 14 m since the 1890s.

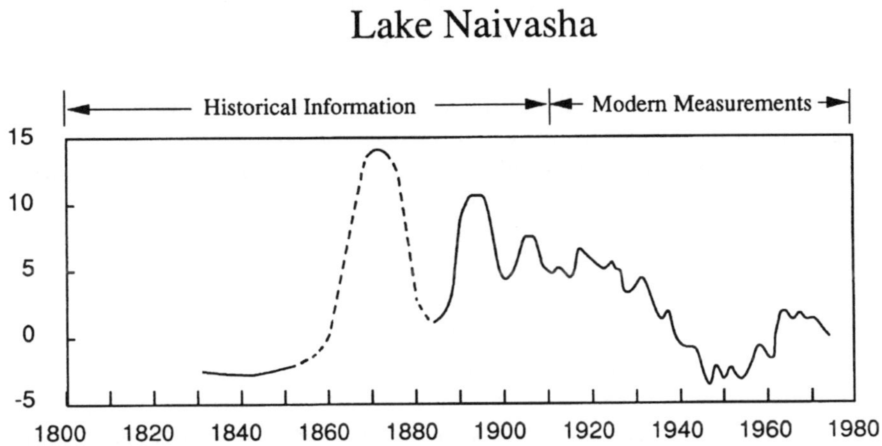

Figure 7. Fluctuations of Lake Naivasha since 1800, modified from Vincent et al. (1979). Years of specific references are indicated at bottom.

The only evidence of a relatively low level in 1883 is the witness of a Masai, who said that Crescent Island could still be reached (Appendix A). Fischer (1885), however, described a large and deep lake and indicated its surface at 1900 msl on his map. This elevation is 20 m above recent stands of the lake. At that time a long island in the southeast of the lake disappeared. If his measurement is accurate, it indicates a high stand had already been reached by 1883 that exceeded the lake maximum of the 1890s.

The assumed rise of the lake between Fischer's time and the visit of Teleki and von Hoehnel in 1888 is inconsistent with the frequent reports of famine or drought in Kenya nearly every year from 1882 to 1892. It is also incongruent with the regressions of other East African lakes that are well documented to have taken place in the 1880s, as noted here and by Nicholson (1998a, 1998b). Furthermore, the droughts of the 1880s were preceded by exceedingly wet conditions in 1878 and by lake level maxima in that year or 1879 for Lakes Victoria, Albert (Mobutu Sese Seko), Tanganyika and Rukwa. Thus, it is more likely that Lake Naivasha had been relatively high during Fischer's visit. In our reconstruction his figure of 1900 msl is accepted, and a subsequent decline to the 1888 levels is indicated.

5. Summary and Conclusions

The fluctuations of Lakes Victoria, Stefanie, Turkana and Naivasha show striking similarities. Evident in all four curves are a) very low lake levels throughout the early nineteenth century, b) high lake levels towards the end of the nineteenth century with a sharp fall towards 1900, c) relatively low levels throughout the first half of the twentieth century, and d) a rapid increase in the early 1960s. The fluctuations are not completely contemporaneous in all four lakes. For example, Lake Victoria reached its maximum in 1878, while that at Rudolf was probably some years earlier Also, Lakes Turkana and Naivasha fell much more gradually than Victoria during the early twentieth century.

It is not clear to what extent the differences can be attributed to various conditions of climate in their catchments or to differences in lake response to hydrologic fluctuations. That the latter is more likely is illustrated by the study of Vincent et al. (1979). They show that there is no statistical correlation between Victoria and Naivasha, pointing to the very gradual fall of the latter and the abrupt fall of the former at the turn of the century. However, individual annual anomalies in lake level are clearly synchronous in the two lakes. Thus, variations in the lake-level curves probably reflect differences in response times of the two lakes.

Long-term fluctuations could be established only for Lake Victoria. These fluctuations tend to be opposite those of both Lake Malawi in southern Africa and Lake Chad in northern Africa. This suggests that a pattern of opposition between subtropical and equatorial latitudes has characterized major climatic fluctuations on a scale of decades to centuries. This pattern is one of the most common in the modern record on time scales of individual years (Nicholson 1986).

Within the last two centuries there have been exceptions to this long term pattern. The arid episode of low lake levels early in the nineteenth century and the wetter conditions suggested by the transgressions of numerous lakes late in the nineteenth century appear to have been roughly synchronous in all areas. The same is true for the decline commencing in the late 1890s. Beginning in the 1950s, however, the major fluctuations of rainfall were out of phase (Nicholson 1995).

The arid episode is evidenced throughout the continent (Nicholson 1978, 1981, 1995) and appears to have climaxed in the 1820s and 1830s. Lakes Ngami, Chilwa, Malawi and others throughout southern Africa experienced very low stands, as did lakes Tanganyika and Rukwa. There was considerable discussion of the "drying up" of southern Africa, as water supplies continually diminished. Severe drought episodes persisted throughout the Sahel and Soudan zones of West Africa. It was a period of continental aridity analogous to the 1980s (Nicholson 1994; Fig. 8).

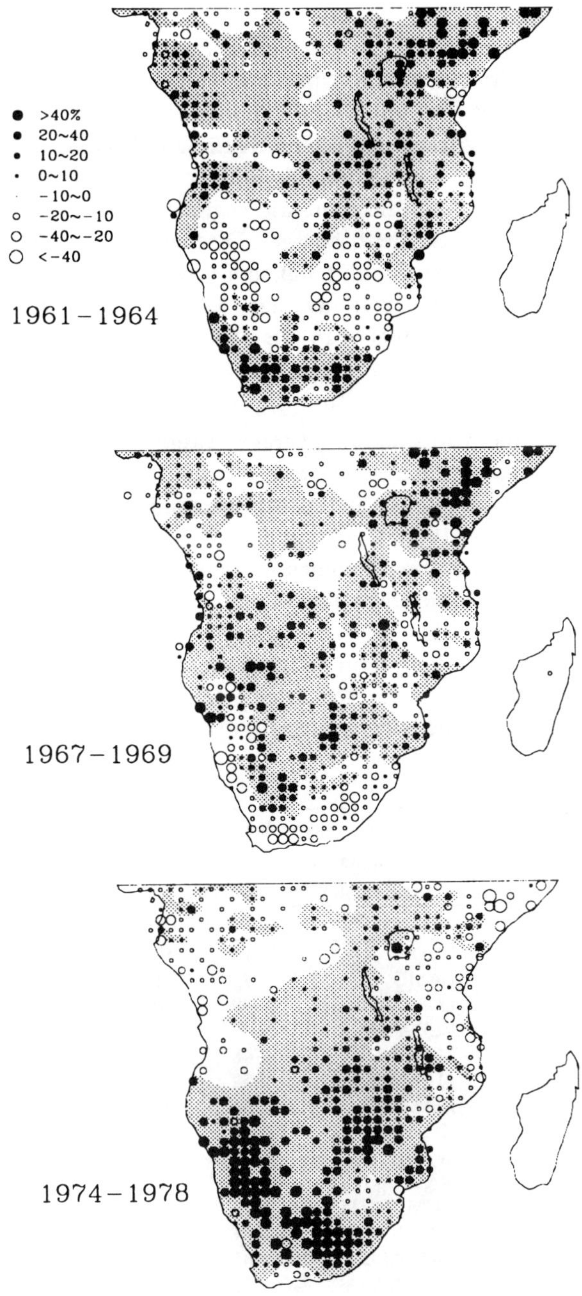

Figure 8. Rainfall anomalies in Southern Africa for 1961-1964, 1967-1969 and 1974-1978, expressed as percent above or below the long-term mean for one-degree grid squares (from Nicholson 1995). Positive anomalies are shaded.

The wetter period in the last few decades of the 18th century also appears to have been nearly continental in extent, although western equatorial sectors appear to have experienced relatively dry conditions at that time (Nicholson 1981). In addition to the lakes surveyed here, Lakes Magadi, Nakuru, Jilor and Baratuma in eastern Africa, Lake Chad and others in West Africa, and Lakes Ngami and Malawi in southern Africa all maintained very high stands, then fell rapidly until around 1914 (Hobley 1914). Wetter conditions prevailed throughout the semi-arid regions of West Africa. The degree of change is illustrated using the Nile discharge. Discharge at Aswan from 1870 to 1899 averaged 109 gigatons, compared with 83 gigatons from 1900 to 1949. Since the early part of this period was drier than average, the discharge must have decreased tremendously after the mid-1890s. In the year 1899-1900, discharge totaled 64 gigatons, (Grove 1972), but it was even lower (42 gigatons) in 1913-1914, in the middle of a continental period of drought.

During the 1960s there was an abrupt transgression of lakes throughout West and East Africa. Aside from the lakes already described, Lakes Nakuru, Manyara, Mobutu Sese Seko, Baringo, Tanganyika, and Chad maintained high stands for about one decade (Grove 1972; Lamb 1966; Butzer et al. 1972). The cause was an extraordinary increase of rainfall in the low latitudes (Flohn 1987; Nicholson 1996; Fig. 9). Examples of this episode in northern Kenya are given in Table 2. During November 1961, the stations Wajir, Eldama and Lokitaung received 612, 402 and 302 mm respectively, compared with long-term monthly means of 58, 48 and 39. This rainfall was coupled with extremely warm sea-surface temperatures along the East African coast. By the 1970s, when the lakes again fell, drier conditions had returned to eastern equatorial Africa (Fig. 9).

TABLE 2. November rainfall (mm) at Kenya stations in 1961, compared to mean rainfall for that month.

Station	Mean	1961
LODWAR	16	190
LAMU	35	212
LOKITAUNG	39	302
ELDAMA	48	402
MAKINDU	48	478
MANDERA	52	193
MALINDI	54	238
KITALE	57	365
WAJIR	58	612
NAKURU	60	280
RUMURUTI	79	350
GARISSA	81	412
MOYALE	86	362
MOMBASA	99	217
MARSABIT	143	612
MURANGO	187	610
EMBU	189	605
MACHAKOS	189	600
KITUI	307	682
MERU	317	875

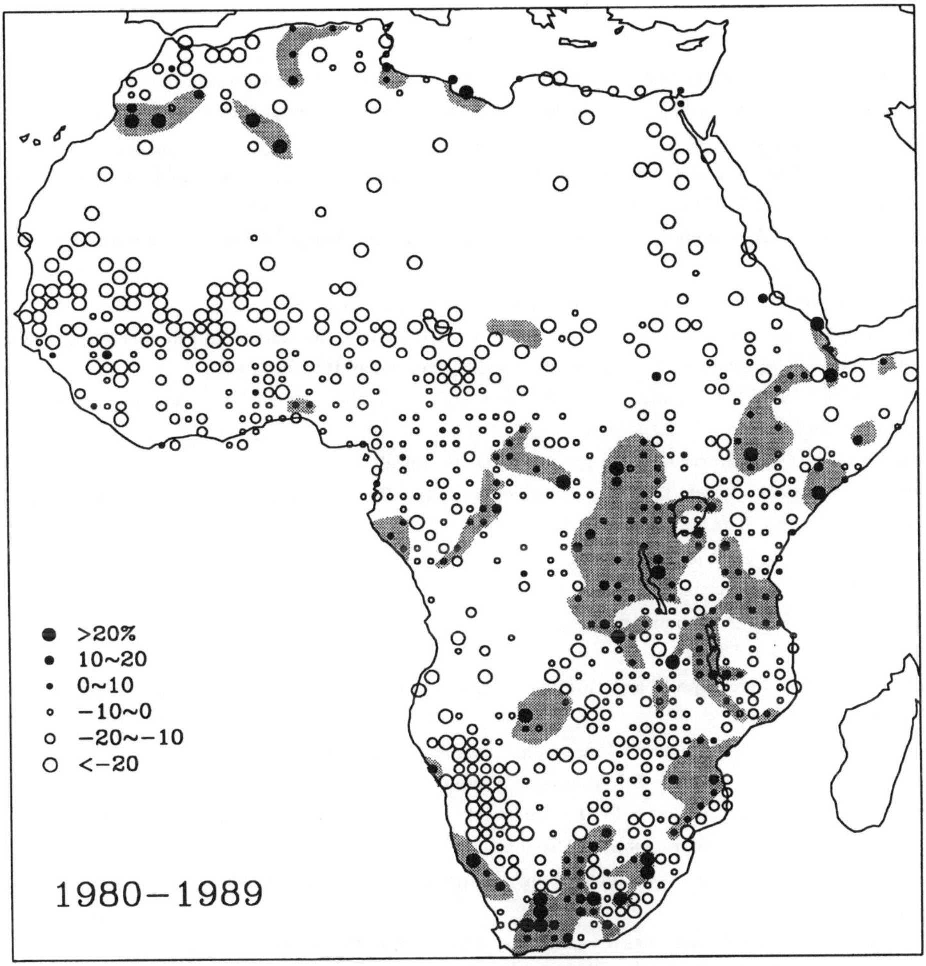

Figure 9. Mean rainfall for 1980-89. Rainfall expressed as percent departure from long-term mean (i.e., the mean for the entire length of record, commencing in 1901 or as soon thereafter as rainfall observations begin). Station data are averaged to facilitate presentation (from Nicholson 1995).

This study has extended the records of the East African lakes and climate back in time and has added detail to the records. The results further underscore the continental scale of the major rainfall fluctuations in Africa. Several anomalous climate intervals have been identified that warrant further study in the context of the meteorological causes of global change.

6. Acknowledgments

This work was supported by grants from NOAA (NA46GP0285) and the National Science Foundation (ATM9417063).

7. References

Aloo, P. A. 1996. Anthropogenic impact on fisheries resources of Lake Naivasha, p. 325-336. *In* T. C. Johnson and E. Odada [eds.], The limnology, climatology and paleoclimatology of the East African lakes. Gordon and Breach.

Åse, L.-E., K. Sernbo, and P. Syrén. 1986. Studies of Lake Naivasha, Kenya, and its drainage area. Forskningsrapport från Naturgeografiska Institutionen Stockholms Universitet. V. 63.

Åse, L.-E. 1987. A note on the water budget of Lake Naivasha, Kenya. Geografiska Annaler **89A**: 415-429.

Beadle, L. C. 1974. The Inland Waters of Tropical Africa. Longman.

Bonnefille, R., and U. Mohammed. 1994. Pollen-inferred climatic fluctuations in Ethiopia during the last 3000 years. Palaeogeography, Palaeoclimatology, Palaeoecology **109**: 331-343.

Brooks, C. E. P. 1923. Variations in the levels of the central African lakes, Victoria and Albert. Geophysical Memoirs, No. 20. (Meteorological Office).

Brueckner, E. 1890. Klimaschwankungen seit 1700. E. Holzel.

Butzer, K. W. 1971. Recent history of an Ethiopian delta. The Omo River and the level of Lake Rudolf. Univ. of Chicago, Dept. of Geography Research Paper 136.

Butzer, K. W., G. L. Isaac, J. L. Richardson, and C. Washbourn-Kamau. 1972. Radiocarbon dating of East African lake levels. Science **175**: 1069-1076.

Cohen, D. W. 1974. A preliminary study of the climatic trends in the lakes' region of East Africa. ASA Conference, Chicago.

Dale, I. R. 1954. Forest spread and climate change in Uganda during the Christian era. Empire Forestry Review 33.

Datta, R. R. 1981. Certain aspects of monsoonal precipitation dynamics over Lake Victoria, p. 333-349. *In* J. Lighthill and R. P. Pearce [eds.], Monsoon Dynamics. Cambridge University Press.

deBaulny, H. L., and D. Baker. 1970. The water balance of Lake Victoria. Technical Note, Water Development Department, Republic of Uganda.

Donaldson Smith, A. 1896. Expedition through Somaliland to Lake Rudolf. Geographical Journal **8**: 221-239.

Edmondson, R. N. 1977. Lake was dry land once. Daily Nation 1977 03 03.

Fischer, G. A. 1885. Das Masai-Land. L. Riedericksen and Co.

Flohn, H., and K. Fraedrich. 1966. Tagesperiodische Zirkulation und Niederschlagsverteilung am Victoria-See (Ostafrika). Meteorologische Rundschau **19**: 157-165.

Flohn, H., and T. Burkhardt. 1985. Nile runoff at Aswan and Lake Victoria: a case of a discontinuous climate time series. Zeitschrift fuer Gletscherkunde und Glazialgeologie **21**: 125-130.

Flohn, H. 1987. East African rains of 1961/62 and the abrupt change of the White Nile discharge. Palaeoecology of Africa **18**: 3-18.

Fraedrich, K. 1968. Das Land- und Seewindsystem des Victoria-Sees nach aerologischen Daten. Archives fuer Meteorologie, Bioklimatologie, und Geophysik, Series A. **17**: 186-206.

Fraedrich, K. 1971. Modell einer lokalen atmosphaerischen Zirkulation mit Anwendung auf den Victoria-See. Beitrage zur Physik der Atmosphaere **44**: 95-114.

Garstin, W. 1901. Report on irrigation projects on the Upper Nile, etc. (In a dispatch from His Majesty's Agent and Consul-General at Cairo). Parliamentary Accounts and Papers 1904. **111**: 1149.

Garstin, W. 1904. Report on the basin of the Upper Nile. Cairo.

Gasse, F. 1975. L'évolution des lacs de l'Afar Central (Ethiopie et T.F.A.I.) du Plio-Pléistocène à l'Actuel: Reconstitution des paleomilieux lacustres à partir de l'étude des Diatomées. D.Sc. thesis, University of Paris VI, 3 vol.

Grove, A. T. 1972. Climatic change in Africa in the last 20,000 years. Les problèmes de developement du Sahara septentrional. Colloque de Ouarghla. 2 vol.

Grove, A. T., and A. S. Goudie. 1971. Secrets of Lake Stefanie's past. Geographical Magazine **43**: 542-547.

Grove, A. T., F. A. Street, and A. S. Goudie. 1975. Former lake levels and climatic change in the Rift Valley of southern Ethiopia. The Geographical Journal **141**: 177-202.

Harlow, V., and E. M. Chilver.[eds.] 1965. The History of East Africa, Vol. 2. Clarendon Press.

Harrison, J. J. 1901. A journey from Zeila to Lake Rudolf. Geographical Journal **18**: 258-275.

Hassan, F. A. 1981. Historical Nile floods and their implications for climatic change. Science **212** :1142-1145.

Hastenrath, S. L., and J. E. Kutzbach. 1983. Paleoclimatic estimates from water and energy budgets of East African lakes. Quaternary Research **5**: 141-153.

Hecky, R. E. 1993. The eutrophication of Lake Victoria. Verhanglungen der Internationale Vereinigung für Limnologie **25**: 39-48.

Herring, R. S. 1979a. Hydrology and Chronology: The Rodah nilometer as an aid to dating interlacustrine history, p. 39-86. *In* J.B. Webster [ed.], Chronology, Migration and Drought in Interlacustrine Africa. Africana Publishing Company.

Herring, R. S. 1979b. The view from Mount Otuke: migrations of the Lango Omiro, p. 283-316. *In* J. B. Webster [ed.], Chronology, Migration and Drought in Interlacustrine Africa. Africana Publishing Company.

Hobley, M. 1914. The alleged desiccation of East Africa. Geographical Journal **44**: 467-470.

von Hoehnel, L. 1894. Discovery of Lakes Rudolf and Stefanie. Longman, Green and Co.

von Hoehnel, L. 1938. The Lake Rudolf region: its discovery and subsequent exploration, 1888-1909. Journal of the Royal African Society **37**: 16-45, 206-226.

Howell, P. P., M. Lock, and S. Cobb. 1988. The Jonglei Canal: Impact and Opportunity. Cambridge University Press.

Johnson, T.C., C. A. Scholz, M. R. Talbot, K. Kelts, G. Ngobi, K. Beuning, I. Ssemmanda, and J. A. McGill. 1996. Lake Pleistocene desiccation of Lake Victoria and rapid evolution of cichlid fishes. Science **273**: 1091-1093.

Lamb, H. H. 1966. Climate in the 1960s. Geographical Journal **132**:183-212.

Lamb, H. H. 1977. Climate: Present, Past and Future. Vol. 2. Methuen.

Lehman, J. T. 1997. Role of climate in the modern condition of Lake Victoria. Copyrighted ms.

Lyons, Sir H. G. 1906. The physiography of the River Nile and its basin. Cairo: Finance Ministry, Survey Dept.

Makin, M. J., T. J. Kingham, A. E. Waddams, C. J. Birchall, and Tamene Teferra. 1974. Development prospects in the southern rift valley, Ethiopia. Draft Land Resour. Stud. Land Resources Division, No.PR/25/74, Ministry of Overseas Development.

Moerth, H. T. 1967. Investigations into the meteorological aspects of the variation in the level of Lake Victoria. East African Meteorological Department Memoirs **4** (2): 1-22.

Nicholson, S. E. 1978. Climatic variations in the Sahel and other African regions during the past five centuries. Journal of Arid Environment **1**: 3-24.

Nicholson, S. E. 1980. Saharan Climates in Historic Times, p 173-200. *In* M. A. J. Williams and H. Faure [eds.], The Sahara and the Nile. A. A. Balkema.

Nicholson, S. E.. 1981. The Historical Climatology of Africa, p. 249-270. *In* T. M. L. Wigley, M. J. Ingram and G. Farmer [eds.], Climate and History: Studies on Past Climates and Their Impact on Man. Cambridge Press.

Nicholson, S. E. 1986a. The spatial coherence of African rainfall anomalies: Interhemispheric teleconnections. Journal of Climate and Applied Meteorology **25**: 1365-1381.

Nicholson, S. E. 1986b. The nature of rainfall variability in Africa south of the equator. Journal of Climatology **6**: 515-530.

Nicholson, S. E. 1994. Recent rainfall fluctuations in Africa and their relationships to past conditions over the continent. The Holocene **4**: 121-131.

Nicholson, S. E. 1995. Environmental Change within the Historical Period, p. 60-75. *In* A. S. Goudie, W. M. Adams and A. Orme [eds.], The Physical Geography of Africa. Oxford University Press.

Nicholson, S. E. 1996. A review of climate dynamics and climate variability in eastern Africa, p. 25-56. *In* T. C. Johnson and E. Odata [eds.], The limnology, climatology and paleoclimatology of the East African lakes. Gordon and Breach.

Nicholson, S. E. 1998a. Fluctuations of Rift Valley Lakes Malawi and Chilwa during historical times: A synthesis of geological, archaeological and historical information, p. 207-231. *In* J. T. Lehman [ed.], Environmental Change and Response in East African Lakes. Kluwar.

Nicholson, S. E. 1998b. Historical and modern fluctuations of Lakes Tanganyika and Rukwa and their relationship to rainfall variability. Climatic Change (in press).

Nicholson, S. E., and H. Flohn. 1980. African environmental and climatic changes and the general atmospheric circulation in late Pleistocene and Holocene. Climatic Change **2**: 313-348.

Nilsson, E. 1940. Ancient changes of climate in British East Africa and Abyssinia. A study of ancient lakes and glaciers. Geografiska Annaler **22**: 1-78.

Ojiambo, B. S., and W. B. Lyons. 1996. Residence times of major ions in Lake Naivasha, Kenya, and their relationship to lake hydrology, p. 267-280. *In* T. C. Johnson and E. Odata [eds.], The limnology, climatology and paleoclimatology of the East African lake. Gordon and Breach.

Oliver, R. 1963. The History of East Africa. Vol. 1. Clarendon Press.

Pankhurst, R. M. 1866. The great Ethiopian famine of 1888-1892. Journal of the History of Medicine **21**: 95-124.

Plote, H. 1974. L'Afrique sahélienne se desseche-t-telle? Bur. Rech. géol. et minières, Service geologique national Orléans, France.

Popper, W. 1951. The Cairo Nilometer. Universitiy of California Press.

Ravenstein, E. G. 1901. Lake-level of the Victoria Nyanza. Geographical Journal **18**: 403-406.

Riehl, H. M., el-Bakry, and J. Meitin. 1979. Nile river discharge. Monthly Weather Review **107**: 1546-1554.

Riehl, H. M., and J. Meitin. 1979. Discharge of the Nile river: A barometer of short-period climate variations. Science **206**: 1178-1179.

Rognon, P., and F. Gasse. 1974. Evolution des lacs depuis le Pléistocène supérieur dans l'Afar central (Ethiopie et Tfai), p. 9-18. *In* Ass. sénégal. Et. Quatern. Afr., Bull. Liaison, Sénégal. No. 42-43.

Sieger, R. 1887. Schwankungen der innerafrikanischen Seen, p. 41-60. *In* Vereinigung der Geographie, Jahresbericht. University of Vienna.

Sikes, H. L. 1936. Notes on the hydrology of Lake Naivasha. Journal of East Africa and Uganda Natural History Society **13**: 74-89.

Spigel, R. H., and G. W. Coulter. 1996. Comparison of hydrology and physical limnology of the East African Great Lakes: Tanganyika, Malawi, Victoria, Kivu and Turkana (with reference to some North American Great Lakes), p. 103-139. *In* T. C. Johnson and E. Odata [eds.], The limnology, climatology and paleoclimatology of the East African lake. Gordon and Breach.

Stager, J. C., B. Cumming, and L. Meeker. 1997. A high-resolution 11,400-year diatom record from Lake Victoria, East Africa. Quaternary Research **47**: 81-89.

Toussoun, O. 1925. Memoire sur l'histoire du Nil. Memoire de l'Institut d'Egypt. Vol. 9. Français de l'Archaéologie Orientale.

Verschuren, D. 1996. Comparative paleolimnology in a system of four shallow tropical lake basins, p. 559-572. *In* T. C. Johnson and E. Odata [eds.], The limnology, climatology and paleoclimatology of the East African lake. Gordon and Breach.

Vincent, C. E., T. D. Davies, and A. R. C. Beresford. 1979. Recent changes in the level of Lake Naivasha, Kenya, as an indicator of equatorial westerlies over East Africa. Climatic Change **2**: 175-189.

Weatherby, J. 1979. The raindrums of the Sor, p. 317-323. *In* J. B. Webster [ed.], Chronology, Migration and Drought in Interlacustrine Africa. Africana Publishing Company.

Webster, J. B. 1979. Noi! Noi! Famines as an aid to interlacustrine chronology, p. 1-38. *In* J. B. Webster [ed.], Chronology, Migration and Drought in Interlacustrine Africa. Africana Publishing Company.

WMO. 1974. Hydrometeorological survey of the catchment of Lakes Victoria, Kyoga and Albert. RAF 66-025, Technical Report 1, 4 volumes.

WMO. 1981. Hydrometeorological survey of the catchment of Lakes Victoria, Kyoga and Mobutu Sese Seko.

Xin, Y. and S. E. Nicholson. 1998. The water balance of Lake Victoria. Journal of the Hydrological Sciences (in press).

Zavattari, E. 1942. La risoluzione del problema del lago Stefania. Bolletino Società Geografica Italiana, Series **7**: 321-344.

ca. 1825-40 Many traditions in northern Uganda indicate drought there between ca. 1825 and 1840, and similar reports exist from Ethiopia and elsewhere in East Africa (Herring 1979a), consistent with other reports that suggest probable low lake level. Howell et al. (1988) indicate that the Nuer, who originally lived near the Bahr el Jebel, suffered severe population pressure at the beginning of the 19th century, including drought and famine. The worst period occurred in the 1820s, which was the time when the Nuer migration began. The drought and famine affected the entire Sudan and adjacent territories.

1830s Intense drought in Kenya dried up wells and forced migrations (Dale 1954), implying a period of low levels of Victoria. The Iminito and the Ngarriama, of eastern Uganda migrated out of the region in response to famine during the drought of the 1830s. This decade the whole of northeastern Uganda suffered badly from drought (Weatherby 1979). A drought some time during the 1830s caused an exodus of the proto-Omiro further west (Herring 1979b).

1836 About this time there was an important famine in Kemba country (east of Lake Magadi, NE of Mt. Kilimajaro, ca. 2.5 S, 37 E; Oliver 1963).

ca. 1840 About 1840 and earlier the flow of the Nile was particularly low. Flow generally appeared to increase from 1850s to 1870s (Sieger 1887, citing an analysis of Fritz in Meteorologische Zeitschrift).

1840-50 According to tribal tradition, about this time Lake Rudolf (Turkana) sank to a very low level and Lake Naivasha was reduced to a puddle (Dale 1954). This would likely mean low levels of Victoria as well. According to Edmonson (1977), there is an indication that around the middle of the 19th century, the Masai were herding cattle on dry land in what is now the middle of Lake Naivasha. A Masai tradition describes a very dry period immediately before the arrival of white men; cattle grazed on Crescent Island, indicating that it was then a peninsula (Sikes 1936; Åse 1986). Appearance of submerged trees during the period of low levels in the 1940s suggests that the lake had previously been very much lower; these trees would originate before the high levels of the 1880s (Åse et al. 1986) and possibly from early in the 19th century.

1847-51 This may have been a period of drought in Uganda, as Herring (1979a) describes low Nile levels during the periods ca. 1847-51, 1872-75, 1882-92 and 1899-1909 and indicates that droughts are reported from different parts of Uganda during most of these periods.

1850s Since about 1850 in various parts of East Africa there was a growth of the forest edges, expansion of riparian forests and growth of woodland, suggesting increasingly wetter conditions (Dale 1954).

1858 Lake Victoria discovered by Speke. Sieger (1887) indicates a recession in this year.

1862 Sieger (1887) indicates a recession of Lake Victoria in this year.

1870s Sieger (1887) indicates that Lake Victoria was rising rapidly in the 1870s.

1875-1900 According to Garstin (1901) "there appears to be no doubt that, for many years previous to 1901, the mean levels of this lake [Victoria] have shown a steady fall, independent of the annual fluctuations due to the rainfall. This general lowering of the water levels was noticed both by Sir. H. Stanley and by Mason Bey at the time of their respective visits" (quoted in Howell et al. 1988).

"Lastly, there is a statement of Mr. McAllister, Sub-Commissioner of the Nile Province ..., that the French Fathers at Muanza, in the south of the lake, had a register showing from marks on the rocks that there has been a fall of some 2 - 4 meters during the last 25 years" (Garstin 1904, quoted in Howell et al. 1988).

Lyons, who accompanied Garstin to the White Nile in 1901 and 1903 and made a special study of Victoria's levels, concluded that the lake was very high in August and September, 1878, fell from 1878 to 1892 and rose in 1892-95 (Howell et al. 1988).

The high levels of Lake Victoria had probably begun already by 1870 (Howell et al. 1988). There are early accounts from ca. 1879-85 of immense herds of Bari cattle grazing above Mongalla.

1872-75 This may have been a period of drought in Uganda, as Herring describes low Nile levels during the periods ca. 1847-51, 1872-75, 1882-92 and 1899-1909 and indicates that droughts are reported from different parts of Uganda during most of these periods.

1874 Size of Victoria became known when Stanley circumnavigated it (Lamb 1966).

1875 Low stand of Victoria, according to Lyons (1906; cited by Cohen 1974).

ca. 1876 Catholic missionaries at Buganda reported that the average water level of Victoria was 8 feet (2.4 m) higher than during that of 1898 (Ravenstein 1901). This suggests (Lamb 1966) that the level around 1876-1880 was slightly (about 0.5 to 0.7 m) above the 1964 peak level. Beadle (1974), presumably based on this information, gives the maximum as 1876, with a level of 14.0 m at the Jinja gauge (see also 1878 entry). Lake Albert also reached a high stand in 1876 (Sieger 1887). High levels at this time are also supported by heights of Nile floods (Sieger 1887).

late 1870s A very steep rise of Lake Victoria (Sieger 1887). Rainfall observations near the lake showed a greater number of rainy days in 1876, 1877 and 1878 than during later years (Sieger 1887).

1878 Lake Victoria was very high in the northeast during August and September; this was associated with very high rainfall (Brooks 1923, quoting Lyons).

1878-79 Lake Victoria was high (Lamb 1966). Garstin's report (see earlier entry) suggests that the maximum occurred in 1878. This is in agreement with a report that 1878 was particularly rainy along the upper Nile (Sieger 1887) and with evidence from the Bahr el Jebel. In 1878 Lado, on the Bahr el Jebel, was flooded and the river was still very high in December 1878, suggesting also an unusually high level of Lake Albert (Howell et al. 1988). The White Nile was also unusually high that year (Howell et al. 1988), with peak flow at Gondokoro occurring in August 1878. The tremendous floods brought enormous quantities of vegetation debris that blocked parts of the river. The Nuer oral history (see Howell et al. chapter 9) also suggested very high Nile floods at the time.

1881 Very dry year in Tanzania, only 729 mm rain in Tabora (5 02 S, 32 49 E), with famine at many places between Tabora and the coast (*Petermann's Mitteilungen* 1881).

1880-90 Lake Victoria was falling (Lamb 1966). Between 1879 and 1884 an island developed in Lake Albert in shallows near Mahagi on the west coast (Sieger 1887). This implies falling lake levels.

1883	Fischer's map from 1883 shows Lake Natron (ca. 2.5 S, 36 E) as 40 km long and 10 km wide, and Lake Naivasha at 1900 m, with no long island in the SE (Fischer 1885). This suggests Crescent Island was either inundated (indicating high lake levels), or had become a peninsula or part of the mainland (indicating low lake levels). Sikes (1936) states that when Fischer reached the lake in June 1883 the water level had already begun to rise. He concludes that Crescent Island was then a peninsula, as a Masai told Fischer it was still possible to get across to Crescent Island via mud. Sike thus estimated the level of Lake Naivasha to be 1880 msl (Åse et al. 1986).
1884	Drought in the area northeast of Lake Victoria (Brooks 1923).
1886	Drought in the area northeast of Lake Victoria (Brooks 1923).
1887	Famine in western lakes area (Oliver 1963), and in Elgeyo district in December (von Hoehnel).
1887	During von Hoehnel's visit to Lake Baringo (ca. 0 40 N, 36 E), the lake was a large expanse of water with one long island and several islets. In December there was famine in Komasia (W of the lake), which was generally the granary of the whole area.
1888	Famine near Mt. Kulall (SE shore of Turkana, ca. 2.5 N, 37 E) and famine everywhere from Lake Baringo (ca. 0 40 N, 36 E) to Ngaboto (2 N, 35 E). According to von Hoehnel, as of May 27 near Lake Turkana (ca. 3 to 4 N, 36 E) there was low rainfall, with only three small showers.
1888	General fall of Lake Victoria on the southern shore (Brooks 1923). von Hoehnel's map of Lake Naivasha indicated that Crescent Island was definitely isolated from the lake shore and his depiction of a bay implied levels approximately 20 feet above the lake's 1959 level, i.e., at 1893 msl (Åse et al. 1886).
1890	General fall of Lake Victoria on the southern shore (Brooks 1923).
1890	Drought in western Kenya (Oliver 1963).
1891	Low levels of Lake Victoria (Brooks 1923)
1892	Famine in western Kenya (Oliver 1963).
1882-92	This may have been drought in Uganda, as Herring (1979a) describes low Nile levels during the periods ca. 1847-51, 1872-75, 1882-92 and 1899-1909 and indicates that droughts are reported from different parts of Uganda during most of these periods.
1892	Lake Victoria was very high, with a tendency to rise. There was heavy rainfall (Brooks 1923).
1892-95	Lake Victoria recovered somewhat to a few years of higher levels (Lamb 1966 - evidently quoting Lyons).
1893	Lake Naivasha was very high during Gregory's visit. Gregory's description suggests a very wet year (Åse et al. 1986).
1894	Lake Naivasha was very high during Hobley's visit (Åse et al. 1986). Based on Hobley's description of the lake level with respect to the railroad, Sikes (1935) concluded that the lake stood at 1899 msl and possibly higher.
1895	Lake Victoria was very high (Brooks 1923 - probably quoting Lyons). Rainfall was also very high at Machachos, Kenya (Åse et al 1986).

1895-1902	Steady fall of Lake Victoria of 2.5 feet (0.76 m) in this seven year period (Lamb 1966). Beadle (1974) indicates on a graph a level of 11.5 m for 1895 or 1896
1890s	Terrible famine and drought in northern Uganda - worst in the memory of tribal elders of that time (Webster 1979). The 1890s were years of extreme deprivation in NE Uganda, with not only droughts, but also wars, locusts and human and animal epidemics (Herring 1979a).
1896-99	Four years of severe drought, culminating in famine in the years 1898 and 1899 (Åse et al 1986). In 1897 to 1899 there was famine from Kikuyu (ca. 0 S, 36 30 E) to Kamba (east of them, ca. 1.5 S, 38 E). In about 1898 there was a famine due to drought in Busoga (ca. 1 N, 33.5 E) and in Bunyoro (Uganda) (Harlow and Chilver 1965). In 1898 and 1899 there was an "unprecedented" drought affecting all of East Africa (Grove et al. 1975; Harrison 1901). There was widespread drought in Uganda in 1899 to 1903, possibly continuing part or most of the period until 1909 (Herring 1979a).
1898	Railway measurements in September and November show that the lake was considerably lower than during 1894 and probably stood at about 1895 msl (Sikes 1935; Åse et al. 1986).
1908	Drought in Kilimanjaro district of Tanzania and in the Busoga region of Uganda (Harlow and Chilver 1965).
1918, 1919	Major famine in Busoga and in the Bukedi district of Uganda. Pestilence and famine ravaged Kenya, particularly the Kukuyu country in 1918 to 1920 (Harlow and Chilver 1965).

1884-88	Rapid fall of the lake (Hobley 1914; Grove et al. 1975). According to Plote (1974) this was a period of moderate regression.
1888	In April 1888 Count Teleki found a broad sheet of water 20 to 25 km wide, but his guide was astonished at how much it had fallen since three years earlier (Grove et al. 1975; Hobley 1914). A strip of lake bed a km wide was exposed, the water was brackish and decomposing dead fish were strewn over the area. It was a saline lake with many crocodiles (Butzer 1971).
1895-97	A strong transgression of the lake and the complete extent of the basin was flooded (Butzer 1971; Plote 1974). In 1895 Donaldson Smith found that the waters had risen since 1888, but none of his soundings were more than 8 m deep (Grove et al. 1975). The overall depth of the lake was probably about 6 m (Grove and Goudie 1971).
1898-99	Lake levels were extremely low and there was tremendous drought. The Sagan River no longer reached the Dulei River at the northern end of the Stefanie trough (Grove et al. 1975). In December 1899 the southern extremity of the lake had dried up and was covered with dead fish (Butzer 1971; Plote 1974). In these two years there was an "unprecedented" drought affecting all of East Africa (Grove et al. 1975; Harrison 1901). There was famine and starvation in the Sagan River valley; the river was so low that it spread out and disappeared before reaching the Dulei River at the north end of Lake Stefanie.
1900	In March 1900 the lake was almost totally desiccated (Butzer 1971; Plote 1974). Grove and Goudie (1971) mention that Lake Stefanie was totally dry at the very end of the last century.
1901	Lake Stefanie was dry (Zavattari 1942).
1902	No absolute level is given, but the channel connecting Lake Chamo with the Sagan River, which flows into Lake Stefanie, was not filled. This channel had been filled for six or seven years (ca. 1963-70) at the time of the Grove et al. (1975) expedition. The traveler Neumann noted in this year that if the water in Lake Chamo were to rise a few inches, a large river would flow from it into the Sagan.
1903	Lake Stefanie was dry mud or dry swamp (Zavattari 1942).
1908	Lake Stefanie was dry mud or dry swamp (Zavattari 1942).
1914	Lake Stefanie was extremely low; according to Hobley (1914) it was "hardly more than a puddle."
1917-18	Lake Stefanie was presumably higher, as the Sagan river was then flowing more strongly than around 1970 (Grove et al. 1975).
1922	Lake Stefanie was dry mud or dry swamp (Zavattari 1942).
1938-39	Lake Stefanie was dry mud or dry swamp (Zavattari 1942).
1959	Lake Stefanie was a marsh (Butzer 1971).
1960s	Lake Stefanie was deep again during the 1960s (Butzer 1971). The rise probably occurred ca. 1961, as with other lakes. It was particularly high in 1968 and 1968, but 1965 Gemini photos showed it to be relatively dry with extensive vegetation along the south shore. The channel connecting Lake Chamo with the Sagan River had been filled for 6 or 7 yr (ca. 1963-70) at the time of the Grove et al. (1975) expedition. This probably had not occurred since ca. 1917-18.

1970s Lake Stefanie was dry again. When Grove et al. (1975) arrived there in 1970 it was dry except near the mouth of the Sagan. After a few rain events it became a shallow but extensive lake occupying much of the dry lake floor.

Other lake evidence from this area

Lake Shallo was a considerable body of water in the 19th century. Even as late as 1901 it was united with Lake Awasa in the rains to form a single lake. In 1940 Nilsson noted continuing desiccation of the lake, such that it was in danger of becoming a papyrus swamp. Aerial photographs from the 1950s shows the lake occupying a 12 km^2 area in the north end of the swamp.

Lakes in the southern Rift Valley of Ethiopia

Lake Chamo exhibited dramatic rise (over 2 m) between 1964 and 1968 that caused marginal vegetation including mature trees to be flooded. Rise was possibly related to the fact that the Lake Abayo has only overflowed continuously into Lake Chamo since 1963 (i.e., to ca. 1973). Lake Chamo has overflowed into the Sagan River and hence into Lake Stefanie only since 1965 (Grove et al. 1975; Makin et al. 1974).

Lakes further north:

Lakes Ziway and Langano attained particularly high levels in the late 1960s and subsequently dropped rapidly. Ziway continually overflowed into Lake Abiyata from 1966 to 1972. The flooding of Abiyata caused the drowning of shoreline vegetation, including mature acacias and fig trees. Abiyata started to fall in 1973, as did Lake Awasa, in between these northern and southern areas.

Lake Awasa was some 3 m higher in 1901 than in 1974. Probably at least a 5 m fall by 1940, when Nilsson suggested Lake Shallo was in danger of desiccation. Again rising levels between the mid-1950s and 1972, then a fall. Surprisingly the rise of the lake corresponded to a period of decreasing rainfall, so probably changes in evaporation or rainfall seasonality and distribution produced the changes of the lake.

Rognon and Gasse (1974) indicate higher levels of Lake Abhe prior to the 20th century and say much in witness to its rapid regression commencing toward the end of the 19th century.

ANCIENT ANALOGUES FOR RECENT ENVIRONMENTAL CHANGES AT LAKE VICTORIA, EAST AFRICA

J. C. STAGER
Math/Science Division, Paul Smith's College
Paul Smiths, NY 12970, USA
and
Climate Change Research Center, University of New Hampshire
Durham, NH 03824, USA

1. Abstract

Fine-interval (~29-35 ^{14}C year) diatom analysis of a sediment core from northern Lake Victoria provides a long-term context for recent environmental changes in the lake, which include increased water column stability, a switch from *Aulacoseira*- to *Nitzschia*-dominated diatom assemblages, and an overall increase in cyanobacteria relative to diatoms. The sediment record indicates that similar events have occurred abruptly and repeatedly in the lake since ~7.2 ^{14}C ka, apparently due to climatically driven disruptions of mixing regimes. Holocene fluctuations in the fossil diatom assemblages display strong periodicities (~500, ~1100, ~1400, and ~2,350-2,550 calendar year cycles) which point to global-scale forcing mechanisms. This paleolimnological record suggests that Lake Victoria and its algal communities are inherently unstable, and that future environmental and fishery management models should consider the lake's sensitivity to climatic changes.

2. Introduction

Paleolimnology has enjoyed tremendous advances in recent years. Spectral analysis of time series now permits detection of periodicities in microfossil records, and quantitative reconstructions of past limnological conditions can now be inferred through the use of transfer functions supported by relatively large sedimentary microfossil assemblage and water chemistry data sets, along with multivariate techniques such as Canonical Correspondence Analysis (Dixit and Smol 1994; Halfman et al. 1994; Gasse et al. 1995; Stager et al. 1997). Calibration of the global radiocarbon record permits development of calendar year chronologies for Holocene sediment records (Bard et al. 1993; Stuiver and Reimer 1993), and fine-interval sampling of sediment cores vastly increases the amount of information obtainable from sediment records (Renberg 1990; Halfman et al. 1994; Stager et al. 1997). These tools are

J.T. Lehman (ed.), Environmental Change and Response in East African Lakes, 37–46.

helping paleolimnology to become one of the primary sources of information about long-term ecological processes and environmental changes on local to global scales.

Over the last 3 decades, phytoplankton productivity has increased dramatically in Lake Victoria (Lehman and Branstrator 1993; Mugidde 1993; Hecky 1993). *Aulacoseira*-dominated diatom assemblages, typical of well-mixed conditions prevalent before the mid-1960s, have been replaced by *Nitzschia*-dominated assemblages as epilimnetic temperatures have risen and water columns have stabilized (Talling 1957 and 1966; Hecky 1993; Hecky et al. 1994). In addition, silica concentrations and N:P ratios have declined, and cyanobacteria far outnumber diatoms for most of the year (Hecky 1993; Ochumba and Kibaara 1993). Causes of these changes, that apparently began soon after the mid-1960s, remain uncertain but have been attributed variously to cultural eutrophication, trophic relations, and climatic changes (Gophen et al. 1993; Hecky 1993; Hecky et al. 1994). Sediment records provide perspectives on recent changes by yielding information about the natural range of environmental variability.

The microfossil record presented here refines a preliminary diatom analysis by Stager (1984) of a 9.9 m core collected from 32 m depth in the Damba Channel, Uganda (0° 05' 0"N, 32° 48' 2"E) by D. Livingstone and colleagues in 1971. Full discussion of the fine-interval diatom record is presented elsewhere (Stager et al. 1997); this paper deals primarily with Holocene records of the genera *Aulacoseira* and *Nitzschia*, and explores what they tell about recent changes in the lake.

In this study, *Aulacoseira* species were used as indicators of stronger or more frequent mixing under storm conditions, and *Nitzschia* species as proxies for increased water column stability under calm conditions. Prior to the late 1960s, fast-sinking *Aulacoseira* species such as *A. granulata* (Ehr.) Simonsen, *A. ambigua* (Grun.) Simonsen, and *A. nyassensis* var. *victoriae* (O. Müller) Simonsen dominated Lake Victoria's diatom communities, especially during seasonal wind-driven mixing events when diatoms were most abundant (Talling 1957, 1966). Today, these taxa have been replaced by various *Nitzschia* spp. (Hecky 1993). Attenuate, slow-sinking *Nitzschia* (referred to here as "long *Nitzschia,*" including *N. acicularis* and *N. lacustris*) tend to be more prevalent in large African lakes when mixing is reduced (Talling 1966; Haberyan and Mhone 1991; Owen and Crossley 1992), and attenuate *Nitzschia* have been found enmeshed in Lake Victoria's cyanobacterial blooms during thermal stratification (Ochumba and Kibaara 1989). Small *N. fonticola* Grun. are often epiphytic on cyanobacteria and are commonly found growing on *Microcystis* in the lake today (Kilham et al. 1986; Haberyan and Hecky 1987; H. Kling personal communication 1995). It is therefore assumed that *Nitzschia* (especially *N. fonticola*) is a possible indicator of cyanobacterial productivity or relative water column stability.

3. Methods

The uppermost 8.3 m of the core was resampled in 1995 at 2.5 cm increments, for a total of 328 subsamples representing about 29 to 35 ^{14}C year intervals (about 30 to 45 calendar years). Three 6 cm sections were lost in core pipe couplings in the field. Two bulk ^{14}C dates were obtained previously (11,710 ± 120 yr B.P., and 7,910 ± 80 yr B.P., centered at 7.85 and 5.27 m, respectively; Stager 1984). An additional bulk date of 3,070 ± 90 yr B.P. (Beta 9134) was obtained for the 2.06-2.16 m level. Because these dates display a linear relation with depth, the ages of the diatom subsamples were interpolated between the direct dates and the core top. A core from Pilkington Bay, dated in 28 places, also documented remarkably steady Holocene sedimentation rates (core P-2, Kendall 1969), as does an offshore core collected in 1995 (Johnson et al. 1996). A 600-year correction was subtracted from the Damba Channel chronology because linear extrapolation of direct dates yielded an apparent age of ca. 600 yr for the core top, as was also found for the P-2 core (Stuiver 1970). Calendar ages relative to 2000 A.D. were determined with the CALIB program (Stuiver and Reimer 1993).

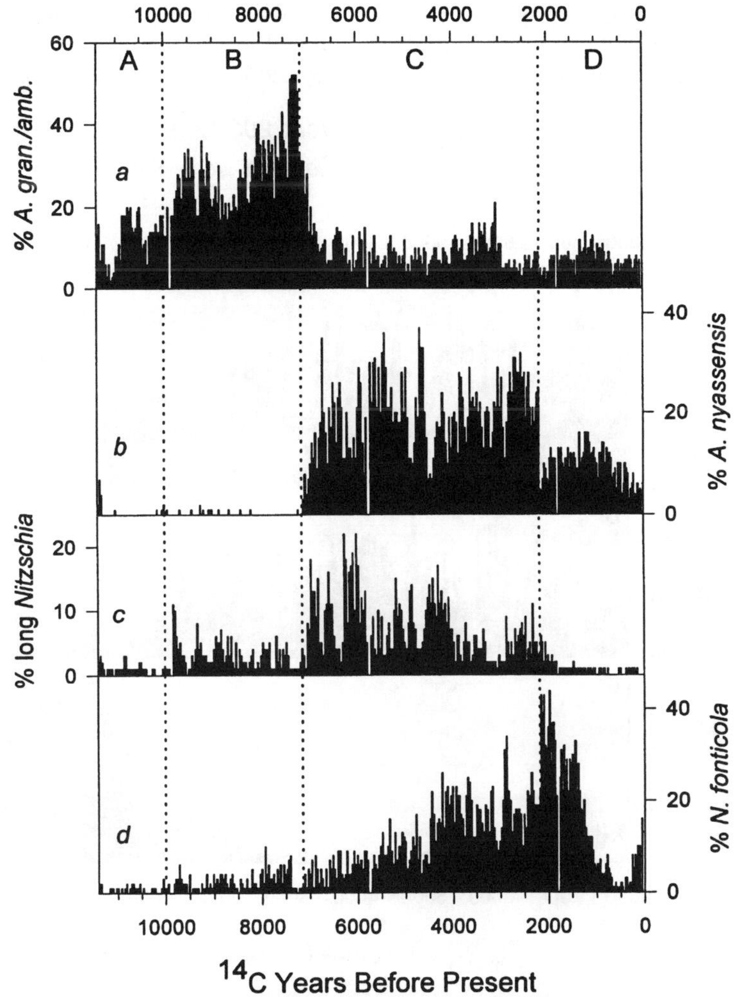

Figure 1. Selected diatom records from Damba Channel core, sampled at 2.5 cm intervals. Uppermost panel (a) combines *Aulacoseira granulata* and *A. ambigua*. Qualitative microfossil assemblage zones A-D after Stager et al. 1997 (see text). Gaps in series represent sediments lost in 3 core pipe couplings (9.8, 5.7, 1.8 ka).

Between 300 and 550 complete diatom valves were counted per subsample, except for the very youngest and oldest samples (n=15) in which diatoms were rare and poorly preserved. Multiple counts of several slides were performed to test counting consistency; percent abundance of major taxa varied by less than 3% between recounts.

4. Results

Below the 5.3 m level (ca. 7.2 ka), *Aulacoseira* populations represented in the Damba Channel core were dominated by *A. granulata* and *A. ambigua* (Fig. 1a, Zone B); these were largely replaced in less than 200 years by *A. nyassensis* var. *victoriae* (Fig. 1b, Zone C). The change coincided with a continental-scale switch from exceptionally humid conditions to more seasonal precipitation and mixing regimes in tropical Africa (Kendall 1969; Nicholson and Flohn 1980; Maley 1982; Stager et al. 1997).

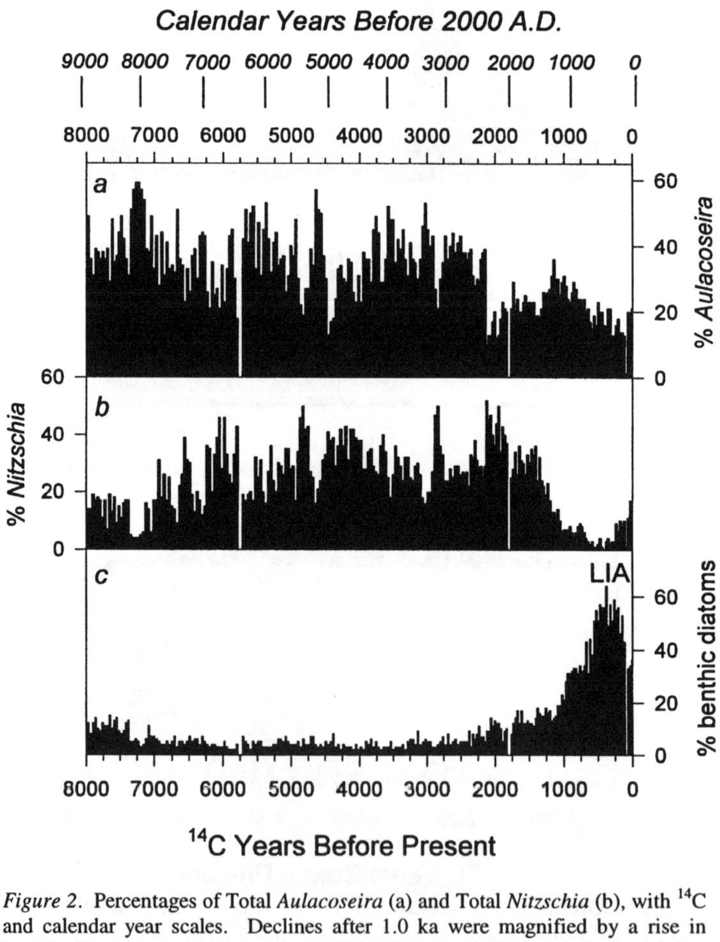

Figure 2. Percentages of Total *Aulacoseira* (a) and Total *Nitzschia* (b), with ^{14}C and calendar year scales. Declines after 1.0 ka were magnified by a rise in benthic, littoral taxa (c) during the arid Little Ice Age (LIA) interval. Sampled at 2.5 cm intervals; gaps in series represent sediments lost in core pipe couplings.

After this climatic switch, the lake supported two abruptly alternating offshore diatom assemblages for most of the following 7.2 ka (Fig. 2a, b). During what appear to have been more thoroughly mixed periods, the dominant taxa were *A. nyassensis* and various euplanktonic *Stephanodiscus* spp. (mostly cf. *S. astraea* varieties, *sensu* Stager 1984). During presumably more stabilized conditions, long *Nitzschia* (Fig. 1c), *N. fonticola* (Fig. 1d) and *Stephanodiscus* spp. predominated.

Major *Nitzschia* peaks averaged 37% (range 30-49%), typically lasted less than a century, and were largest between 4.5 and 1.8 ka. The peaks developed abruptly, with Total *Nitzschia* proportions rising on average by 45% in less than 35 years (range 17-140%) and terminating sharply with average declines of 24% (range 7-56%) in less than 35 years. The largest Total *Nitzschia* peaks in the Damba Channel core are listed in Table 1. The chronology of these fluctuations should be considered tentative until supported by additional high-resolution records.

TABLE 1. Largest peak percentages of Total *Nitzschia* in the Damba Channel record, with sample depths and tentative time intervals. Calendar year dates are relative to 2000 A.D.

Depth (m)	Radiocarbon Year	Calendar Year	Percent *Nitzschia*
3.49-3.44	4490-4420	4960-4880	40
3.36-3.24	4310-4120	4750-4550	42
2.94	ca. 3680	ca. 4050	40
2.41-2.39	2910-2870	3190-3150	49
1.84-1.76	2150-2060	2300-2260	46
1.71-1.69	2000-1970	2200-2160	47
1.64-1.61	1910-1880	2100-2070	42

Over the last 2,000 years, deposition of littoral diatom taxa increased in the Damba Channel, probably by transport from increasingly proximate coastal areas (Zone D in Fig. 1, and Fig. 2c). *Nitzschia* and *Aulacoseira* percentages both declined as benthic diatoms (Fig. 2c; including *Fragilaria pinnata, F. construens, Navicula scutelloides, Cocconeis placentula, Epithemia zebra*), and green algae (*Botryococcus, Pediastrum*) increased. The uppermost portion of the record shows the beginning of an apparent return to *Nitzschia/Aulacoseira*-dominated assemblages during the last century or so (Fig. 2a, b).

5. Discussion

With the evolution of paleolimnological techniques, temporal resolution of African microfossil records has greatly increased. As a result, environmental conditions are known to fluctuate more frequently and abruptly than was once commonly assumed (Anderson 1995). Using *Aulacoseira* percentage as a proxy for environmental change, Figure 3 illustrates this process in the case of Lake Victoria. Kendall's (1969) coarse-interval diatom record from Pilkington Bay provided evidence of general long-term climatic instability in tropical Africa in supplement to his detailed pollen stratigraphy

(Fig. 3a). Preliminary analysis of the Damba Channel core, subsampled at 10 to 30 cm increments, later showed that diatom communities had been even less stable than previously indicated (Fig. 3b; Stager 1984). However, lack of temporal resolution precluded evaluation of the rates and durations of these changes. Fine-interval resampling of the Damba Channel core now shows that phytoplankton community structure fluctuated rapidly and often in the past, on time scales similar to those observed since 1960 (Fig. 3c).

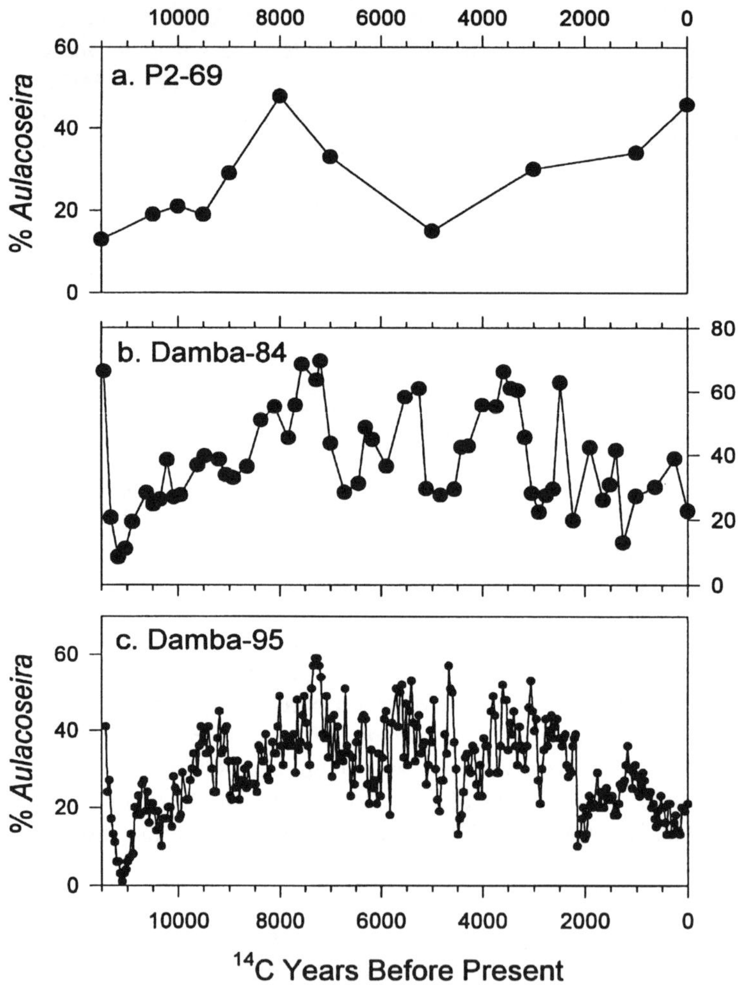

Figure 3. Comparison of sampling resolutions for diatom stratigraphies from Lake Victoria. (a) From core P-2, Pilkington Bay (Kendall 1969), (b) from Damba Channel core (Stager 1984), (c) from Damba Channel core, this study.

Pollen microfossil data from northern Lake Victoria document a long-term pattern of increased rainstorm seasonality since the mid-Holocene (Kendall 1969), and the concurrent increase of *N. fonticola* in the Damba Channel core suggests that cyanobacterial productivity likewise increased during that interval (Fig. 1d, Zones C and D). Diatom valve abundance in the Damba Channel core declined markedly over the last 5,000 yr, as did pollen deposition at Pilkington Bay (Kendall 1969, Stager 1984). Declines in abundances of diatom and pollen microfossils were not by inorganic sediment dilution, as sedimentation rates and organic contents at both sites remained fairly stable. The pattern is consistent with documented transitions from diatom- to cyanobacteria-dominated populations in other lakes which experience reduced mixing effectiveness (Reynolds 1980). Land clearance in Uganda is thought to have had significant impacts on the region only after ca. 2.2 ka (Taylor 1990), so anthropogenic alterations of local vegetation are unlikely to explain the entire post-5.0 ka pollen decline. The microfossil data suggest both diatom productivity and pollen deposition have indeed declined over the long term, possibly reflecting a reduction of lake mixing and offshore pollen transport related to a reduction of wind strength or duration.

This study shows that abrupt changes in phytoplankton community composition, probably reflecting climatic fluctuations, have been common since the mid-Holocene. Around 2.2 ka, for example, *Aulacoseira* populations declined sharply and *N. fonticola* rose to ~40% of total diatom abundance in less than 30 yr (Fig. 1b, d); percentages of *N. fonticola* may have been even higher for brief periods during the 1 to 2 decades represented in each of the 1 cm thick sediment subsamples. Preliminary evidence from an offshore core suggests that this may have been a lake-wide event (Stager, unpublished data). Thus, recent disruption of phytoplankton communities, while unique in particulars, has been foreshadowed repeatedly since the mid-Holocene.

Spectral analyses of time series in the Damba Channel core provide supporting evidence of climatic forcing behind many of Lake Victoria's Holocene diatom community fluctuations (Stager et al. 1997). Strong periodicities significant at the 99.9% statistical test level in *Aulacoseira* and *Nitzschia* profiles explain 53% and 39% of the variation in those records, respectively. Several of the periodicities, such as the ~500, ~1100, ~1400 and ~2,350-2,550 year cycles which also occur in other paleoclimatic records around the world (Damon and Jirikowic 1992; Magny 1993; O'Brien et al. 1995; Mayewski et al. in press), point to global-scale forcing.

Humans played additional roles in the last 2,200 years or more of environmental change in the Victoria basin, although the history of anthropogenic land clearance in the region remains unclear (Taylor 1990). Climate has continued to set the pace of much change in the phytoplankton communities of the Damba Channel, as evidenced by an increase of benthic diatoms (Fig. 2c) in response to a probable lake level decline during the arid Little Ice Age interval ca. 0.6-0.1 ka, which also appears elsewhere in East Africa as a cool or arid signal (Hamilton 1982; Bonnefille and Mohammed 1994).

Climatic events may likewise have initiated the recent switch from *Aulacoseira*-dominated to *Nitzschia*/cyanobacteria-dominated communities. East African weather during the mid-1960s was unusually wet and was followed by a pronounced reduction in precipitation (Nicholson 1994) that presumably reduced the strength or duration of storm-driven mixing in Lake Victoria. Recent anthropogenic nutrient inputs and trophic effects may have amplified and prolonged weather-induced water column stabilization. For example, thermal absorption by dense cyanobacterial blooms might have helped to stabilize water columns by raising epilimnetic temperatures, and hypolimnetic anoxia might have stimulated phytoplankton productivity by enhancing nutrient release from sediments.

6. Conclusions

This paleolimnological record shows that Lake Victoria's water column structure and phytoplankton communities have been dynamic for thousands of years, shifting abruptly between well-mixed and more stabilized states in response to climatic fluctuations. Recent anthropogenic nutrient inputs and trophic effects may amplify and prolong the lake's responses to natural climatic variability. It will be interesting to note the relative strengths of anthropogenic and climatic influences on the lake and its communities should local weather conditions again shift to favor greater water column mixing. Environmental scientists and fishery professionals should consider effects of inevitable climatic variability on the lake ecosystem when developing future models.

7. Acknowledgments

Special thanks to S. Grimm, R. Hecky, K. Kelts, H. Kling, T. Johnson, K. Laird, J. Lehman, D. Livingstone, P. Mayewski, L. Meeker, Paul Smith's College and J. Smol for technical support and valuable discussions.

8. References

Anderson, N. J. 1995. Temporal scale, phytoplankton ecology and paleolimnology. Freshwater Biology **34**: 367-378.

Bard, E., M. Arnold, R. G. Fairbanks, and B. Hamelin. 1993. ^{230}Th-^{234}U and ^{14}C ages obtained by mass spectrometry on corals. Radiocarbon **35**: 191-199.

Bonnefille, R., and U. Mohammed. 1994. Pollen-inferred climatic fluctuations in Ethiopia during the last 3000 years. Palaeogeography, Palaeoclimatology, Palaeoecology **109**: 331-343.

Damon, P. E., and J. L. Jirikowic. 1992. Solar forcing of global climate change?, Chapter 9. *In* R. E. Taylor, A. Long, and R. S. Kra [eds.], Radiocarbon after four decades: an interdisciplinary perspective. Springer-Verlag.

Dixit, S. S., and J. P. Smol. 1994. Diatoms as indicators in the Environmental Monitoring and Assessment Program - Surface Waters (EMAP-SW). Environmental Monitoring and Assessment **31**: 275-306.

Gasse, F., S. Juggins, and B. Khelifa. 1995. Diatom-based transfer functions for inferring past hydrochemical characteristics of African Lakes. Palaeogeography, Palaeoclimatology, Palaeoecology **117**: 31-54.

Gophen, M., P. B. O. Ochumba, U. Pollinger, and L. S. Kaufman. 1993. Nile perch (*Lates niloticus*) invasion of Lake Victoria (East Africa). Verhandlungen der Internationale Vereinigung für Limnologie **25**: 856-859.

Haberyan, K. A., and R. E. Hecky. 1987. The late Pleistocene stratigraphy and paleolimnology of Lakes Kivu and Tanganyika. Palaeogeography, Palaeoclimatology, Palaeoecology **61**: 169-197.

Haberyan, K. A., and O. K. Mhone. 1991. Algal communities near Cape Maclear, southern Lake Malawi, Africa. Hydrobiologia **215**: 175-188.

Halfman, J. D., T. C. Johnson, and B. P. Finney. 1994. New AMS dates, stratigraphic correlations and decadal climatic cycles for the past 4 ka at Lake Turkana, Kenya. Palaeogeography, Palaeoclimatology, Palaeoecology **111**: 83-98.

Hamilton, A. C. 1982. Environmental History of East Africa. A study of the Quaternary. Academic Press.

Hecky, R. E. 1993. The eutrophication of Lake Victoria. Verhandlungen der Internationale Vereinigung für Limnologie **25**: 39-48.

Hecky, R. E., F. W. B. Bugenyi, P. Ochumba, J. F. Talling, R. Mugidde, M. Gophen, and L. Kaufman. 1994. Deoxygenation of the deep water of Lake Victoria, East Africa. Limnology and Oceanography **39**: 1476-1481.

Johnson, T. C., C. A. Scholz, M. R. Talbot, K. Kelts, G. Ngobi, K. Beuning, I. Ssemmanda, and J. A. McGill. 1996. Late Pleistocene desiccation of Lake Victoria and rapid evolution of cichlid fishes. Science **273**: 1091-1093.

Kendall, R. L. 1969. An ecological history of the Lake Victoria basin. Ecological Monographs **39**: 121-176.

Kilham, P., S. S. Kilham, and R. E. Hecky. 1986. Hypothesized resource relationships among African planktonic diatoms. Limnology and Oceanography **31**: 1169-1181.

Lehman, J. T., and D. K. Branstrator. 1993. Effects of nutrients and grazing on the phytoplankton of Lake Victoria. Verhandlungen der Internationale Vereinigung für Limnologie **25**: 850-855.

Magny, M. 1993. Solar influences on Holocene climatic changes illustrated by correlations between past lake-level fluctuations and the atmospheric 14C record. Quaternary Research **40**: 1-9.

Maley, J. 1982. Dust, clouds, rain types, and climatic variations in tropical North Africa. Quaternary Research **18**:1-16.

Mayewski, P. A., L. D. Meeker, M. S. Twickler, S. I. Whitlow, Q. Yang, and M. Prentice. in press. Major features and forcing of high latitude northern hemisphere atmospheric circulation over the last 110,000 years. Journal of Geophysical Research.

Mugidde, R. 1993. The increase in phytoplankton primary productivity and biomass in Lake Victoria (Uganda). Verhandlungen der Internationale Vereinigung für Limnologie **25**: 846-849.

Nicholson, S. E. 1994. Recent rainfall fluctuations in Africa and their relationship to past conditions over the continent. The Holocene **4**:121-131.

Nicholson, S. E., and H. Flohn, 1980. African environmental and climatic changes and the general atmospheric circulation in late Pleistocene and Holocene. Climatic Change **2**: 313-348.

O'Brien, S. R., P. A. Mayewski, L. D. Meeker, D. A. Meese, M. S. Twickler, and S. I. Whitlow. 1995. Complexity of Holocene climate as reconstructed from a Greenland ice core. Science **270**:1962-1964.

Ochumba, P.B.O., and D. I. Kibaara. 1989. Observations on blue-green algal blooms in the open waters of Lake Victoria, Kenya. African Journal of Ecology **27**:23-34.

Owen, R. B., and R. Crossley. 1992. Spatial and temporal distribution of diatoms in sediments of Lake Malawi, Central Africa, and ecological implications. Journal of Paleolimnology **7**: 55-71.

Renberg, I. 1990. A 12,600 year perspective of the acidification of Lilla Öresjön, southwest Sweden. Philosophical Transactions of the Royal Society of London B **327**: 357-361.

Reynolds, C.S. 1980. Phytoplankton assemblages and their periodicity in stratifying lake systems. Holarctic Ecology **3**: 141-159.

Stager, J.C. 1984. The diatom record of Lake Victoria (East Africa): the last 17,000 years, p. 455-476. *In* D. G. Mann [ed.], Proceedings of the Seventh International Diatom Symposium. Strauss & Cramer.

Stager, J. C., B. Cumming, and L. Meeker. 1997. A high-resolution 11,400-year diatom record from Lake Victoria, East Africa. Quaternary Research **47**: 81-89.

Stuiver, M. 1970. Long-term C14 variations, p. 197-213. *In* I. U. Olsson [ed.], Radiocarbon variations and absolute chronology. Proceedings of the Twelfth Nobel Symp. Wiley.

Stuiver, M., and P. J. Reimer. 1993. Extended 14-C data base and revised CALIB 3.0 14-C age calibration program. Radiocarbon **35**: 215-230.

Talling, J.F. 1966. The annual cycle of stratification and phytoplankton growth in Lake Victoria (East Africa). Internationale Revue der gesamten Hydrobiologie **51**: 545-621.

Talling, J.F. 1957. Some observations on the stratification of Lake Victoria. Limnology and Oceanography **2**: 213-221.

Taylor, D. M. 1990. Late Quaternary pollen records from two Ugandan mires: evidence for environmental change in the Rukiga Highlands of southwest Uganda. Palaeogeography, Palaeoclimatology, Palaeoecology **80**: 283-300.

INITIAL RESULTS OF 1995 IDEAL SEISMIC REFLECTION SURVEY OF LAKE VICTORIA, UGANDA AND TANZANIA

C. A. SCHOLZ
Rosenstiel School of Marine and Atmospheric Science
University of Miami, Miami, FL 33149-1098, USA

T. C. JOHNSON
Large Lake Observatory, University of Minnesota-Duluth
Duluth, MN 55812, USA

P. CATTANEO
Rosenstiel School of Marine and Atmospheric Science
University of Miami, Miami, FL 33149-1098, USA

H. MALINGA
Uganda Petroleum Exploration and Production Department
P.O. Box 9, Entebbe, Uganda

S. SHANA
Tanzania Petroleum Development Corporation
P.O. Box 5233, Dar es Salaam, Tanzania

1. Abstract

Digital airgun seismic reflection data acquired on Lake Victoria during the 1995 IDEAL field program suggest evidence of major lake level fluctuations in the mid-late Quaternary. At least five discrete depositional packages can be clearly delineated in the upper 40-50 m of the sedimentary section, and are defined on the basis of stratal relationships and marked variations in acoustic character. The uppermost sequence boundary is observed lake-wide as a high-amplitude reflection, and is overlain by a very low-amplitude 3-9 m thick sequence. The high-amplitude reflection correlates with a soil horizon observed in sediment cores, which together suggest complete desiccation of the lake near the end of the Pleistocene. The lowermost sequence observed on the single-channel data is thickest on the far western margin of the lake, and thins dramatically to the east, the site of the modern depocenter. The timing of deposition of the older sequences and of the shift of the basin depocenter from west to east is unconstrained, but likely coincided with the uplift of the western rift shoulder that occurred during the mid-late Pleistocene. Multichannel seismic data show that unlike the extremely smooth, flat, and continuous nature of the lake floor, the basement

J.T. Lehman (ed.), Environmental Change and Response in East African Lakes, 47–57.

topography underlying the lacustrine sediments is highly variable and irregular. Basement is typically observed between about 200 and 600 m below the modern lake surface on most seismic profiles. During this survey, approximately 1950 km of single channel and multichannel seismic reflection data were acquired in a series of transects across broad reaches of the lake basin.

2. Introduction

Lake Victoria is the largest lake in Africa, and is the world's third largest lake by area. Despite its great size, it ranks only eighth in total volume (Herdendorf 1990), on account of its shallow maximum depth of about 70 m (Fig. 1). It is distinct among the large lakes of Africa because it occupies a broad cratonic sag, rather than a narrow rifted trough such as is the case with Lakes Malawi and Tanganyika. It is renown for its extensive endemic lacustrine faunas (e.g., Meyer et al. 1996), and ranks behind only Lakes Baikal, Malawi and Tanganyika in terms of number and diversity of endemic species. The lake straddles the equator (Fig. 1), and thus experiences an "endless summer", resulting in high year-round primary productivity (Wetzel 1983). Because of its high primary production and its great expanse, the Victoria lake floor probably receives more organic carbon annually than any other large lake system. The lake is the source of the White Nile, and climatic and tectonic perturbations in the lake system probably had far reaching effects elsewhere in the Nile drainage basin.

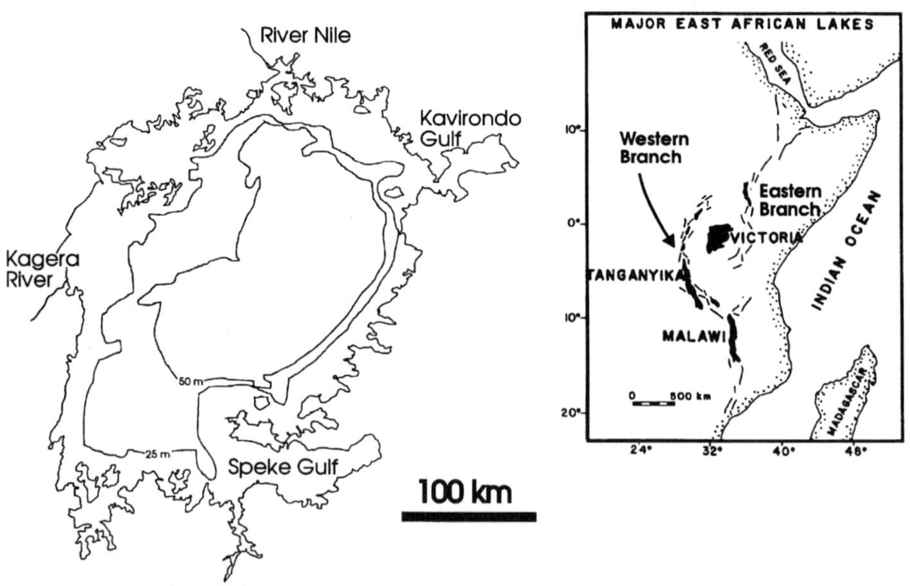

Figure 1. Bathymetric map of Lake Victoria. Contour interval = 25 m. Locations of major East African lakes, and the Western and Eastern Branches of the East African Rift System are shown at right.

The formation of the lake has long been attributed to the uplift along the margins of the Eastern and Western Branches of the East African Rift System (Bishop and Trendall 1967). Consequently the history of the lake is also the history of drainage development on the margins of these rift branches. This paper presents the initial results of the IDEAL 1995 seismic reflection survey on the Ugandan and Tanzanian parts of Lake Victoria, and discusses these results in the context of regional tectonic and climatic events.

3. Geological and Tectonic Framework

Lake Victoria is located in a broad topographic depression, atop part of the Tanzania craton, between the Eastern and Western Branches of the East African Rift System (Fig. 1). It has been postulated that East African rifting has followed zones of weaker crust in several Proterozoic Mobile Belts, and has largely circumvented the Archean Tanzania craton (e.g. Nyblade et al. 1996). Numerous authors report evidence for an ancient east-to-west flowing drainage system across East Africa, possibly dating back to the Mesozoic (Bishop and Trendall 1967). The drainage divide of this system is reported to be near the present day Kenya rift. The west-flowing drainage may pre-date the broad topographic super-swell of eastern Africa, which formed when the African plate came to rest atop a series of mantle hot spots about 30 million yr before present (Burke 1996).

Stratigraphic work in the middle part of this century documented the complex development of a series of regional, and in some cases continent-wide, peneplains or erosion surfaces (e.g., Dixey 1943; Bishop and Trendall 1967; Temple and Doornkamp 1970), which are related to the evolution of the broad continental swell. Unfortunately, the age constraints on these surfaces are weak, except for the latest Pleistocene and Holocene events. Of these, the Buganda (oldest, pre-Miocene or Mesozoic) and Kyoga or Tanganyika (early-mid Tertiary) surfaces are most widespread and universally accepted (Bishop and Trendall 1967; Temple and Doornkamp 1970).

The bedrock geology surrounding Lake Victoria is dominated by Proterozoic and early Paleozoic rocks mainly affiliated with the Bukoban, Irumide, Ruzizian, and Mozambique orogenic belts. Archean-age rocks are also found mostly in the southeast, and form part of the Tanzania craton. Karroo-age sediments are also observed in many localities within the drainage basin, mainly associated with Mesozoic-age rifts. In the eastern part of the drainage basin are Miocene-age and younger volcanics and volcano-clastics associated with rifting and volcanism of the Eastern Branch of the East African rift (Kent 1944).

The shoreline of Lake Victoria is highly crenulated, especially in the north, east, and south, suggesting drowning in the recent geological past. The Kavirondo (Winam) and Speke Gulfs display widespread evidence of basement faulting such as is seen in the Western and Eastern Branch lakes (Darracott 1974; Mboya 1984; Scholz et al. 1990). The river systems on the western margin of the lake show a distinctive "barbed arrow" drainage pattern, and are documented by many workers (see summary by Bishop and Trendall 1967). This pattern suggests the reversal of these drainages, and combined with the drowned shoreline morphology of much of Lake Victoria, indicates a major uplift of about 700 m of the western rift margin and an eastward shift in the location of Lake Victoria in the Pleistocene (Temple and Doornkamp 1970). The initial uplift was slow, with the west-flowing rivers simply incising into the uplifting terrain. In the mid-late Pleistocene more rapid uplift caused the actual reversals of the river systems, resulting in their present-day eastward flows.

Superimposed on this tectonically-induced shift are climatically driven fluctuations in lake level. Previous workers have documented falls in the level of the lake at the end of the Pleistocene and during the Holocene (Kendall 1969; Stager et al. 1986). Analyses of short sediment cores indicate that the lake level fell to 26 m depth or less between 14,750 and 13,700 yr B.P. (Stager et al. 1986).

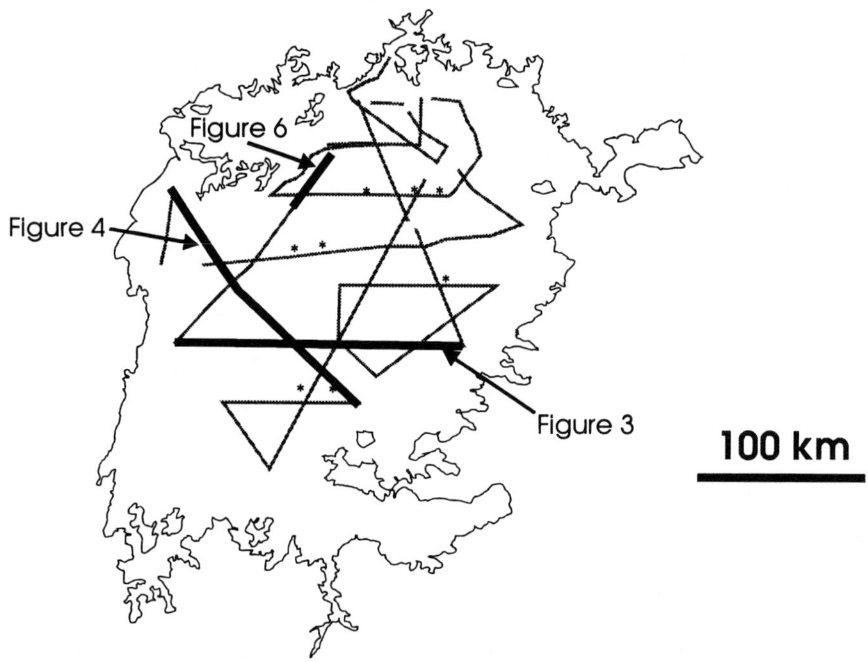

Figure 2. Trackline map from 1995 seismic reflection survey of Lake Victoria. Lines presented in this chapter are in bold. * = locations of sediment cores.

4. Methods

About 1950 km of single-fold and 1050 km of multi-fold seismic reflection data were acquired during the 1995 IDEAL cruise on Lake Victoria aboard the *R/V Ibis*, operated by Uganda Fisheries Research Institute (Fig. 2). We used a small air gun (1 in^3, Bolt 600B) for the seismic source. Two seismic receiver and acquisition systems were deployed during the cruise. One system consisted of a single-channel ITI ST-5 streamer coupled to an ELICS Delph2 acquisition computer. The ITI streamer contained 10 hydrophones with a 1-m spacing. The other system consisted of a 150 m,

24-channel Teledyne towed array with 6.25 m group interval and 50 m lead-in cable, for a total offset of 200 m. Acoustic signals were digitally acquired using an OYO DAS-1 recording instrument. The two different streamer-acquisition instrument systems were optimized for different imaging objectives, although they used the same seismic source system. The single-channel data were processed to image the details of the shallow subsurface stratigraphy, whereas the multichannel data were best suited to deeper imaging of the base of the lacustrine section, interpreted to be a crystalline basement erosive surface. Shot spacing was 12.5 m on lines with coincident SCS and MCS acquisition. MCS data are nominally 6-fold. Position data were acquired about every 10 m using a Global Positioning System navigation unit operating in autonomous mode.

Single-channel data processing consisted of gain manipulation, 10-trace horizontal stack, and time-variant filtering of the data over a total record length of 250 ms. The multichannel processing flow consisted of trace editing, sort to 30-trace supergather, velocity analysis, pre-stack mute, normal moveout correction and stack, post-stack predictive deconvolution, automatic gain control application, and time-variant filter. Total record length of the multichannel data was 750 ms. Thirty seven lines of SCS data were acquired, totaling 1950 km, and 18 lines (1050 km) of MCS were collected (Fig. 2).

Profiles are presented in two-way travel time of sound, rather than in depth. We assume that the velocity of sound is about 1500 m s^{-1} in the uppermost part of the stratigraphic section. Thus on the high resolution single-channel records presented here, 1 ms of two-way travel time in the sedimentary section equates to approximately 0.75 m. In order to discern subtle variations in the subsurface stratigraphy, profiles are displayed here with considerable vertical exaggeration.

5. Results of Seismic Reflection Profiling

The airgun seismic records collected from Lake Victoria in 1995 reveal several episodes of sedimentation, each correlating to a discrete eustatic or tectonic event in the lake's recent past. Initial interpretations of the seismic data were made taking a sequence stratigraphic approach (e.g., Vail 1987; van Wagoner et al. 1987). At least five depositional sequences have been identified in the single channel seismic records, and a clear crystalline basement arrival is observed on multichannel seismic profiles.

This is the most comprehensive survey to date of the lake's offshore regions. The best prior bathymetric information is from the original British Admiralty survey, collected during the early part of the 20th century. One of the most notable aspects of the new data set is that it reveals the Victoria lake floor to be extremely level and smooth over most of the lake (Fig. 3). Few faults cross the main body of the lake, and the only significant relief on the lake floor is associated with submerged bedrock highs or inselbergs. These are primarily observed in the southern part of the seismic grid, north of Ukerewe Island, and in the northern part of the lake.

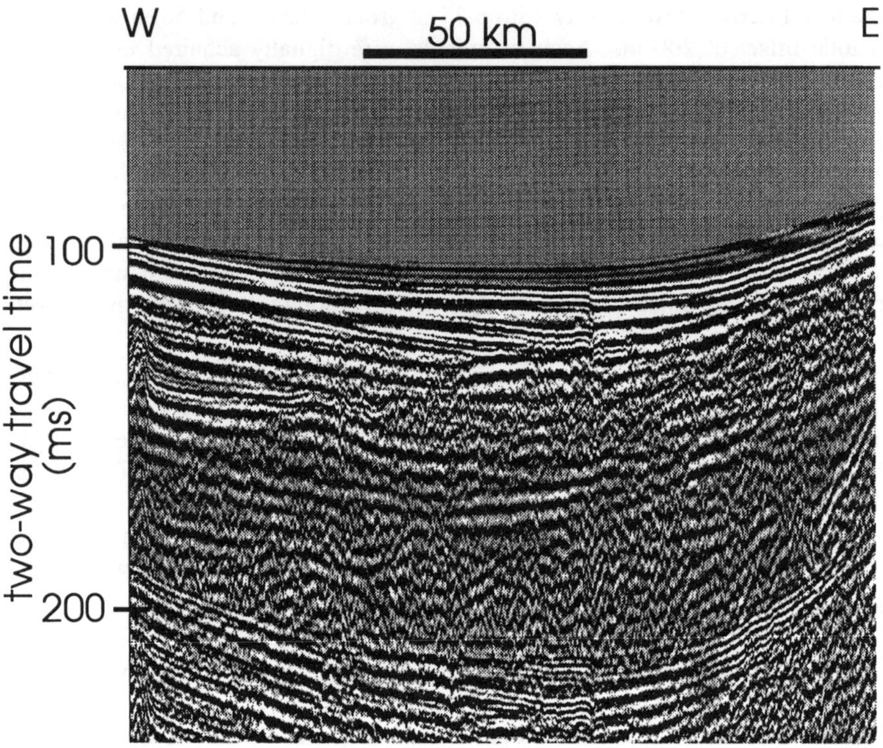

Figure 3. Single-channel seismic profile V14A/B across the central part of Lake Victoria.

A low-amplitude "acoustically transparent" sequence (Sequence A) is observed above a high-amplitude reflection in many areas (Figs. 3, 4). This distinctive sequence averages 6 to 12 ms (ca. 4.5 to 9 m) in thickness, with greater accumulations in shallow depressions between islands, and with maximum accumulations in the northeast part of the lake (Fig. 5). Several sediment cores recovered during the 1995 field program (Fig. 2) penetrate through the upper sequence. Magnetic susceptibility profiles from several of these sediment cores indicate a marked increase in magnetic susceptibility near the base of the cores (Johnson et al. 1996; Ngobi et al. 1998). Additionally, water content decreases and wet bulk density values are higher by a factor of 2 or more below the A-B interface. The lithologic observations (Johnson et al. 1998) and physical property data are consistent with the interpretation of the high-amplitude reflector as a soil horizon and desiccation surface. This high-amplitude reflection is observed on virtually all seismic lines. Whereas maximum accumulations of Sequence A are observed in the northeast part of the lake, the sequence thins and disappears in the southwestern corner of the basin.

Underlying sequences display significantly different character than Sequence A described above. All lower sequences consist of high-amplitude, high-continuity reflections, which show relatively little variability across the basin. The subtle variations in the external form of the individual sequences become clearly evident only on profiles with extreme vertical exaggeration (e.g., Figs. 3, 4). Sequences B and C follow similar patterns as in Sequence A, thickening toward the northeast part of basin. Sequence D shows a markedly different character and external geometry, however. This sequence has a higher frequency content than the other sequences, possibly indicating a section of alternating sands and muds. Most significantly, the sequence is thickest on the western margin of the lake (Fig. 4). On lines 9A and 16A the sequence is more than 35 ms (ca. 26 m) thick in the west, thinning to zero in the southeast.

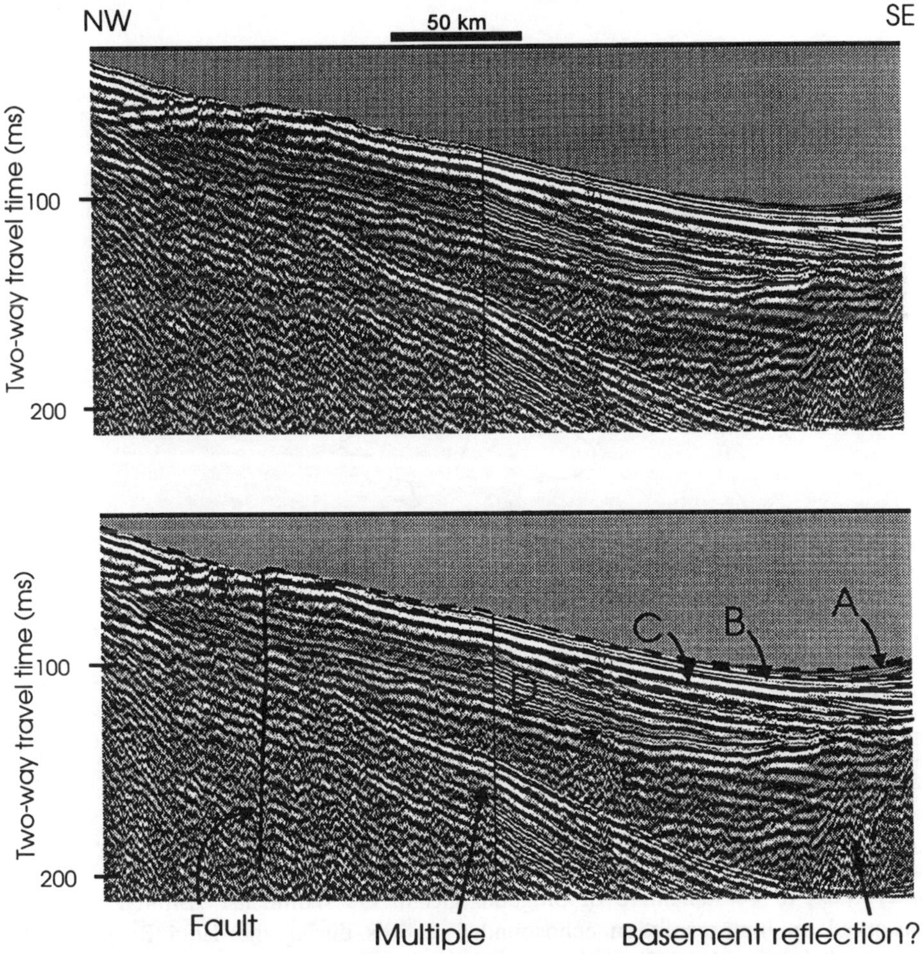

Figure 4. Seismic profiles V95-9A and -16A across Lake Victoria from NW to SE. Letters correspond to depositional sequences identified on single channel seismic data.

Multichannel seismic profiles were acquired and processed in order to optimize the images of pre-lacustrine crystalline basement. This approach was successful (e.g., Fig. 6), although the fine stratigraphic details seen in the single-channel profiles were obscured in these records. The most important initial result of the analysis of the multichannel records is the observation of the considerable relief and highly irregular topography of the basement underlying the horizontal lacustrine sediments. Basement is observed between 200 and 600 ms (ca. 200 to 550 m) on the sections, and the considerable relief suggests an ancient erosive surface and drainage network. Assuming an average velocity of sound in the sediments of 1800 m s^{-1} for the bulk of the lacustrine sediments, the total lacustrine section thickness is estimated to vary between 100 and 450 m over much of the lake, and locally disappears in the vicinity of submerged inselbergs. Examinations of the MCS continue, and are focused on identifying detailed trends of incised basement valleys across the Lake Victoria basin.

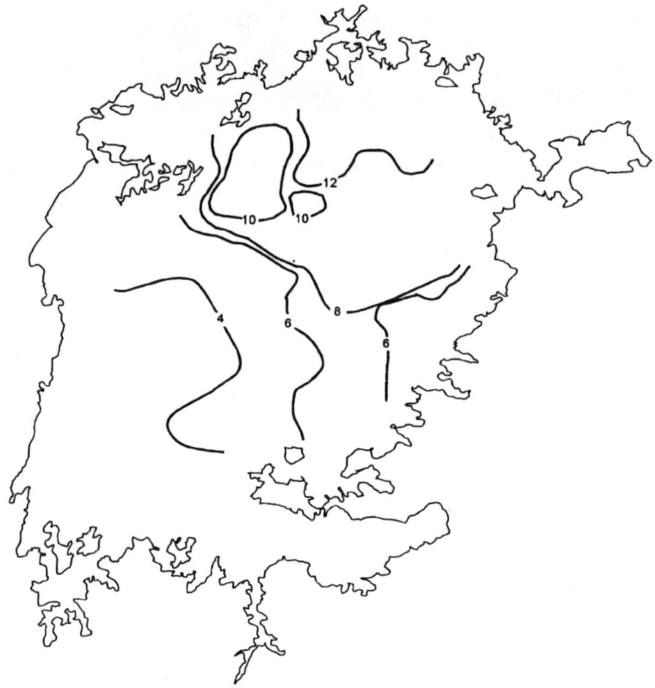

Figure 5. Lake Victoria showing thickness of Sequence A (latest Pleistocene and Holocene sequence). Contours are in ms (two-way travel time). Sound velocity in the sediment is assumed to be 1500 m s^{-1} (10 ms = 7.5 m).

Single-channel seismic data show several regions of elevated noise which are interpreted to be indicative of biogenic gas in the sediments. Similar zones were observed on high-resolution echosounder records during the 1985 Project PROBE survey (Scholz et al. 1990).

6. Discussion

The high-amplitude reflection observed lakewide in this seismic data set correlates well with the soil horizon identified in deep-water sediment cores (Johnson et al. 1996; Ngobi et al. 1998) and with a similar reflection observed in high-resolution echo-sounder records (Scholz et al. 1990). Although the track line spacing of the existing seismic grid is still wide (more than 40 km in some parts of the lake), we interpret these data to be an indication of a lake-wide desiccation surface which developed at the end of the Pleistocene. Dates on the sediment cores indicate the desiccation event occurred before 12,400 ^{14}C yr B.P. (Johnson et al. 1996). Whereas the exact duration of this event is still poorly constrained, it appears to be coincident with the cooler and dryer climatic conditions which were widespread at that time (e.g. Kutzbach and Street-Perrott 1985). Thus this particular sequence boundary is interpreted to be climatically induced.

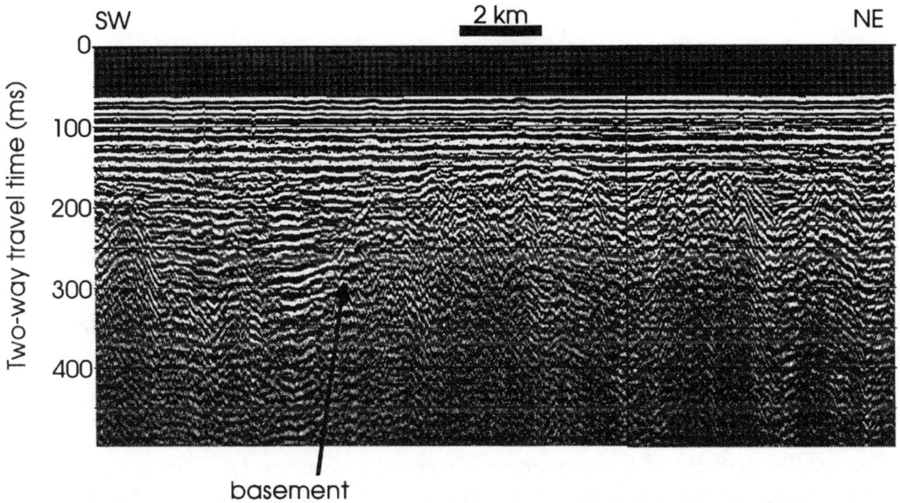

Figure 6. Multichannel seismic profile V95-10C from the northwestern part of Lake Victoria. Note the considerable relief on the reflections, which are interpreted to be crystalline basement.

The evolution of the endemic faunas of the African lakes have been the focus of considerable study and controversy. Lake Victoria contains over 300 endemic species of haplochromine cichlids, more than any lake except Lake Malawi. If our assertion of a long-lived dry phase in the late Pleistocene history of Lake Victoria is correct, then explanations are required for the presence of endemic cichlids and other endemic fishes with open-water or deep-water affinities. Since we see no evidence in the geophysical data for any deep-water refugia at the end of the Pleistocene, many of these species must have evolved since 12,400 yr B.P. (Johnson et al. 1996). DNA analyses of the Victoria species flocks suggest that three or four ancestral lineages may have survived the desiccation event (A. Meyer, personal communication).

The lower sequences observed in the seismic data (Fig. 4) appear to correlate to an earlier phase of Lake Victoria, when the basin depocenter was situated in the western part of the modern drainage system. We tentatively correlate Sequences D and E with the sediments observed in the Katonga Valley, described by Temple and Doornkamp (1970). If this correlation is correct, then the seismic records display evidence for the initial development of Lake Victoria in the west, and the gradual migration of the depocenter to the northeast part of the system. This depocenter shift is interpreted to be a consequence of increased uplift of the western rift margin during the mid-late

Pleistocene (Temple and Doornkamp 1970). Sequence D is more than 35 m thick in the west, thinning to zero in the east. The seismic sequences are separated by erosional truncation surfaces in several areas (Fig. 4). It seems likely therefore that the desiccation surface described between the A and B sequences may be just the most recent episode of several lake level lowerings to have occurred during the mid-late Pleistocene.

Several faults cross the seismic lines in a number of localities. In some cases there is actually surficial relief on the faults (Fig. 3); in most instances offsets of 1 to 2 m are observed on the late-Pleistocene unconformity surface beneath Sequence A.

7. Summary

a) The Victoria lake floor is remarkably smooth, flat, and continuous, with very little relief except for isolated bedrock inselbergs.

b) Single-channel, intermediate resolution, air gun seismic reflection records reveal the presence of a thin, low amplitude depositional sequence over a high-amplitude reflector interpreted to be a lake-wide desiccation surface. This reflector correlates with a soil horizon observed in sediment cores recovered from the deep part of the lake. The desiccation event is thought to have occurred prior to 12,400 yr before present.

c) Five discrete depositional sequences are observed on the single-channel seismic data, each separated by erosional unconformities in marginal areas.

d) One of the deepest of the sequences observed (sequence D) is thickest on the far western margin of the lake, and thins dramatically toward the northeast where the Holocene sediments are thickest. This pattern is interpreted to reflect uplift of the western branch rift shoulder and the shift of the Lake Victoria depocenter to the northeast.

e) Multichannel seismic reflection records indicate that there is considerable relief on the basal lacustrine crystalline basement. Basement reflections are typically observed between about 200-550 m below the modern lake surface. The highly variable relief of the basement reflections are interpreted to indicate an ancient, pre-lake drainage system.

8. Acknowledgments

We thank the governments of Uganda and Tanzania for permission to conduct this research, and for their assistance in carrying out the work. We thank Dr. F. Bugenyi, the staff of the Uganda Fisheries Research Institute, and especially the crew of the *R/V Ibis* for their assistance in conducting the field program. D. Rea and an anonymous reviewer provided constructive reviews. This research was funded by the Earth Sciences Program of the U.S. National Science Foundation, (NSF-EAR 9421565 to CAS).

9. References

Bishop, W. W., and A. F. Trendall. 1967. Erosion surfaces, tectonics, and volcanic activity in Uganda. Quarterly Journal of the Geological Society of London **122**: 385-420.

Burke, K. 1996. The African Plate. South African Journal of Geology **99**: 339-410.

Darracott, B. W. 1974. The structure of Speke Gulf, Tanzania and its relation to the East African rift system. Tectonophysics **23**: 155-175.

Dixey, F. 1946. Erosion and tectonics in the East African rift system. Quarterly Journal of the Geological Society of London **102**: 339-388.

Herdendorf, C. E. 1990. Distribution of the world's large lakes, p. 3-38. *In* M. Tilzer and C. Serruya, [eds.], Large Lakes, Springer-Verlag.

Kendall, R. L. 1969. An ecological history of the Lake Victoria Basin. Ecological Monographs **39**: 121-176.

Kent, P. E. 1944. The Miocene beds of Kavirondo, Kenya. Quarterly Journal of the Geological Society of London **100**: 85-118.

Kutzbach, J. E., and F. A. Street-Perrot. 1985. Milankovitch forcing of fluctuations in the level of tropical lakes from 18 to 0 kyr BP. Nature **317**: 130-134.

Johnson, T. C., C. A. Scholz, M. R. Talbot, K. Kelts, R. Ricketts, G. Ngobi, K. Beuning, I. Ssemmanda, and J. A. McGill. 1996. Late Pleistocene desiccation of Lake Victoria and rapid evolution of cichlid fishes. Science **273**: 1091-1093.

Johnson, T. C., Y. Chan, K. Beuning, K. Kelts, G. Ngobi, and D. Verschuren. 1998. Biogenic silica profiles in Holocene cores from Lake Victoria: Implications for lake level history and initiation of the Victoria Nile, p. 75-88. *In* J. T. Lehman [ed.], Environmental change and response in East African lakes. Kluwer.

Mboya, B. 1984. The genesis and tectonics of the N.E. Nyanza rift valley, Kenya. Journal of African Earth Sciences **1**: 315-320.

Meyer, A., C. M. Montero, and A. Spreinat. 1996. Molecular phylogenetic inferences about the evolutionary history of East African cichlid fish radiations, p. 303-323. *In* T. C. Johnson, and E. Odata [eds.], The limnology, climatology, and paleoclimatology of the East African lakes. Gordon and Breach.

Ngobi, G. N., K. Kelts, T. C. Johnson, and P. A. Solheid. 1998. Environmental magnetism of Late Pleistocene/Holocene sequences from Lake Victoria, East Africa, p. 59-73. *In* J. T. Lehman [ed.], Environmental change and response in East African lakes. Kluwer.

Nyblade, A. A., C. Birt, C. A. Langston, T. J. Owens, and R. J. Last. 1996. Seismic experiment reveals rifting of craton in Tanzania. EOS **77**: 517, 521.

Scholz, C. A, B. R. Rosendahl, J. Versfelt, and N. Rach. 1990. Results of echo-sounding of Lake Victoria. Journal of African Earth Sciences **11**: 25-32.

Stager, J. C., P. N. Reinthal, and D. A. Livingstone. 1986. A 25,000 year history for Lake Victoria, East Africa, and some comments on its significance for the evolution of cichlid fishes. Freshwater Biology **16**: 15-19.

Temple, P. H., and J. C. Doornkamp. 1970. Influences controlling lacustrine overlap along the northwestern margins of Lake Victoria. Zeitschrift für Geomorphologie **14**: 301-317.

Vail, P. R.. 1987. Seismic Stratigraphy Interpretation using Sequence Stratigraphy, p. 1-10. *In* A. W. Bally [ed.], Atlas of Seismic Stratigraphy. American Association of Petroleum Geologists Studies in Geology #27, v. 1.

Van Wagoner, J. C., R. M. Mitchum, Jr., H. W. Posamentier, and P. R. Vail. 1987. Key definitions of sequence stratigraphy, p. 11-14. *In* A. W. Bally [ed.], American Association of Petroleum Geologists Studies in Geology #27, v.1.

Wetzel, R. G. 1983. Limnology, 2nd edition. Saunders.

ENVIRONMENTAL MAGNETISM OF LATE PLEISTOCENE / HOLOCENE SEQUENCES FROM LAKE VICTORIA, EAST AFRICA

GIDEON N. NGOBI
KERRY KELTS
Limnological Research Center, University of Minnesota
Minneapolis, MN 55455, USA

THOMAS C. JOHNSON
Large Lakes Observatory, University of Minnesota
Duluth, MN 55812, USA

PETER A. SOLHEID
Institute of Rock Magnetism, University of Minnesota
Minneapolis, MN 55455, USA

1. Abstract

The East African region has undergone significant landscape modification since the last glacial maximum, and this history is archived in sediments from Lake Victoria. Comparison of systematic variations in sediment magnetic properties with changes in lithic grain size, water content and other proxies provides a sensitive proxy record for evaluation of transport energy and weathering in the drainage basin. Seven cores were recovered from Lake Victoria in 1995 as part of an International Decade for East African Lakes (IDEAL) multidisciplinary study of the lake. Whole core magnetic susceptibility of these seven cores which extend to more than 14,000 years B.P. show a general intra-basin correlation of stratigraphy overlying a firm soil. One core was selected as a representative type core. Volume magnetic susceptibility identified soil horizons below the Holocene muds, and mass normalized susceptibility determined that this soil zone derives from the precursor lake muds with some evidence of diagenesis of the magnetic carrier. Prior to 12,700 yr B.P. the proxies reflect very low water levels with some periods of complete desiccation adequate to form thick crumbly soils, oxidize organic matter and form hematite. During the filling stage to ca. 7,500 yr B.P., magnetic mineral sources remain stable, although lake level fluctuated. There is a general increase in magnetic susceptibility of sediments through the Holocene once the lake overflows. This suggests that although the pre-Holocene magnetic mineral components are soils derived from precursor lake sediments, the Holocene magnetic mineral components are derived from the drainage basin. An anomalous magnetic spike at 180 to 160 cm depth has been identified as increased amounts of the same source material, possibly a major wash-in event. Increase in magnetic properties and lithic grain size in the uppermost 30 cm is probably due to increased soil erosion, a

J.T. Lehman (ed.), Environmental Change and Response in East African Lakes, 59–73.

reflection of human forest clearance activity. Although the Holocene magnetic record has diamagnetic and possibly diagenetic overprint, a 2,000-year oscillation of concentration and composition shows that the record is sensitive to major changes in water balance history.

1. Introduction

Tropical Africa experienced significant climate change within the last glacial maximum, which probably had significant impact on the regional landscape. The Lake Victoria basin was dry, probably covered with savanna grasslands and shrubs, when it began to refill with water around 11,900 radiocarbon years before present (11.9 ka) (Johnson et al. 1996). Use of environmental magnetism in landscape studies has shown that in many different settings the abundance, types and magnetic grain size of magnetic minerals incorporated in sediments have varied in response to climate change (Verosub and Roberts 1995). The glacial - interglacial record in Lake Baikal, southeast Siberia, was inferred from a pattern of magnetic property variations that indicates dilution of magnetite by biogenic production during warm intervals and enhanced eolian input during cold periods (Peck et al. 1994). Rosenbaum and others (1997) have related changes in magnetic properties of sediments from Buck Lake, Cascade range, Oregon, with pollen data and trace element geochemistry to provide sensitivity proxy record of climate change.

Paleomagnetic results obtained by Mothersill (1996) from Pilkington Bay of Lake Victoria indicate that the sediments have good enough magnetic stability to preserve a record of geomagnetic secular variation. The work of Barton and Torgersen (1988) for Lake Turkana showed that the 500-year sediment record was quite uniform magnetically, with little down-core variation in natural remanent magnetization (NRM). Lake Tanganyika records span the last 25 to 5 ka, and show that in spite of low paleomagnetic intensity of the sediments and high variability of sedimentation rate, inter-basin magneto-stratigraphic correlation is possible (Williamson et al. 1991). However, none of these regional records have used variations in magnetic mineral properties to study past climate change.

This study uses rock magnetic properties supported by other proxies including lithic grain size and water content to interpret paleoclimate processes such as weathering, erosion, soil formation, and anthropogenic effects in the Lake Victoria drainage basin.

2. The region and sites

The Lake Victoria basin is a shallow tectonic crustal sag, formed between the eastern and western arm of the Great East African Rift Valley. The system between the East African rift shoulders is regarded as tectonically stable for the reference time scale (13,000 yr B.P. to present). Primary inflows to the basin are from Kagera and Katonga rivers in the west and numerous small rivers in the east. However, over 90 % of the annual water input to the basin is from 1470 mm yr^{-1} of precipitation (Crul 1995). The shoreline along the northern and southern margin of the lake is highly irregular and appears morphologically a drowned landscape, with numerous near-shore islands. The eastern shore has a shallow littoral gradient, this leading to less erosive power and wider swamps filled with recent sediments. The drainage basin on the western part of the lake has a dense drainage pattern, and the lake has a more irregular shoreline and steeper littoral gradient than the eastern shoreline. The western part of the lake is probably the main source of sediments.

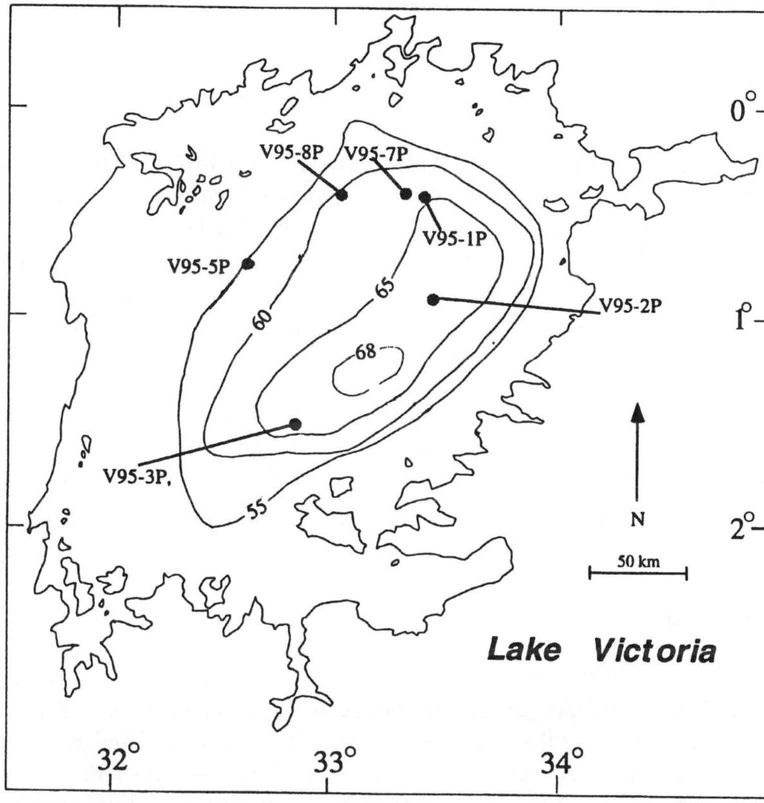

Figure 1. Location map of Lake Victoria, East Africa, showing core sites of the V-1995 IDEAL expeditions. V95-1P refers to a piston core (P) obtained in 1995 at site 1. Bathymetry was updated during a 1996 expedition.

Modern Lake Victoria appears much younger than other rift lakes such as Albert and Tanganyika, and its origin is probably Mid-Pleistocene, some 500,000 years ago. Upwarping of the western rift caused crustal sagging and drainage reversal of the head waters of the previously westward flowing drainage system to the east, filling up the present Lake Victoria basin (Kendall 1969). The basin (68 m) trends NE to SW, which tracks the flexure axis of the crustal sag (Fig. 1). Acoustic stratigraphy does not clearly indicate faults, but on-lap structure of the sediments includes evidence of changing shorelines due to basin downwarping (Johnson et al. 1997 expedition results, in preparation). The underlying Precambrian rocks include massive to weakly foliated granites, quartzite, phyllites, slates and undifferentiated acid gneiss (Geologic map of Uganda, 1963). The geology of the western drainage basin is dominantly rhyolitic tuffs, siltstone, shale and granites (Schluter, unpublished ms.). The northern end of the Lake Victoria basin, in Busoga and Buganda, has exposed red lateritic hills and deeply weathered soils. Laterite is a rigid network of goethite and hematite crystals cemented together into a hard pan. The basement rocks made up of granites and granite gneiss are easily weathered by high rainfall and warm tropical temperatures. Dehydration for hardening may derive from several episodes of aridity before the last glacial maximum.

3. Chronology

Chronology is based on preliminary AMS radiocarbon dates of macro fossils or 40 to 80 μm pollen, and a 1.3 mm size fraction of charcoal from the sediment of core V95-2P. A date of 500 radiocarbon years on the pollen fraction extracted from surface sediments at site V95-2P indicates that a certain amount of older material mixes in with modern sediment and yields radiocarbon dates that are older than the actual age. Eleven dates obtained are corrected by 500 yr (Beuning 1997). The upper 9 dates, obtained from a sporopollen - lignin - charcoal concentrate using a 20 to 70 μm fraction, are subject to residual carbon effect correction. The basal dates are charcoal and emergent sedge, which has terrestrial origin, and are not corrected. The dates from the core 2P indicate a uniform sedimentation rate of 0.69 m kyr^{-1} from 13,240 to 10,400 ^{14}C yr B.P. and a rate of 0.66 m kyr^{-1} from 7,680 to 950 ^{14}C yr B.P. We have a paradoxical offset in the dates between 466 and 487 cm (10,040 to 7,680 ^{14}C yr B.P.).

4. Methodology

Seven piston cores were recovered in 1995 during a one week Uganda Fisheries Research Institute - IDEAL expedition cruise of Lake Victoria aboard the R/V Ibis. Cores of varying lengths (Table 1) were recovered using a modified Kullenberg corer (Kelts et al. 1986) and were retained in their plastic liners and cut into 1 m sections for multidisciplinary study. These cores are now archived at the Limnological Research Center, University of Minnesota.

TABLE 1. Location of cores. Core 4P was lost during coring operations

Core Name	Latitude (S)	Longitude (E)	Total Length (m)	Estimated Age of Base of Core (yr B.P.)
V95 - 1P	00° 27.63'	33° 24.09'	9.09	>13,000
V95 - 2P	00° 58.67'	33° 27.32'	7.51	13,240 ± 80
V95 - 3P	01° 38.58'	32° 51.22'	4.48	>12,000
V95 - 5P	00° 48.05'	32° 37.97'	3.81	~5,000
V95 - 6P	00° 46.04'	32° 49.71'	6.03	>12,000
V95 - 7P	00° 27.70'	33° 21.10'	9.21	8,000
V95 - 8P	00° 28.78'	33° 04.87'	0.85	>13,000 ± 80

The 1-m sections were logged for whole core magnetic susceptibility (MS) using a pass-through Bartington Instruments susceptibility meter. Logging for each section requires about 20 minutes and is accomplished by passing the core in 1 cm steps through the loop sensor; the process provides near-continuous, non-destructive rapid measurements. Volume magnetic susceptibility is obtained in relative units. After the whole core magnetic susceptibility was measured, the cores were split longitudinally, visually described, photographed and immediately sampled for water content and organic carbon content. Microscopic visual core descriptions have been used to define lithologic boundaries. Smear slide samples were taken at regular, 10 cm intervals, using a small amount of sediment and allowed to dry for petrographic aided description of lithology.

Whole core magnetic susceptibility profiles of the 7 cores generally show similar overall intra-basin trends and were used as a basis for selecting core V95-2P as a representative core for detailed rock magnetic and non-magnetic study. Water content, lithic grain size, biogenic silica, carbonate-carbon, and total sulfur were determined for the seven cores. Water content was determined at regular intervals of 10 cm by mass of tared vials filled with 1 cc sediment. Samples were freeze-dried and re-weighed. Grain size analysis samples from the type core were pretreated to remove biogenic silica ($b\mathrm{SiO_2}$) and organic matter, and then measured using a Lasentenc back-scatter analyzer. Detailed descriptions of methodology can be obtained from the LRC core laboratory web site (http://www.lrc.geo.umn.edu; K. Kelts 1997).

A combination of magnetic, petrographic and geochemical methods was used to determine amounts, types, and magnetic grain sizes of magnetic minerals to decipher origins and possible post-depositional alterations of the minerals. The concentration of magnetite was estimated using volume- and mass-normalized magnetic susceptibility. Discrete samples in 5 cm^3 plastic boxes were taken from core V95-2P with a sampling interval of 10 cm. Mass-normalized magnetic susceptibility (Khi), anhysteretic remanent magnetization (ARM), saturation remanent magnetization (SRM), saturation isothermal remanent magnetization (SIRM) and hysteresis measurements were made on core V95-2P. Samples were measured at room temperature for low frequency (430 Hz) magnetic susceptibility using a Geophysica KLY-2 susceptibility bridge. Mass-normalized susceptibility is reported in $m^3\ kg^{-1}$.

Magnetic mineral content was established from isothermal remanent magnetization (IRM; Thompson and Oldfield 1986). IRM was imparted in forward induction of 0.3 T, 1T and a back field of 0.3 T ($IRM_{-0.3T}$) in a magnetizer and measured using a 2 G cryogenic superconducting rock magnetometer. Magnetite saturates below 0.3 T, so that the differences in IRM between 0.3T and 1.0 T are caused by hematite. The ratio ($IRM_{-0.3T}$ /IRM_{1T}) called the S-ratio is a measure of the proportion of magnetite to hematite. High S-ratios indicate large amounts of magnetite relative to hematite and decreasing values indicate increasing proportions of hematite.

Information about the relative magnetic grain size of magnetite was obtained from the magnitude of anhysteretic remanent magnetization (ARM), normalized by a measure of magnetite content (Banerjee et al. 1989). ARM was imparted on discrete samples of core V95-2P with a Schonstedt GSD-1 demagnetizer operating at peak AF field of 50 μT and a steady bias field of 0.1 mT, and measured on a 2 G cryogenic superconducting rock magnetometer. The ratio ARM/X indicates changes in magnetic grain size if the magnetic mineralogy is dominantly magnetite. The ratio ARM/X increases as magnetic grain size decreases and is particularly sensitive to single domain (SD, 30 nm < d < 0.5 μm) and pseudo-single domain (PSD, 0.5 μm < d < 8 μm) size.

Magnetic hysteresis properties were measured for selected intervals of core V95-2P using an alternating gradient force magnetometer to provide information about magnetic grain size distribution down-core. The induced magnetization is sensed by a set of pick-up coils, then amplified and displayed on an X-Y recorder such that the Y-axis displays the magnetization and the X-axis displays the magnetic field. Ratios of hysteresis parameters: saturation remanent magnetization (Mr) to saturation magnetization (Ms), Mr / Ms and coercivity of remanence (Hc) to coercivity (Hc), Hcr / Hc can be used to estimate grain size. Fine SD grains are characterized by Mr/Ms > 0.5 and Hcr /Hc < 1.5, and coarse MD grains are characterized by Mr/Ms < 0.05 and Hcr /Hc > 1.5 (Banerjee et al. 1989). Discrete samples were taken from core V95-2P and air dried to obtain a sample chip (>50 mg) in order to obtain Ms, Mr, Hc and Hcr which are read off the hysteresis loop.

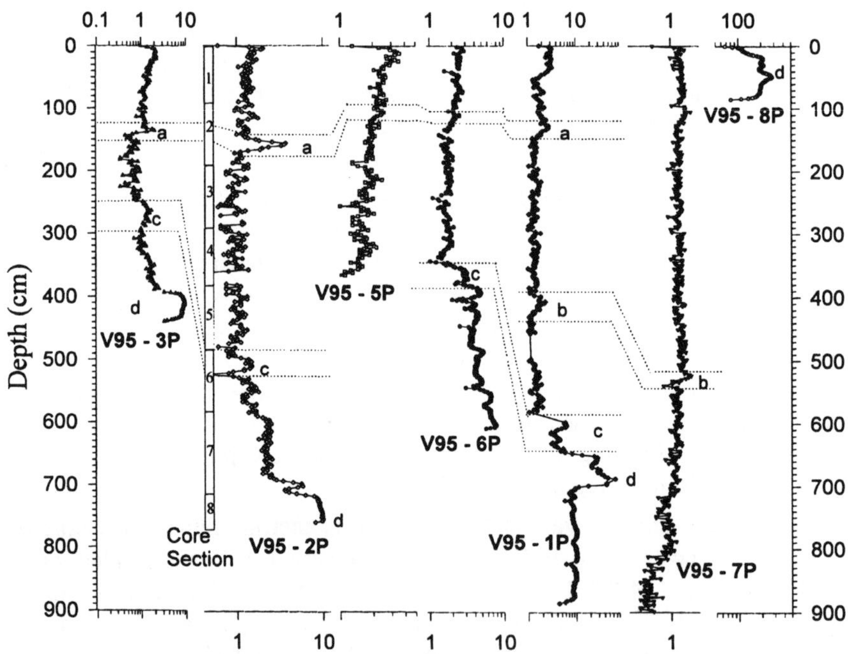

Figure 2. Whole core volume magnetic susceptibility (×10^{-7} relative units). Transect V95-3P to V95-2P cuts across the deepest part of the basin, and represents an almost complete lacustrine stratigraphy.

5. Results and Discussion

Whole core magnetic susceptibility profiles of the 7 cores show similar overall intra-basin trends. Details vary depending on core location (Fig. 2). Cores 5P and 7P are the most difficult to correlate because the long Holocene record possibly has some gaps. Core 8P apparently penetrated directly to pre-Holocene sediments. There is broad coherence among magnetic units which suggests consistent sediment sources throughout the Holocene. This coherence is the basis for selecting core V95-2P for the detailed evaluations presented below. Five magnetic zones have been identified based on the assumption that variations in magnetic mineral concentration (volume MS and mass-normalized MS), composition (S-ratio) and magnetic grain size (ARM / khi) is influenced by climate change.

Zones 5, 4 and 3 represent the late Pleistocene and have about 15% total organic matter (%TOM, estimated as 2.5× %TOC), 15% $b\mathrm{SiO_2}$, and 70% terrigenous sediment content.

TABLE 2. Radiocarbon AMS dates for Core V95 - 2P

Core Section	Absolute depth (cm)	^{14}C age (yr)	Corrected age (yr)	Sample material
2P 1	19	1,450 ± 61	950	Pollen
2P 2	141	3,050 ± 51	2,550	Pollen
2P 3	217	4,110 ± 61	3,610	Pollen
2P 4	330	6,290 ± 61	5,790	Pollen
2P 5	466	8,180 ± 51	7,680	Pollen
2P 6	487	10,540 ± 61	10,040	Pollen
2P 6	508	11,140 ± 61	10,640	Pollen
2P 7	618	12,910 ± 71	12,410	Pollen
2P 8	687	12,400 ± 71	12,400 ± 71	Charcoal
2P 8	709	13,240 ± 82	13,240 ± 82	Charcoal

5.1. ZONE 5

Zone 5 (760-710 cm in the type core), >13,000 ^{14}C yr B.P (Table 2, Fig. 3), shows the highest volume MS, greater than 6×10^{-7} units, in all cores except V95-5P and 7P which do not extend back past 8,000 ^{14}C yr B.P., and may represent the last glacial maximum (LGM) (Fig. 2). The zone implies an episode of sub-aerial exposure and is accompanied by soil formation. Interpretations of sub-aerial exposure is indicated by presence of highest magnetic susceptibility, supported by 50% drop in porosity, vertically oriented rootlets, and prismatic cracking of vertisols; total organic carbon drops to less than 5% with resulting poor pollen preservation. However, mass-normalized magnetic susceptibility does not show a sharp increase from depths 760 to 685 cm. The enhanced whole core magnetic susceptibility (Fig. 3) in this zone is mainly due to compaction and de-watering of lake sediments from pre-Holocene Lake Victoria, and is interpreted as a concentration of the same source material with only minor magnetic enhancement due to soil formation. Magnetic enhancement occurs by formation of micro-cystalline maghemite and magnetite from weakly magnetic oxides and hydroxides via reduction-oxidation cycles under normal pedogenic processes (Mullins 1977). S-ratio is >0.85, and is interpreted to be caused by lower coercivity mineral content such as magnetite or maghemite in this zone. The soil forming processes are influenced by regional climate conditions.

The ARM/X ratio is a concentration-independent proxy, and low values correspond to larger magnetic grain size. However, the interpretation of this ratio may be complicated by significant amounts of superparamagnetic (SP) or paramagnetic material. Zone 5 shows the lowest ARM/X ratio, but magnetic granulometry results do not fall within the standard classification of SD, PSD or MD for depths 760 to 680 cm (samples from section 2P8, Table 3).

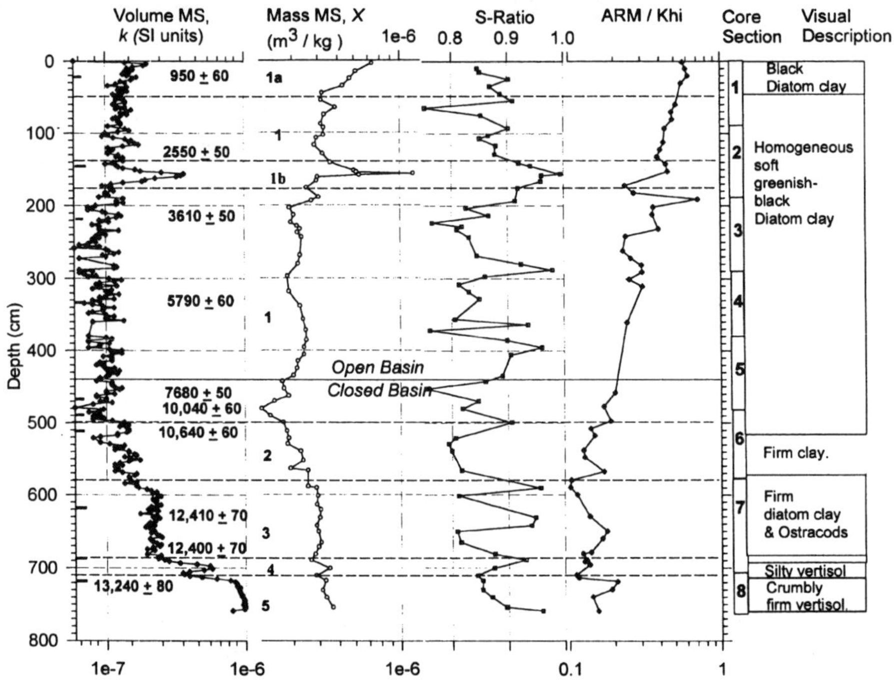

Figure 3. Magnetic properties and visual description of core V95-2P plotted against the core section labels and depth. The numbers 1a, 1b, 2, 3, 4, 5 indicate zones based on volume- and mass-normalized magnetic susceptibility.

5.2. ZONE 4

Zone 4 (710-685 cm), about 13,200 ^{14}C yr B.P. to about 12,000 ^{14}C yr B.P. has high volume MS, ~5.2×10^{-7} units, but almost the same mass normalized MS and no change in ARM/Khi ratio as Zone 5. The S-ratio values peak at this interval (> 0.85). This zone has a porosity increase to about 40%, an increase in lithic grain size from 10 μm to 13 μm, and increases in $b\mathrm{SiO_2}$ from >5% to 15% (Fig. 4). The zone is firm silty greenish black vertisol with extensive shell debris and organic matter including plant cuticle. The layer has an appearance of a matted peat lens and the pollen classification corresponds to typhus swamp, a period probably when the soil horizon in the deepest site is gradually covered with wetland (Beuning 1997). The terrigenous sediment content in this zone is relatively coarse grained magnetite-maghemite interpreted to be caused by detrital reworking of material in a shallow lake as refilling began.

TABLE 3. Hysteresis parameters for core C95-2P. Mr: saturation remanence, Ms: saturation magnetization, Hcr: remanent coercivity, Hc: coercive force. Magnetic zones are those listed in Fig. 3.

Section	core depth (cm)	absolute depth (cm)	Mr /Ms	Hcr/Hc	Magnetic Zone
2P1	6	6	0.20	2.7	1a
	16	16	0.22	2.7	1
	26	26	0.20	2.7	1
	36	36	0.22	9.1	1
	46	46	0.21	3.1	1
2P2	70	162	0.19	2.8	1 b
2P3	9	200.5	0.24	3.4	1
	39	230.5	0.22	3.8	1
2P4	10	293.2	0.01	8.9	1
	50	334.2	0.09	3.1	1
2P5	30	412.7	0.11	47.4	1
2P6	10	488.3	0.11	5.8	1
	30	508.3	0.10	67.7	2
	70	548.3	0.14	54.5	2
2P7	30	609.2	0.19	5.9	3
	70	649.2	0.14	8.2	3
2P8	5	684.2	0.11	5.4	4
	44	732.2	0.17	5.2	5
	80	759.2	0.10	13.0	5

5.3. ZONE 3

Zone 3 (685-580 cm), about 12,000 to about 11,500 ^{14}C yr B.P. has dates that show ^{14}C reversal between 12,910 and 12,400 ^{14}C yr B.P. Although this zone has a significantly lower volume susceptibility value of 1.8×10^{-7} units than zones 5 and 4, mass-normalized susceptibility values do not show a significant change above depths of 685 cm. S-ratio values for Zones 5, 4 and 3 oscillate between high coercivity composition (hematite-goethite) and low coercivity composition (magnetite-maghemite) at regular intervals of 50 cm between peaks. Lithic grain size for these firm, fine-grained ostracod-containing clays drops sharply to about 6 μm and total sulfur concentrations drop to a minimum. The zone represents closed basin sedimentation with minimum clastic influxes into a shallow, swampy, polymictic lake. The landscape was probably semi-arid savanna with gentle hills and minimum runoff. Marginal swamps decrease stream velocity and trap coarse components. The period from about 14,000 ^{14}C yr B.P. to about 11,500 ^{14}C yr B.P. is marked by low %TOC and increasing concentrations of biogenic silica. Decreased volume MS, but the same mass-normalized MS, between

Zone 4 and Zone 3 indicates the same source material, but a 40% higher water content in Zone 3 contributes to decreased magnetization. Although the sediment clastic influx represented by lithic grain size was at a minimum, the oscillation of magnetic composition record may reflect high variability of sediment flux and hydrology prior to Pleistocene - Holocene transition. Sediment composition is about 40% TOM, 20% $b\mathrm{SiO_2}$, and 40% terrigenous sediment.

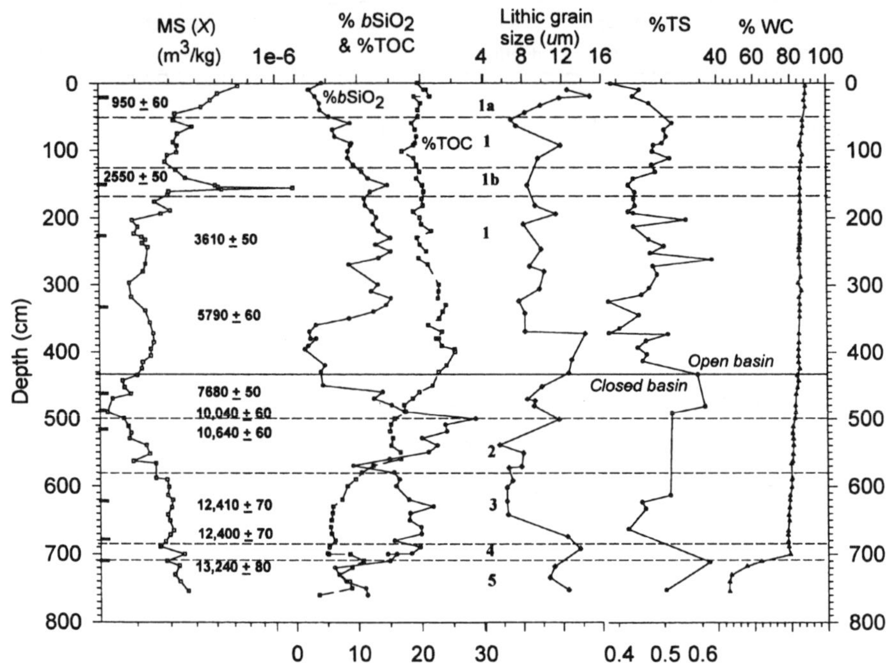

Figure 4. Non-magnetic properties: total organic carbon (%TOC), water content (%WC), total sulfur (%TS), and lithic grain size. Lithic grain size is represented as the mean value of measured distributions.

5.4. ZONE 2

Zone 2 (580-500 cm), about 11,500 ^{14}C yr B.P to 10,040 ^{14}C yr B.P., is marked by the smallest values of volume MS and has a gradual decrease in mass-normalized MS to the lowest value of the record (1.03×10^{-7} $m^3\,kg^{-1}$). This progressive decrease in mass MS corresponds to the gradual increase in $b\mathrm{SiO_2}$ from >5 % to a peak value of 28% at about 10,500 ^{14}C yr B.P. The increase in $b\mathrm{SiO_2}$ coincides with increased %TOC, which dilutes the magnetic susceptibility. S-ratio is less than 0.85, interpreted as high coercivity mineral composition (goethite-hematite). The ratio ARM/Khi does not show a significant change from the values in the lower zones, and granulometry results do not fall within the standard classification of SD, PSD or MD for depths 580 to 500 cm. Water and $b\mathrm{SiO_2}$ are diamagnetic; when exposed to a magnetic field, they produce a negative magnetization and thus measured susceptibility is negative. Hematite and goethite have low magnetizability. This zone with low lithic grain size is homogeneous, firm, fine-grained, greenish-black clay containing few diatoms. The period exhibits very low clastic input, and transition from Zone 2 to Zone 1 corresponds to the northern latitude Young Dryas interval (cold and dry), as recognized by increased productivity in the Lake Victoria diatom assemblage (Stager et al. 1997).

Holocene sediment is mainly biogenic in composition, with about 50% TOM, $b\mathrm{SiO_2}$ variable but greater than 10 % (Johnson et al. 1998), and about 35% terrigenous sediment. The sediment is homogeneous, soft, greenish-black clay.

5.5. ZONE 1

Zone 1 (500-30 cm), about 10,040 ^{14}C yr B.P. to ~1,500 ^{14}C yr B.P. covering most of the Holocene, has relatively low, uniform volume magnetic susceptibility except for a strong magnetic susceptibility event between depths 170 cm to 140 cm (Fig. 3). Mass-normalized susceptibility gradually increases from around 2×10^{-6} m^3 kg^{-1} in the early Holocene to 7.5×10^{-6} m^3 kg^{-1} in the late Holocene sediments. Mass-normalized MS record has low amplitude oscillations until about 3 ka and Holocene S-ratio oscillates between high coercivity composition (hematite-goethite) and low coercivity composition (magnetite-maghemite) at intervals of about 2,000 years between peaks, until about 3 ka in Zone 1b. The amplitude gradually increases from 0.2 at about 10 ka to 0.3 at about 3 ka, which probably represents a transition in the mode of forcing mechanism for the erosional input.

Early Holocene sediments at depths of about 500 cm to 380 cm (about 10 ka to 7 ka) mark the start of progressive increases in mass-normalized MS, magnetite content, and lithic grain size. This is a period when biogenic silica drops to the lowest Holocene levels of about 2%, and %TOC increases towards the highest Holocene levels. S-ratio oscillates in phase with lithic grain size, with low sediment flux zones corresponding to hematite zones (Fig. 4). The peak magnetite content at 380 cm (about 7 ka) corresponds to the peak low amplitude mass-normalized MS values, interpreted as a period of maximum erosion and clastic input. It is likely that detrital magnetic components correspondingly increase due to increased precipitation which results in landscape outwash. This erosion maximum appears to lag the transition from closed to overflowing lake basin at depths of 440 cm, at about 7,500 ^{14}C yr B.P. as interpreted from ^{16}O/^{18}O isotope data (Beuning 1997). When the lake overflowed there was continued erosion of lake shore for at least 400 years.

The depths 380 cm to 300 cm is a zone of stable %TOC content of greater than 20%, gradually decreasing %$b\mathrm{SiO_2}$ and constant lithic grain size. After the detrital flux maximum at about 7 ka to 3 ka, magnetic concentration and composition fluctuate but lithic grain size does not. The oscillations probably represent source sediments nearby, such lake margins that are alternately submerged by lake level fluctuations. This exposure would produce alternating redox environments conducive to composition changes between magnetite and hematite.

The ratio ARM/Khi does not change significantly from the late Pleistocene to early Holocene, but after about 7,000 yr B.P. (about 360 cm absolute core depth) the ratio progressively increases from 2.5 to 10, with a curvilinear shape toward the sediment-water interface. Magnetic granulometry (Fig. 5) plots the Holocene sediments mainly in the PSD region. Magnetic grain size decreases progressively only after the lake basin opens, possibly due to a light diagenesis overprint in the upper Holocene. With a positive water balance, the lake filled to overflowing, and may have stratified to form an anoxic zone above the sediment-water interface. Sediment sulfide content greater than 0.5%, and the curvilinear shape of ARM/Khi toward the sediment-water interface may imply diagenetic magnetite dissolution. We propose that low magnetic susceptibility in the Holocene thus reflects dilution by diamagnetic input and diagenesis. The modern water budget has a positive water P-E, with minimum sediment input (0.69 m kyr^{-1}), a situation that has persisted throughout the late Holocene.

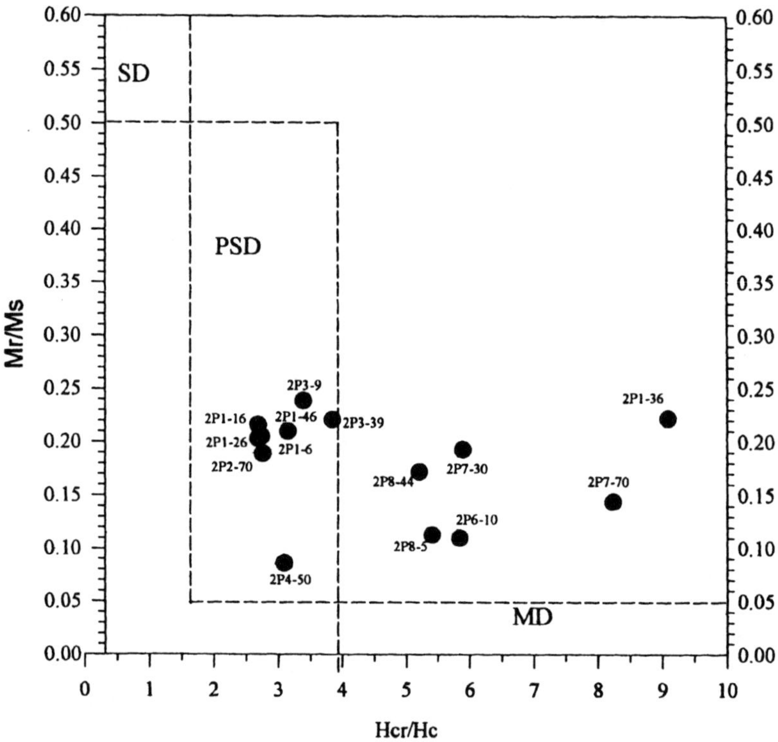

Figure 5. Hysteresis parameters for Lake Victoria core V95-2P. Single domain (SD), pseudo-single domain (PSD), and multidomain (MD) fields after Banerjee et al. (1989) .

5.6. ZONE 1b

Zone 1b (170-140 cm), about 3,000 ^{14}C yr B.P. is marked by a sharp peak in volume MS to more than 3.0×10^{-7} units, by mass-normalized MS of about 1.2×10^{-5} m^3 kg^{-1}, and by a corresponding decrease in the ARM/ Khi ratio (Fig. 3). This zone shows the highest peak in S-ratio, which represents magnetite. This peak is also traced in cores 1P, 3P and 6P with smaller amplitude (Fig. 2). The ARM/Khi ratio sharply decreases, which is interpreted as a sudden increase in magnetic grain size or a sudden inflow of outwash material depleted in more coercive magnetic mineral such as hematite. The latter is probable because a plot of Mr/Ms versus Hcr/Hc (Fig. 5) shows that the magnetic grain size distribution is pseudo single domain (PSD) with no anomalous grain size classification. The peak is interpreted as a sudden influx and distribution of clastic material from the weathered lake margin in a catastrophic geomorphologic event. This could have been triggered by spectacular regional floods, such as those of 1962-1963. Microscopic and mineral studies showed that the event was not caused by volcanic ash or debris. It remains a challenge to correlate this event with other evidence from the region.

5.7. ZONE 1a

Zone 1a (30-0 cm), about 1,500 ^{14}C yr B.P. to present, has a steady increase in mass-normalized MS from 3.0 to 5×10^{-6} m^3 kg^{-1}, lithic grain size from 5 to 15 μm, and ARM / Khi ratio of about 10 (Figs. 3 and 4). S-ratio does not show the previous oscillation pattern in the upper 150 cm after the peak at 3 ka, probably because of diagenetic dissolution of magnetite in the uppermost sediment. Lithic grain size increases but other non-magnetic proxies do not increase correspondingly. These patterns are interpreted as increased lithic input of detrital magnetic particles from soil erosion. Stager et al. (1997) refer to the period after 3,000 years as arid, based on diatoms. However, the magnetic signal may be related less to climate change than to occupation of East Africa by humans, and to deforestation for agricultural practices resulting in widespread soil depletion. Deforestation exposes magnetic carrier soils in lateritic hills located at the northern end of the Lake Victoria basin, in Busoga and Buganda.

Previous regional paleomagnetic studies have shown that the sediments have good magnetic stability to preserve a record of geomagnetic secular variation (Mothersill 1996; Williamson et al. 1991; Barton and Torgerson 1988). However, none of these regional records have used variations in magnetic mineral properties to study past climate change. The zones assigned by magnetic properties illustrate how climate affected catchment processes. Zone 5 represents a paleosol of lacustrine origin. Zone 4 is a silty vertisol deposited when the lake began refilling. Zone 3 represents low lake levels and high variability of detrital influx which may be linked to high variability in climate change. Zone 2 is a dry and cold period prior to Pleistocene/Holocene transition. Zone 1 represents the Holocene with over 65% biogenic sediment content. A 2,000-year oscillation pattern of concentration and composition indicates alternating redox environments conducive to composition changes between magnetite and hematite due to lake level fluctuations.

6. Conclusion

Correlation among six different core sites indicates a high degree of lateral continuity of Holocene sedimentation in Lake Victoria. At first approximation these correlation features could be used as stratigraphic markers within the basin and they have been used as a basis for selecting sections for more detailed studies. Volume magnetic susceptibility identifies soil horizons below Holocene clays (>12,400 ^{14}C yr B.P.), and mass-normalized susceptibility suggests that this soil zone derives from precursor lake clays of the same mineral type.

There is a general increase in Khi through the Holocene once the filling lake overflows. This suggests that although the pre-Holocene magnetic mineral components are soils derived from precursor lake sediments, the Holocene magnetic mineral components are derived from the drainage basin.

Mass-normalized MS has low amplitude oscillations from about 10 ka to 3 ka and Holocene S-value oscillates between hematite and magnetite at regular intervals of about 2,000 yr. The amplitude gradually increases from about 10 ka to about 3 ka which probably represents a transition in the mode of forcing mechanism for the erosional input. This erosion maximum appears to lag the transition from closed to overflowing lake at depths of 440 cm (7.5 ka) by at least 400 yr.

Non-magnetic properties delimit a transition from closed to open basin conditions but magnetic properties are not sensitive to this lake level change. The Holocene Lake Victoria is productive, and diamagnetic biogenic silica and water content are covariant with magnetic susceptibility.

A Holocene event at core depths of 175 to 140 cm at about 3,000 ^{14}C yr B.P. caused briefly enhanced volume and mass-normalized MS and a corresponding decrease in the ARM/Khi ratio. The event is interpreted as a flooding event that re-deposited marginal outwash.

The uppermost 30 cm of enhanced magnetic susceptibility, increase in ARM, and increase in bulk sediment grain size is interpreted as reflecting erosion influenced late Holocene aridity and human forest clearance activity in sediment source areas.

The magnetic susceptibility record basin-wide was smoothened out by the presence of diagenetic organic material in the sediment and it appears to be relatively insensitive to major changes in water balance history. This study shows that mass-normalized magnetic susceptibility trends may be useful for tracking basin sediment source variations, if the interpretations are constrained for mass-balanced diamagnetic components and diagenetic dissolution of the magnetic component.

7. Acknowledgments

This work has been supported by IDEAL project NSF grant EAR-9304962 to K. Kelts. The analytical results used are a collaboration of various research scientists at LLO, LRC, and IRM at the University of Minnesota. We thank Kristina Beuning and Mike Jackson for helpful reviews of the manuscript.

8. References

Barton, C. E., and M. Torgersen. 1988. Paleomagnetic and 210 Pb estimates of sedimentation of Lake Turkana, East Africa. Paleogeography, Paleoclimatology, and Paleoecology **68**: 53-60.

Banerjee, S. K., J. King, and J. Marvin. 1989. A rapid method for magnetic granulometry with applications to environmental studies: Geophysical Research Letters **8**: 333-336.

Beuning, K. R. 1997. Late-glacial and Holocene vegetation, climate, and hydrology of Lakes Albert and Victoria, East Africa. Ph. D. dissertation, University of Minnesota.

Crul, R. C. M. 1995. Conservation and managenment of African Great Lakes Victoria, Tanganyika and Malawi. UNESCO. IHP-IV project M-5.1. Draft Report.

Johnson, T. C., Y. Chan, K. Beuning, K. Kelts, G. Ngobi, and D. Verschuren. 1998. Biogenic silica profiles in holocene cores from Lake Victoria: implications for lake level history and initiation of the Victorian Nile, p. 75-88. *In* J. T. Lehman [ed.], Environmental Change and Response in East African Lakes. Kluwer.

Johnson, T. C ., C. A. Scholz, M. R. Talbot, K. Kelts, G. Ngobi, K. Beuning, I. Ssemmanda, and J. A. McGill. 1996. Late pleistocene desiccation of Lake Victoria and rapid evolution of cichlid fishes. Science **273**: 1091-1093.

Kelts, K., U. Briegel, K. Ghilardi, and K. Hsu. 1986. The Limnogeology - ETH coring system. Schweizerische Zeitschrift für Hydrologie **48**: 104-115.

Kendall, R. L. 1969. An ecological history of Lake Victoria basin. Ecological Monographs **39**: 121-176.

Mothersill, J. S. 1996. Paleomagnetic results from Lakes Victoria and Albert, Uganda. Studia Geophysica et Geodaetica **40**: 25-35.

Mullins, C. E. 1977. Magnetic susceptibility of soil and its significance in soil science. A Review Journal of Soil Science **2**: 223-246.

Peck, J. A., J. King, S. M. Colman, and V. V. Kravchinsky. 1994. A rock-magnetic record from Lake Baikal, Siberia: evidence for late quaternary climate change. Earth and Planetary Science Letters **122**: 221-238.

Rosennbaum, J. G., R. Reynolds, P. D. Adam, J. Drexler, A. M. Sarna-Wojjcicki, G. C. Whitney. 1997. A middle pleistocene climate record from Buck Lake, Cascade Range, southern Oregon - Evidence from sediment magnetism, trace - element geochemistry, and pollen. Geological Society of America Bulletin **108**: 1328-1341.

Stager, J. C., B. Cumming, L. Meeker. 1997. A high resolution diatom record from Lake Victoria, East Africa. Quaternary Research **47**: 81-89.

Thompson, R., and F. Oldfield. 1986. Environmental magnetism. Allen and Unwin.

Verosub, K. L., and A. P. Roberts. 1995. Environmental magnetism: past, present and future. Journal of Geophysical Research **100**: 2175-2192.

Williamson, D., N. Thouvey, C. Hillaire, A. Mondequere, M. Taieb, J. Tercelin, and A. Vincens. 1991. A chronological potential of paleomagnetic oscillations recorded in late quaternary sediments from Lake Tanganyika. Quaternary Science Reviews **10**: 351-361.

BIOGENIC SILICA PROFILES IN HOLOCENE CORES FROM LAKE VICTORIA: IMPLICATIONS FOR LAKE LEVEL HISTORY AND INITIATION OF THE VICTORIA NILE

THOMAS C. JOHNSON
YVONNE CHAN
Large Lakes Observatory, University of Minnesota
Duluth, MN 55812, USA

KRISTINA BEUNING
KERRY KELTS
GIDEON NGOBI
Limnological Research Center, University of Minnesota
Minneapolis, MN 55455, USA

DIRK VERSCHUREN
Large Lakes Observatory, University of Minnesota
Duluth, MN 55812, USA

1. Abstract

Six piston cores from offshore Lake Victoria were analyzed for abundance of biogenic Si. Lake Victoria completely dried up during the last glacial maximum and began to refill around 12,400 radiocarbon years ago (12.4 ka). The abundance of biogenic Si rose immediately in the lacustrine sediments that accumulated over the paleosol, to the highest values observed in Holocene sediments, until about 7.4 ka when they dropped abruptly and remained low until 5.9 ka. They rose to intermediate values by 5.5 ka, after which time the profiles in the cores diverge, depending on their location in the lake. The simultaneous rise and fall in biogenic silica values observed across the entire lake basin between 12.4 and 5.5 ka cannot be attributed solely to changes in primary productivity. A model of the geochemical mass balance of Si in this system indicates that it could not support such widespread change for more than two to three decades. We interpret the biogenic Si profiles to be a record of changing supply rate of dissolved Si early in the new lake's history, and to changes in lake level. The lake remained a closed basin until 7.4 ka, when oxygen isotopic data on aquatic cellulose indicate that it rose to its highest level, 18 m above present, and outflow was initiated. This resulted in an abrupt drop in biogenic Si accumulation at core sites throughout the lake because much of the Si could exit with the outflow. Downcutting at the Nile outlet resulted in a drop in lake level and focusing of diatom deposition in offshore basins. The lake finally stabilized around 5.5 ka. Subsequent shifts in biogenic Si profiles in the cores

J.T. Lehman (ed.), Environmental Change and Response in East African Lakes, 75–88.

are attributed either to a spatially heterogeneous pattern of diatom productivity or to sediment focusing effects.

2. Introduction

Tropical Africa was significantly drier and cooler during the last ice age than it is today. The Lake Victoria basin was dry, probably covered with savanna grasslands and shrubs, and began to re-fill with water around 12,400 radiocarbon years before present (12.4 ka) (Johnson et al. 1996). Seismic reflection profiles indicate that lakes occupied the basin for much of the last 400,000 years as a result of tectonic warping and back-ponding of rivers that previously flowed westward across the region, but on at least four occasions these lakes shrank or disappeared in response to climatic change (Johnson et al. 1996; Scholz et al. 1998).

Major ecological transformation has taken place in Lake Victoria within the last 12,000 yr. The lake had more than 300 endemic species of haplochromid cichlid fish at the beginning of this century (Greenwood 1980; Kaufman and Ochumba 1993), and most of these species must have evolved after the lake formed (Johnson et al. 1996). Other changes undoubtedly occurred as the basin filled with water and adjusted to a new hydrological regime. Cycling of organic C shifted from terrestrial to predominantly lacustrine processes (Talbot and Livingstone 1989), and nutrient dynamics, water chemistry and circulation dynamics evolved as the lake deepened and the climate settled into conditions characteristic of the Holocene.

Here we examine the abundance of biogenic Si in six piston cores from the lake to determine the nature of change that the lake underwent with respect to diatom production and accumulation. How extensive was the change, how rapid was the transition to "typical" Holocene conditions, if such conditions exist, and to what extent were the speed, direction and extent of the change influenced by climate or other physical factors?

3. Methods

Nine piston cores were recovered from Lake Victoria on two cruises in 1995 and 1996 (Fig. 1). These cruises were part of the International Decade for the East African Lakes (IDEAL) multidisciplinary study of the lake. A modified Kullenberg corer (Kelts et al. 1986) with a 10-m barrel recovered cores varying in length between 4.5 and 9.5 m. Core sites were selected on the basis of seismic reflection profiles, focusing on offshore areas where the Holocene sediments were relatively thick. A GPS satellite navigation system was used to determine ship position. Sediment cores were retained in their plastic liners, cut into sections 1 m long, capped, taped and air freighted to the Limnological Research Center at the University of Minnesota in Minneapolis for subsequent analyses. After whole core magnetic susceptibility was measured with a pass-through magnetometer, they were split, visually described, photographed and immediately sampled for water and organic C content.

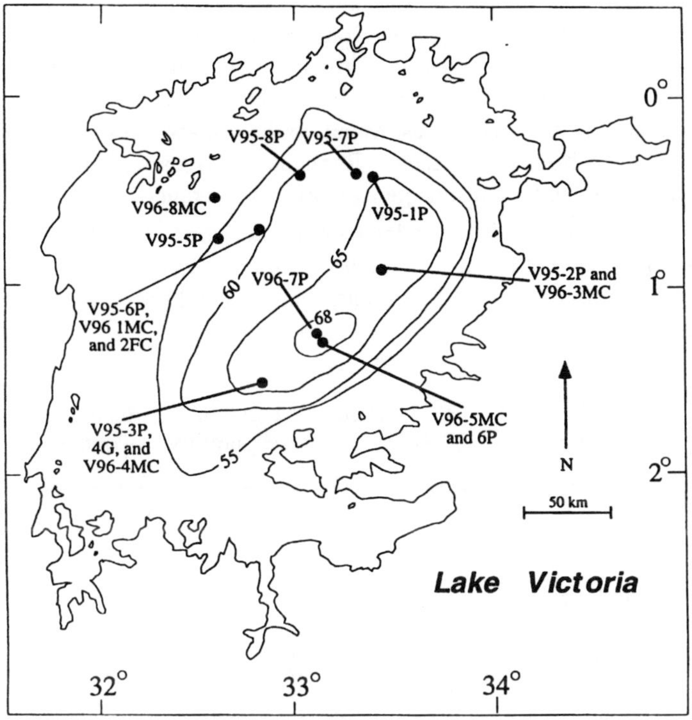

Figure 1. Bathymetric map of Lake Victoria, showing all the cores recovered by IDEAL in 1995 and 1996. Prefixes refer to year the core was obtained, and suffixes refer to core identification number and type: P = piston core, MC = multi-core, and G = gravity core. Contours are in meters.

TABLE 1. Accelerator mass spectrometer (AMS) radiocarbon dates obtained from all the cores in this study.

Core	Depth (cm)	^{14}C Age	SD	Corrected Age	Material
V95-2P	19	1450	60	950	pollen
V95-2P	140	3050	50	2550	pollen
V95-2P	215	4110	60	3610	pollen
V95-2P	321	6290	60	5790	pollen
V95-2P	432	9260	60	8760	pollen
V95-2P	457	8180	50	7680	pollen
V95-2P	480	10540	70	10040	pollen
V95-2P	499	11140	60	10640	pollen
V95-2P	608	12910	70	12410	pollen
V95-2P	677	12400	70	12400	charcoal
V95-2P	699	13240	80	13240	charcoal
V95-1P	831	16760	140	16260	pollen
V95-3P	275	9530	160	9030	pollen
V95-6P	373	12910	230	12410	pollen
V95-7P	915	10190	110	9690	pollen
V96-7P	418	13200	70	12700	pollen

Six of the cores were analyzed for abundance of biogenic Si by wet chemical digestion (modified after DeMaster 1979; Krause et al. 1983). The cores were sampled at 10 cm intervals, and the samples were freeze-dried and ground prior to analysis. Approximately 0.3 g of the ground dry sediment was reacted with 200 ml of 0.5 N NaOH solution at 85° C. This was found to be the weakest base solution that effectively dissolved all diatoms present. Replicate analyses were made on about 5% of the samples, and showed the results generally to have a precision of 1% by weight.

AMS radiocarbon dates were obtained on 11 horizons in V95-2P, and at a few depths in some other cores (Table 1). The upper nine dates were obtained from the 20 to 70 μm treated fraction of pollen-lignin-charcoal and the lowest two dates from grass charcoal (Beuning et al. 1997).

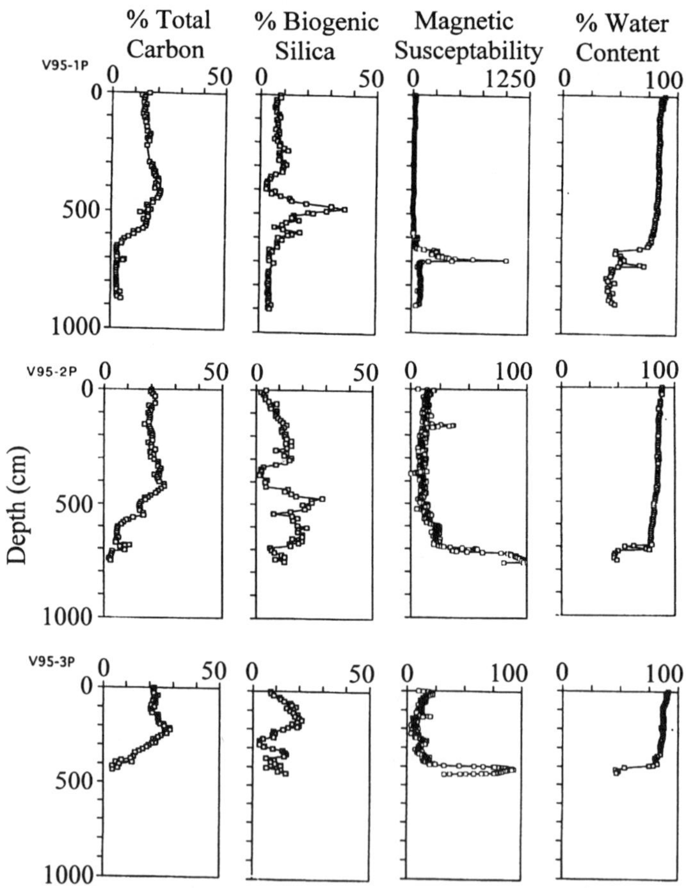

Figure 2. Profiles of total (organic) C, biogenic SiO_2, magnetic susceptibility and water content in 3 piston cores recovered from Lake Victoria. Top of the paleosol recovered in these cores is clearly indicated by the spike in magnetic susceptibility and the abrupt change in water content. Note how profiles of total organic C and biogenic Si, both often interpreted as proxies of paleoproductivity, do not parallel each other.

4. Results

Holocene sediments in offshore Lake Victoria are underlain by a paleosol that formed when the lake basin dried out completely in the late Pleistocene (Johnson et al. 1996). Most of the piston cores penetrated to or through the paleosol, typically just 4 to 7 m below the lake floor. The top of the paleosol is clearly indicated by a peak in magnetic susceptibility and by a significant drop in water content in cores V95-1P, 2P, 3P, and 6P, and V96-7P (Fig. 2). The sediments of Lake Victoria are carbonate free, therefore the percent by weight of total C reflects the abundance of organic C. Carbon is in very low abundance in the paleosol and rises gradually to values near 20% by weight in the overlying Holocene sediments.

Biogenic silica concentrations in the paleosol are low, ranging from about 4 to 5% by weight in core V95-1P up to approximately 10% in cores V95-2P and V95-3P. With the exception of V95-6P, biogenic Si profiles in all cores show a consistent pattern of high concentration in the early Holocene and lower values in the late Holocene (Fig. 3). The details of this pattern are a two-step rise in biogenic Si to maximum values within 2 m above the paleosol, followed by a drop to the lowest values measured in the core. Above this horizon of very low Si concentrations, the profiles in all cores rise and then diverge; some show decreasing values up core (e.g., V95-2P, 3P), some show steady values above this horizon (e.g., V95-1P, 6P, 7P), and one shows rising values with time (V96-7P). V95-6P is from the shallowest site, and may frequently experience sediment winnowing by wave activity (Johnson 1980).

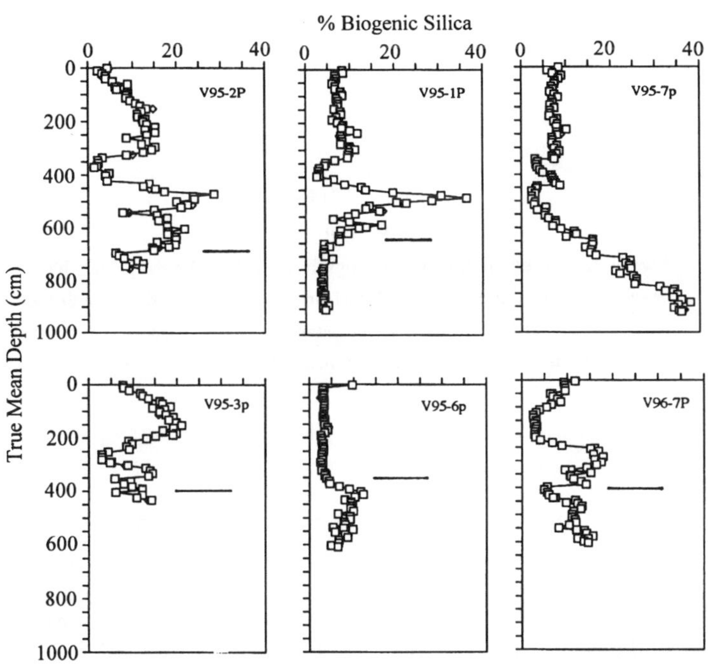

Figure 3. Profiles of biogenic SiO_2 abundance in all 6 cores analyzed in this study. Core locations are shown in Fig. 1. The top of the paleosol is indicated, where recovered, by horizontal line.

A date of 500 yr on the 20 to 70 μm treated pollen-lignin-charcoal fraction extracted from surface sediments at site V95-1P indicates that a certain amount of older material mixes in with modern sediment and yields radiocarbon dates older than the actual age. We have subtracted 500 yr from all dates, except the bottom two dates on charcoal, to reflect this effect. This correction is close to the 400 yr correction applied by Kendall (1969), subsequently modified to 600 yr by Stuiver (1970), to the core they studied from Pilkington Bay at the northern periphery of the lake. The corrected dates in V95-2P are reasonably consistent down-core, with the exception of the four samples at burial depths between 430 and 600 cm (Fig. 4). All but these four dates fall nearly on a straight line. We interpret the ages to reflect a reasonably constant sedimentation rate defined by the linear fit through seven of the dates (Fig. 4). If this is true, the four dates that fall off the line must reflect substantial deposition of reworked older sediment at the site between about 12 and 8.7 ka. Beuning et al. (1997) interpret the dates somewhat differently, invoking a hiatus of about 2500 yr at 466 to 487 cm to explain the age offset that we attribute to reworking. These differences in age assignment do not affect the chronology of the major events or the conclusions of this paper. Dates obtained from the other piston cores (Table 1) are consistent with core V95-2P when stratigraphically matched according to their profiles of biogenic Si (Table 1).

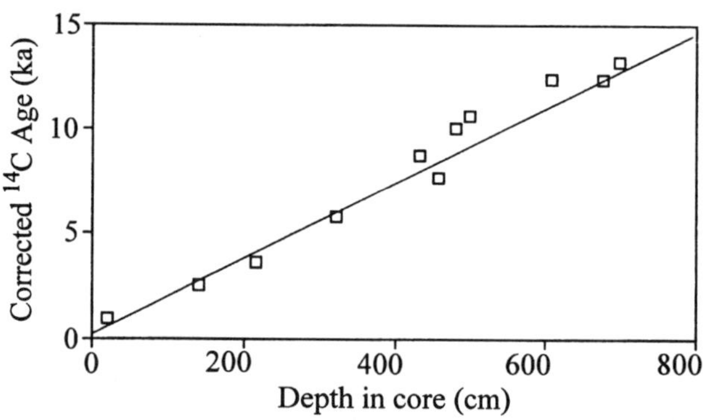

Figure 4. Radiocarbon geochronology in Core V95-2P. All dates have been corrected by subtracting 500 yr from the accelerator dates, with the exception of the bottom two, which were based on charcoal (see text). The straight line defines the ages assigned to the core depths.

5. Discussion

Profiles of biogenic Si and of total organic C in the sediments of large lakes and oceans are often considered to be proxies of past primary productivity. In such environments the biogenic Si consists almost entirely of diatom frustules and the organic C is derived almost exclusively from the plankton. This is the case for Lake Victoria. Microscopic examination of the sediments reveals that the vast majority of the biogenic Si is diatoms, and pyrolysis analyses of the organic matter in the upper two-thirds of the core yield Hydrogen Index values that indicate a predominantly algal origin (Laerdal 1997).

But the biogenic Si and total organic C profiles do not parallel one another (Fig. 2), indicating that at least one of them cannot be considered as a valid proxy for paleo-productivity in the lake.

The consistency of biogenic Si profiles among cores and very high early Holocene concentrations suggest that diatom production and biogenic Si accumulation were influenced more by the geochemical mass balance of silica in the lake system than by primary productivity, at least in the 2 to 4 m immediately overlying the paleosol. The total quantity of biogenic Si that can accumulate in the sediments of Lake Victoria is limited by the net supply of dissolved Si (inflow minus outflow) to the system. There cannot be a simultaneous increase or decrease in biogenic Si deposition across the entire lake basin unless the net supply of dissolved Si changes accordingly.

A simple mass balance model for silica in Lake Victoria consists of river input of dissolved Si as the sole source, $(SiO_2)_R$, and two major sinks consisting of burial of diatom Si, $(SiO_2)_D$ and Nile River outflow of dissolved Si $(SiO_2)_{NO}$:

$$(SiO_2)_R = (SiO_2)_D + (SiO_2)_{NO} \tag{1}$$

This equation holds true regardless of the importance of diatoms as primary producers in the ecosystem. If diatoms are outcompeted by other phytoplankton, less Si will be removed from the water column, thereby increasing the outflow term.

Values for the 3 terms in Eq. 1 can be estimated as follows. River input of dissolved Si is the product of annual river inflow times dissolved concentration in the rivers. Annual river inflow is estimated to be 19×10^9 m^3 yr^{-1} (Sene and Plinston 1994). The mean concentration of dissolved Si in rivers flowing into Lake Victoria is not known, but Kilham (1971) measured a concentration of 0.28 mM in the Kagera River, the largest river flowing into Lake Victoria. Assuming this to be representative of all rivers flowing into the lake, annual input of dissolved Si from the rivers is about 3.2×10^{11} g SiO_2 yr^{-1}.

For Nile outflow, we use the dissolved Si concentration in lake waters that was measured in the early 1960s by Talling (1966). This may be high, for Verschuren et al. (1998) conclude that Si depletion in the lake may already have been underway, starting in the 1930s. Lake Victoria has undergone major ecosystem change since then, and dissolved Si concentrations in the lake are presently about an order of magnitude lower than what Talling measured (Hecky, 1993). Dissolved Si in surface waters of Lake Victoria was about 75 μM in 1960. Multiplying this times the outflow of the Nile (34×10^9 m^3 yr^{-1}) (Sene and Plinston 1994) yields an annual loss of 1.5×10^{11} g SiO_2 yr^{-1}.

The amount of biogenic Si that accumulates annually on the floor of Lake Victoria is determined by:

$$(SiO_2)_D = A \times (BSiO_2/100) \times (1 - \Theta) \times \rho \times LSR \tag{2}$$

where: A = surface area of diatomaceous clay accumulating in the modern lake
$BSiO_2$ = per cent biogenic silica in the sediment
Θ = porosity of the sediment

ρ = dry sediment density
LSR = linear sedimentation rate

The areal extent of diatomaceous clay presently accumulating on the floor of Lake Victoria is equal to about 40% of the modern lake floor, according to the seismic reflection profiles (Scholz et al. 1998), or 34,000 km^2. The average amount of biogenic silica in the upper Holocene sediments that we analyzed is about 8% by weight (Fig. 3). The porosity of the upper Holocene sediments is about 0.93 and the dry sediment density is about 2.3 g cm^{-3}. The linear sedimentation rate is based on a mean thickness of Holocene sediment equal to 6 m, assuming that it has accumulated for 12,400 yr since the lake began to refill, or 0.5 m kyr^{-1}. Multiplying these values together yields a value for total biogenic silica accumulation rate in the lake of 1.6×10^{11} g SiO_2 yr^{-1}. Mass balance is nearly achieved by these rough calculations; it is certainly within the range of uncertainty of the values that were used.

Based on the values above, a mean residence time (T) for dissolved silica in the lake basin can be determined from:

$$T = (SiO_2)aq / (SiO_2)_R \qquad (3)$$

where $(SiO_2)aq$ is the total amount of dissolved silica in the lake. In 1960 the average concentration of dissolved Si throughout the water column was about 80 μM SiO_2 (Talling 1966). Multiplying this times the volume of Lake Victoria (2700 km^3) yields a value of 13×10^{12} g SiO_2 for $(SiO_2)aq$. Dividing this by the annual river influx yields a residence time of about 41 yr. This indicates that within just four decades of a major simultaneous increase in biogenic Si accumulation across the entire lake floor, as was observed in the early Holocene, diatom production and burial could have depleted the water column of dissolved Si. The high biogenic Si accumulation rates observed in all of the cores following re-flooding of the basin lasted for about 4,000 yr, from 12.4 to 8.4 ka. Based on estimates of current silica fluxes, these rates could not have been sustained for more than about four decades if they were supported solely from the reservoir of dissolved Si already in the lake.

The basin-wide increase in biogenic Si accumulation rates observed in piston cores could have been caused by two factors. First, there probably was an abundance of readily mobilized silica in the drainage basin when the lake first re-filled after the desiccation event, around 12,400 years ago. Seasonal river inflow to the low, grassy plain probably carried dissolved Si to this region, where it may have been taken up in the formation of grass phytoliths or other unstable siliceous solids. These would have accumulated in the soil, and been readily dissolved when water began to flow and accumulate as conditions became wetter around 12 ka. Thus, the concentration of dissolved Si in the rivers may have been higher for a few decades or centuries than they are now. Second, if lake level did not immediately rise to overflowing, it would have remained in a closed-basin configuration for several millenia. The biogenic Si accumulation rate would have been higher in deep-water sites because there would have been no Nile outflow removing dissolved Si from the lake (in 1960, Nile outflow removed about 50% of the river input of dissolved Si), and also the area of biogenic Si accumulation would have been less extensive than it presently is.

Changing lake level can also explain the basin-wide decrease in biogenic Si around 7.4 ka. Recent paleohydrological reconstruction of Lake Victoria based on the oxygen isotopic composition of aquatic cellulose suggests that lake level did not rise to overflowing conditions until about that time (Beuning et al. 1998). This is a controversial hypothesis because Kendall (1969) suggested that overflow occurred around 10 ka, coincident with rising lake levels throughout north Africa. But the hypothesis of Beuning et al. (1998) is based on an 8 per mil drop in the cellulose-inferred isotopic composition of the lake water at 7.4 ka, a phenomenon that is difficult to explain any other way. This isotopic shift coincides with the abrupt drop in biogenic silica concentration to about 2%, the lowest Holocene values observed in all of the cores. Beach terraces at 18 m above the present lake level (Temple 1964) and isotopic modeling (Beuning et al. 1997) indicate the height to which the lake may have risen at this time. The 18 m terrace has not been dated, but since it is found at the same elevation around the lake, suggesting that tectonic uplift has not had time to introduce tilt, it must be relatively young (Temple 1964). Kendall (1969) suggested that it may represent an early Holocene highstand.

Inception of significant outflow would have affected potential diatom production and biogenic Si accumulation because it would have carried dissolved Si out of the lake. The areal extent of accumulation of diatomaceous clays may have been more widespread in the deeper lake than it is today. The drop in biogenic Si accumulation rate could reflect lowered inflow of dissolved Si to Lake Victoria, either by a drop in river discharge or by a drop in the dissolved Si concentration in the river water. However river discharge could not have dropped at this time when the ^{18}O of the lake water was at its lowest value in the Holocene (Beuning et al. 1998). If anything, river discharge to the lake was at a maximum at this time. Given this situation, it is difficult to imagine a drop in the rate of chemical weathering in the basin and a corresponding drop in the concentration of dissolved Si in the inflowing rivers. The subsequent rise to Holocene values that typically fall in the range of 5 to 10% by weight is interpreted to reflect the downcutting of the Nile to its present sill depth, with reductions in both the areal extent of diatomaceous clay and the proportion of dissolved Si removed from the lake by Nile discharge.

Explanation for the major peaks and valleys of the biogenic Si curve in terms of lake level variations is based on two assumptions: first, that diatoms were the dominant phytoplankton group throughout the Holocene and their productivity was limited by the supply of dissolved Si, and second, the amount of dissolved Si flowing into the lake remained fairly steady after the high values immediately following the flooding of the Victoria basin. We do not have independent evidence to prove that diatoms dominated the phytoplankton throughout the Holocene, and it is conceivable that the period of lowest biogenic Si accumulation rates (e.g., 3.3 to 4.2 m in core V95-2P) resulted from diatoms being outcompeted by other phytoplankton groups. Neither ^{13}C nor ^{15}N of the bulk organic matter show significant shifts in value that coincide with this interval of low biogenic Si accumulation, but Hydrogen Index and C/N values are slightly higher in this interval than anywhere else in core V95-2P, perhaps reflecting stratification and anoxia in the bottom waters (Laerdal 1997), or perhaps a subtle indication that diatoms were not the major component of the phytoplankton. Microscopic examination of sediments shows fewer diatoms relative to other sedimentary components, but this only confirms the low abundance of biogenic Si, and not its cause.

Timing of events hypothesized above is best determined from the chronology in V95-2P (Fig. 4). High accumulation rates of biogenic Si in the lake occurred between 12 and 7.4 ka, with a brief interlude of lower accumulation rate around 9 ka (Fig. 5). We do not know the cause for this short period of lower accumulation rate but it is seen in all cores. It may reflect a brief wet period when lake level rose almost to overflowing and the extent of diatom clay deposition was greater than just before or afterwards. Based on the V95-2P chronology, the event coincides with lake highstands throughout much of north Africa (Street-Perrott and Roberts 1983). Alternatively, this may have been a dry period, with the river input of dissolved Si lower than usual, but on the basis of the silica profiles from Lake Victoria alone, we cannot distinguish between two explanations that are diametrically opposite in terms of climatic implications. But calling upon less river inflow at this time would be in conflict with paleoclimate data from many other lakes in Africa that recorded wet conditions at 9 ka.

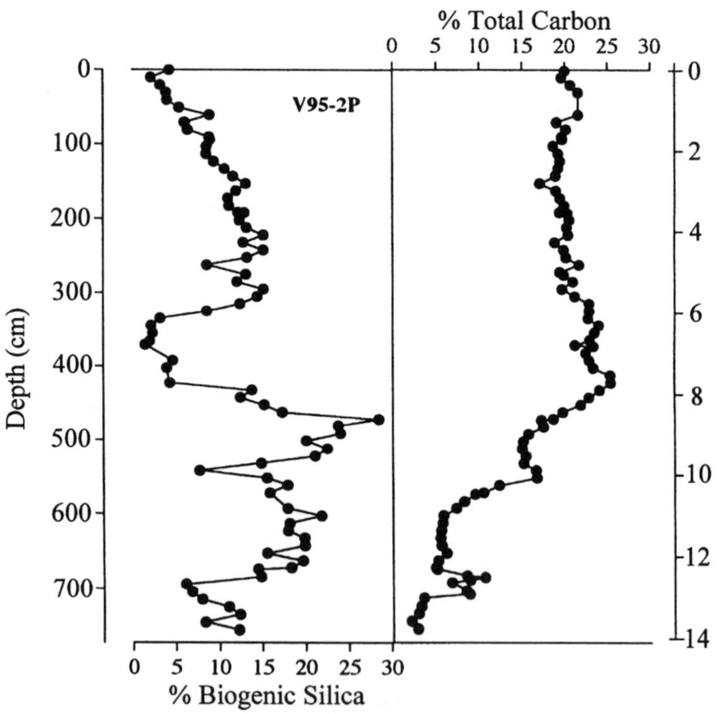

Figure 5. Profiles of biogenic Si and total C in V95-2P, plotted against depth in core and radiocarbon age, based on the straight line fit to the corrected dates in Fig. 4.

The drop to low biogenic Si abundance lasted from about 7.4 to 5.8 ka, and is interpreted to reflect the highest Holocene level of Lake Victoria at 18 m above present. There is evidence to suggest that at this elevation the lake initially drained to the south, to the Manonga and Malagarasi basins (Hurst 1957; Temple 1964). If true, breaching and downcutting at the Nile outlet must have happened soon afterwards, resulting in capture of the lake's outflow, and initiation of the Victoria Nile. Progressive and rapid downcutting of the Nile outlet was accomplished by about 5.5 ka, resulting in a reduced lake area, a more restricted area of deposition of diatomaceous clays, and consequently a higher concentration of biogenic Si in the sediments younger than 5.5 ka. Alternatively, diatoms regained their dominance in the phytoplankton community, and less Si was lost from the lake to Nile outflow.

The trend in biogenic Si profiles since 5.5 ka is not consistent among the cores (Fig. 3). The concentration increases with time in V96-7P (although this may be a result of over-penetration of this core, thereby bypassing the latest Holocene record), it decreases over the same interval in V95-2P and V95-3P, and it remains fairly stable in V95-1P, 6P and 7P. These trends suggest an areal shift in the locus of high diatom accumulation in the lake towards the deepest part of the basin from 5.5 ka to the present. Whether this is a signal of a shift in the region of high primary production or of sediment refocusing cannot be determined.

Relative abundances of diatom species in the Lake Victoria cores are presently being determined by J. C. Stager. The total abundance of diatom valves in a core he analyzed from Damba Channel, at a depth of 32 m (Stager 1984), appears to track the total biogenic Si profile of V95-2P. If we apply our 500 yr correction to Stager's radiocarbon dates, the site appears to have been flooded shortly before 11 ka, and remained a relatively shallow water site until about 9.5 ka when the abundance of sponge spicules and black spherules declined and the water content and organic matter content increased (Stager 1984). The record between about 9.5 and 7.5 ka is dominated by *Stephanodiscus astrea* and *Aulacoseira* (*Melosira*) *granulata* and *A. ambigua*, with the relative abundance of *Aulacoseira* to *Stephanodiscus* steadily increasing through this period (Stager et al. 1997). Around 7.5 ka, the two species of *Aulacoseira* were replaced rather abruptly by *A. nyassensis* (Stager 1984; Stager et al. 1997). Stager (1984) found this puzzling because *A. granulata* and *A. ambigua* are found primarily in the shallower portions of Lake Victoria (Talling 1966) whereas *A. nyassensis* is more abundant in the deeper stratified waters of the lake. Kendall (1969) suggested that Lake Victoria had rapidly risen to overflowing conditions by about 10 ka so this transition in *Aulacosira* species was difficult to explain.

Stager et al. (1997) speculated that the change in *Aulacoseira* species may have resulted from decreased turbulence in the lake after 7.4 ka. However if the lake in fact remained a closed basin lake until around this time, then the transition in *Aulacoseira* species may have been caused by changing salinity. Salinity would have risen steadily as evaporation concentrated dissolved ions from the rivers. Assuming river concentrations, based on Kilham's (1971) analyses of Kagera River water, of Ca = 0.13 mM, Mg = 0.16 mM, K = 0.09 mM, Na = 0.31 mM and Cl = 0.11 mM, and multiplying these by the modern river influx of 19×10^9 m^3 yr^{-1} for 5,000 yr (12.4 to 7.4 ka) permits an estimate of the total mass accumulation over time. Assuming the ions dissolved in a lake volume of 2700 km^3 results in accumulated concentrations of Ca = 4.6 mM, Mg = 5.6 mM, K = 3.3 mM, Na = 11 mM, and Cl = 3.7 mM. These are not unusually saline conditions, and the values plot in the range of lakes labeled IIID of fig. 4 by Gasse et al. (1983), where *Aulacoseira granulata* (especially var. *angustissima*) and *A. ambigua* are found to be abundant.

Stager (1984) also remarks on the decline in diatom abundance in the upper part of the Damba Channel core, and suggests that sediment refocusing may be the cause. Our results, showing a rise in biogenic Si with time at the deepest water site, confirms his suggestion.

6. Conclusions

The sediments of Lake Victoria record major changes in the lake ecosystem in the first several thousand years after the basin began to refill with water at 12,400 radiocarbon years before present (12.4 ka). Abundance of total organic C rose steadily for about 5000 yr until 7.5 ka, and stabilized at about 20% by weight for the remainder of the Holocene. Abundance of biogenic Si rose more rapidly throughout the lake basin to maximum values between 9.5 and 7.4 ka, reflecting a high initial influx of dissolved Si that probably was readily available when tropical rains inundated previously dry savanna and mobilized silica in the soils. But the initial high flux of dissolved Si probably did not last longer than a few centuries. High biogenic Si concentrations observed in the cores in this time period could have been maintained for only several thousand years if the lake basin were smaller than present. Under these circumstances no Si would have been lost from the lake by outflow, and the areal extent of diatomaceous sediments would have been less than today.

The dramatic drop in biogenic Si concentration at 7.4 to 5.8 ka marks a highstand in Lake Victoria, probably to the + 18 m level that is reflected in strandlines around the entire basin. River outflow, at first perhaps out of the southern end of the lake and into the Lake Eyasi basin, but shortly thereafter through the Nile at the north, suddenly would have carried away a significant portion (ca. 25%) of the dissolved Si supplied to the lake. During this highstand the areal extent of biogenic Si deposition would have been larger than at present. Both factors would have contributed to declines in abundance of biogenic Si in offshore cores. It is possible that diatoms were outcompeted by other phytoplankton during this period, but there is no strong, independent evidence to support this. Erosional downcutting of the Nile outlet between about 5.8 and 5.4 ka brought the lake level to near its present elevation, and biogenic Si concentrations rose during this time as a result of reduced loss via Nile outflow and more focused accumulation of diatoms on the lake floor. Since 5.4 ka the locus of biogenic Si accumulation has shifted to the deepest part of the basin, perhaps reflecting as much the effects of sediment focusing as of primary productivity in the overlying waters.

Two important conclusions can be drawn from this study. First, biogenic Si profiles from Lake Victoria support the conclusion of Beuning et al. (1998) based on oxygen isotope data that the Victoria Nile did not begin to flow until some time between 7.4

and 5.8 thousand years ago. This has important implications for downstream basins in the Nile system, including Lake Kyoga and Lake Albert. Second, profiles of biogenic Si in the tropical lakes of East Africa are not necessarily good recorders of past productivity. The geochemical mass balance of Si in each lake system must be assessed before conclusions are drawn regarding changes in trophic dynamics, particularly if shifts in biogenic Si occur simultaneously basin-wide.

7. Acknowledgments

We thank the Fisheries Institute of Uganda under the direction of Dr. F. Y. B. Bugenyi for providing logistical support and the *R/V Ibis* for our field work on Lake Victoria, and the governments of Uganda and Tanzania for permission to conduct research on Lake Victoria. While the conclusions of this paper are the sole responsibility of the authors, the manuscript was significantly improved by comments on an earlier draft provided by Michael Talbot, J. Curt Stager, and an anonymous reviewer. Financial support was provided by the U.S. National Science Foundation grants EAR-9421566 to TCJ and EAR-9304962 to KK.

8. References

Beuning, K., K. Kelts, and T. C. Johnson. 1998. Paleohydrology of Lake Victoria, East Africa inferred from $^{18}O/^{16}O$) ratios in sediment cellulose. Geology (in press).

DeMaster, D. J. 1979. The marine budgets of silica and Si32. Ph.D. Dissertation. Yale University.

Gasse, F., J. F. Talling, and P. Kilham. 1983. Diatom assemblages in East Africa: classification, distribution and ecology. Revue D'Hydrobiologie Tropicale **16**: 3-34.

Greenwood, P. H. 1980. Towards a phyletic classification of the "genus" Haplochromis (Pisces, Cichlidae) and related taxa. Part II; The species from Lakes Victoria, Nabugabo, Edward, George and Kivu. Bulletin of the British Museum of Natural History (Zoology) **45**: 209-231.

Hecky, R. E. 1993. The eutrophication of Lake Victoria. Mitteilungen Vereinigung Internationale Limnologie **25**: 39-48.

Hurst, H. E. 1957. The Nile. Constable.

Johnson, T. C. 1980. Sediment redistribution by waves in lakes, reservoirs and embayments, p. 1307-1317. *In* H. Stefan [ed.], Proceedings of Symposium on Surface Water Impoundments. American Society of Civil Engineering.

Johnson, T. C., C. A. Scholz, M. R. Talbot, K. Kelts, R. D. Ricketts, G. Ngobi, K. Beuning, I. Ssemmanda, and J. W. McGill. 1996. Late Pleistocene desiccation of Lake Victoria and rapid evolution of cichlid fishes. Science **273**: 1091-1093.

Kaufman, L., and P. Ochumba. 1993. Evolutionary and conservation biology of cichlid fishes as revealed by faunal remnants in northern Lake Victoria. Conservation Biology **7**: 719-730.

Kelts, K., U. Briegel, K. Ghilardi, and K. Hsu. 1986. The limnogeology-ETH coring system. Schweizerische Zeitschrift für Hydrologie **48**: 104-115.

Kendall, R. L. 1969. An ecological history of the Lake Victoria basin. Ecological Monographs **39**: 121-176.

Kilham, P. 1971. Biogeochemistry of African Lakes and Rivers. Ph.D. Dissertation. Duke University.

Krause, G. L., C. L. Schelske, and C. O. Davis. 1983. Comparison of three wet-alkaline methods of digestion of biogenic silica in water. Freshwater Biology **13**: 73-81.

Laerdal, T. 1997. A Reconstruction of the Sedimentology and Paleolimnology of Lake Victoria During the Last 15 Thousand Years, Using Stable Isotopes (Carbon and Nitrogen) and Rock-Eval Pyrolysis. Candidate Scient. Thesis, University of Bergen, Norway.

Scholz, C. A., T. C. Johnson, P. Cattaneo, H. Malinga, and S. Shana. 1998. Initial results of the 1995 IDEAL seismic reflection survey of Lake Victoria, Uganda and Tanzania, p. 47-57. *In* J. T. Lehman [ed.], Environmental Change and Response in East African Lakes. Kluwar.

Sene, K. J., and D. T. Plinston. 1994. A review and update of the hydrology of Lake Victoria in East Africa. Hydrological Sciences Journal **39**: 47-63.

Stager, J. C. 1984. The diatom record of Lake Victoria (East Africa): the last 17,000 years, p.455-476, *In* Proceedings of the 7th Diatom Symposium. Koeltz, Koenigstein.

Stager, J. C., B. Cuming, and L. Meeker. 1997. A high-resolution 11,400 year diatom record from Lake Victoria, East Africa. Quaternary Research **47**: 81-89.

Street-Perrott, F. A., and N. Roberts. 1983. Fluctuations in closed-basin lakes as an indicator of past atmospheric circulation patterns, p. 331-345. *In* A. Street-Perrott, et al. [eds.], Variations in the global water budget. D. Reidel Publishing Company.

Stuiver, M. 1970. Long-term C14 variations, p. 197-213. *In* I. U. Olsson {ed.], Radiocarbon Variations and Absolute Chronology. Proceedings of the Twelfth Nobel Symposium. John Wiley and Sons.

Talbot, M. R., and D. A. Livingstone. 1989. Hydrogen index and carbon isotopes of lacustrine organic matter as lake level indicators. Palaeogeogaphy, Palaeoclimatology, Palaeoecology **70**: 121-137.

Talling, J. F. 1966. The annual cycle of stratification and phytoplankton growth in Lake Victoria (East Africa). Internationale Revue der gesamten Hydrobiologie **51**: 545-621.

Temple, P. H. 1964. Raised features along the southern shoreline of Lake Victoria. Proceedings of the East African Academy **1**: 13-22.

Verschuren, D., D. N. Edgington, H. J. Kling, and T. C. Johnson. 1998. The eutrophication of Lake Victoria: sedimentary signals at offshore stations. Journal of Great Lakes Research (in press).

ROLE OF APATITE WEATHERING IN THE EUTROPHICATION OF LAKE VICTORIA

JOSH HOLTZMAN
JOHN T. LEHMAN
Department of Biology, University of Michigan
Natural Science Building, Ann Arbor, MI 48109, USA

1. Abstract

We investigated biogenic and geochemical constituents of a short sediment core obtained during 1992 from a nearshore region in northern Lake Victoria. Diatom stratigraphy reveals disappearance of *Aulacoseira* from near surface layers, consistent with neolimnological observation of the lake. X-ray fluorescence, dry combustion, and wet chemical methods were used to measure bulk composition, as well as C, N, H, non-apatite P, and biogenic Si. Organically-bound P has increased somewhat in the top 10 cm, but the main Total-P increase appears to have been caused by increased amounts of apatite weathered from carbonatite minerals that are abundant in the region. Increases in non-biogenic Si and Ti in the top 10 cm indicate increased importance of terrigenous inputs during the 20th Century. Our data are consistent with the explanation that modern changes in land use practices, coupled with rainfall and lake level changes, have accelerated erosion of phosphate rich mineral soils into the lake. Weathering of these phosphate rich minerals are probably the source of persistent and elevated offshore P concentrations in Lake Victoria and other historically P rich lakes in East Africa. The main removal mechanism for lake P is sedimentation and burial, resulting in a residence time for P of 25 years. Loss rates of P parallel those of biogenic Si, indicating that elevated P concentrations are partly a consequence of floristic changes, especially the loss of fast-sinking diatoms.

2. Introduction

Lake Victoria has experienced radical changes in its flora, fauna, temperature regime, and water chemistry since the 1950s and 1960s (Hecky 1993). These changes have been interpreted as a rapid eutrophication of the lake resulting from human population growth, agriculture, biomass burning, and livestock herding (Lipiatou et al. 1996). In addition, there has been a profound change in lake water level (Bugenyi and Magumba 1996), apparently caused by changes in the balance of precipitation, runoff, and evaporation (Kite 1981; Piper et al. 1986). The lake is now both warmer and apparently more stably stratified than in historical records. The increased stability of

J.T. Lehman (ed.), Environmental Change and Response in East African Lakes, 89–98.

thermal stratification and associated increased primary productivity has contributed to extensive hypolimnetic oxygen depletion (Ochumba and Kibara 1989; Hecky 1993; Bugenyi and Magumba 1996), with consequent reduction of fish habitat. Lehman and Branstrator (1993), Lehman (1996), and Ochumba (1996) have pointed out that the changing conditions of the lake are caused in part by climatic variations that have reduced the magnitudes and temporal extents of vertical mixing and eddy diffusion. Retrospective studies of the lake that document changes in its condition are relevant for understanding the development of its current state.

Hypotheses for the altered modern condition of Lake Victoria were recently reviewed by Lehman (1996). The changes have been regarded to result either from (1) elevated nutrient income, (2) climate change, or (3) food web manipulation by introduction of an exotic piscivore, the Nile perch (*Lates niloticus*). Hecky (1993) examined a sediment core taken from deep water at the eastern margin of the lake in Kenya. He found that sediment P and N have increased during the past 30 years, and effectively refuted Hypothesis 3 by demonstrating that the changes preceded the rise of the introduced piscivore. Lipiatou et al. (1996) reinforced the chronology of change through further study of the same core. Instructive and provocative, these reports are nonetheless based on analysis of a single core from a single site. The objective of this study was to test Hecky's observations at a different site and to investigate certain chemical changes in greater detail.

Johnson et al. (1998) report analyses of six short cores from deep water sites in central Lake Victoria. We report a coring site that is shallower and closer to shore than either Hecky or Johnson et al. in an effort to identify land-watershed interactions that may have influenced the lake in recent times.

3. Methods

Our sediment cores were obtained in October 1992 from Ugandan territorial waters of northern Lake Victoria in Mubaale Bay east of Bugaia Island at 23 m depth. Cores were extruded in the field at 1 cm intervals, then freeze-dried. Organic Matter content (O.M.) was determined by loss on ignition (Bengtsson and Magnus 1986). Major elemental analyses were performed by X-ray fluorescence (XRF) for Al, Si, P, Ca, Mg, Na, K, Ti, and Fe (XRAL Laboratories, Ann Arbor, Michigan). XRF analyses for P were compared with wet oxidation by acid persulfate. C, N, and H were measured by dry combustion and gas chromatography (University of Michigan Department of Chemisty Instrumentation Services, CHN/AA Laboratory). Biogenic Si (BSi) was measured by time-series extraction in 1% Na_2CO_3. Non-apatite P, called operationally biogenic P (BP), was measured by extraction in 0.1 N NaOH with mechanical agitation overnight. Concentrations are expressed per unit of dry mass of the original sediment samples. For comparisons of elemental composition of the mineral fraction of the sediment, results are expressed as percentage composition of the organic-free sediment ash. Pearson correlation coefficients were estimated among the sediment constituents (SYSTAT 5.0), and significance levels of correlations were judged from Bonferroni-adjusted probabilities.

Diatom stratigraphy was assessed from freeze-dried samples. Diatoms were prepared by M. Julius at the University of Michigan Phytoplankton Laboratory (M. Julius, personal communication): Sediment subsamples were boiled for 30 min in 30% H_2O_2 at 110 °C, followed by addition of 25 ml of concentrated HNO_3. After subsidence of the resulting exothermic reaction, the samples were heated to 120 °C for 1 h, then rinsed six times with distilled water. Each entire subsample was settled on 18 mm circular cover slips in Battarbee chambers, and was mounted using Hyrax™ mounting medium.

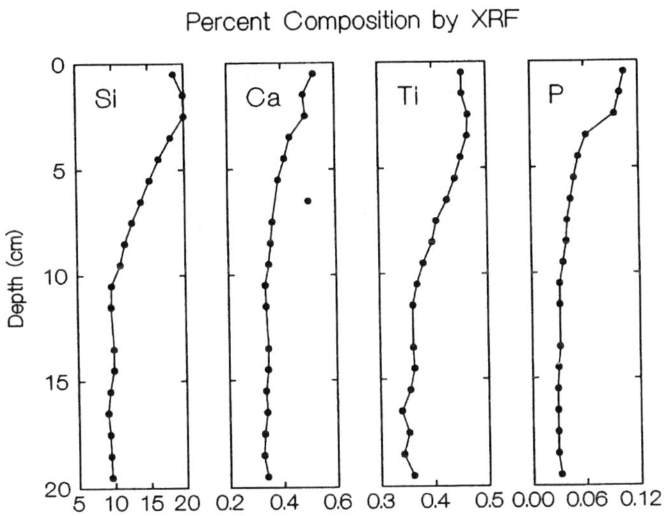

Figure 1. Bulk composition of sediment determined by X-ray fluorescence, expressed as percentage composition of dry sediment.

4. Results

Our core showed the transition from *Aulacoseira* to *Nitzchia* described by Hecky (1993). We also recorded the same increase in P near the surface (Fig. 1), but the increased P content did not appear to originate with sedimenting biological material. In our core it was associated with increasing contents of Si, Ca, and Ti. The increase in titanium is considered indicative of elevated contributions of terrigenous source materials. Increases in terrigenous mineral sediments outpaced some of the increases in organic matter (O.M.), total C, and biogenic P. Moreover, the modern terrigenous inputs were great enough to dilute some of the clay-associated elements present in the mineral (i.e., organic-free ash) fraction of the sediments, particularly the elements Al, K, and Mg (Fig. 2).

Most of the increased sediment P was inorganic, crystalline mineral P (Fig. 3), and the mineral content increased toward the surface of the core. Biogenic Si constituted only a few percent of the total Si, but the biogenic content increased in modern sediments. Increased biogenic Si in the surface sediment is consistent with the observed depletion of dissolved Si from the lake during its recent eutrophication (Hecky 1993).

Accelerated inputs of terrigenous materials, primarily non-biogenic Si, resulted in a shift from Al, K, and Mg to Si in the mineral fraction, as well as a dilution of the C and N content of the upper sediments (Fig. 4). Non-mineral P (BP) and Biogenic-Si increased, however, despite the terrigenous inputs. We consequently examined the stoichiometries of C, N, Si, and P, using BP rather than total P as the reference for comparison. The analysis (Fig. 5) showed a progressive decline in the C:P and N:P ratios toward the top of the core, whereas Si:P ratios increased at the surface. Strong positive correlations exist among total organic matter, Al, K, and Mg (Table 1). Total Si and P also correlate positively with each other, but both Si and P are strongly negatively correlated with Al, K, Mg, and total organic matter.

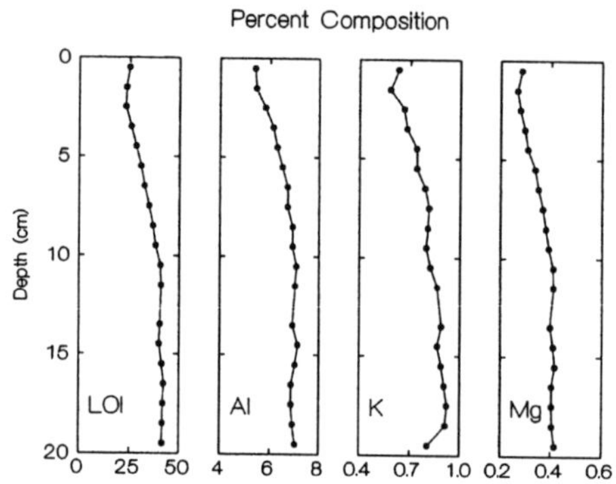

Figure 2. Loss-on-ignition (LOI), and contents of Al, K, and Mg in mineral ash, determined by X-ray fluorescence. Constituents reported as percentage composition of dry sediment.

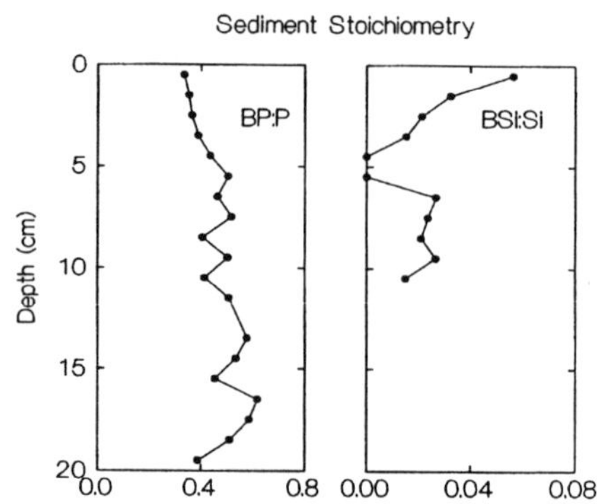

Figure 3. Proportions of total sediment P and Si that are non-crystalline minerals, and putatively are of biological origin. BP and BSi were determined by alkaline digestion.

TABLE 1. Pearson product moment correlations ($P < 0.001$ in all cases, $n = 19$) among sediment core constituents. Elemental constituents are expressed as percent composition of the mineral fraction based on XRF analysis. O.M. is percent dry mass.

Constituent	Al	K	Mg	O.M.	P
K	+0.901				
Mg	+0.953	+0.927			
LOI	+0.924	+0.941	+0.991		
P	-0.910	-0.913	-0.910	-0.895	
Si	-0.994	-0.944	-0.994	-0.999	+0.904

Bulk Ca:P stoichiometric ratios by moles decrease toward the top of the core during the period of accelerated terrigenous inputs (Fig. 6). The changes imply that apatite is the primary source of P accumulating in the modern sediments at our nearshore site. We used the stoichiometry of Ca:P in apatite (5:3) to calculate the increase in sedimentary Ca that must be expected from the stratigraphy of crystalline mineral P (putatively apatite: Fig. 7). Calculations showed that almost all of the modern increases in accumulation of Ca in the sediments results from an increased income of apatite.

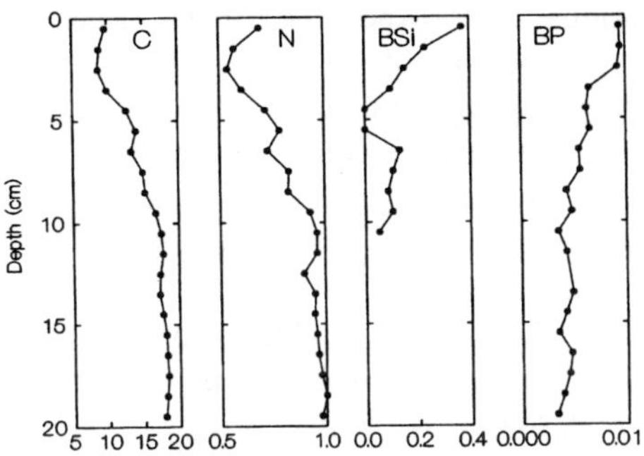

Figure 4. Elemental composition of biogenic components, mmol g^{-1} sediment.

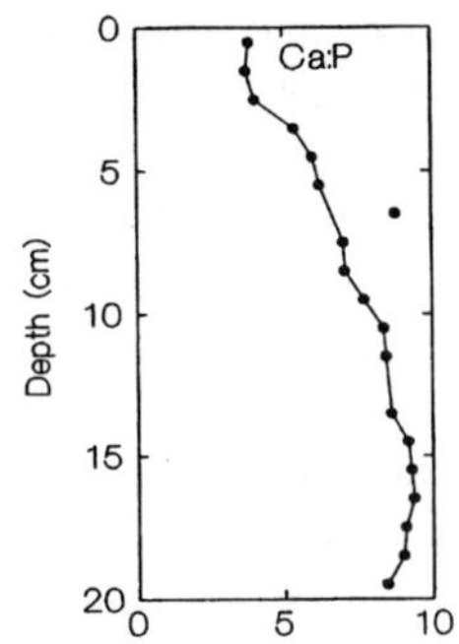

Figure 6. Molar ratios of Ca to P.

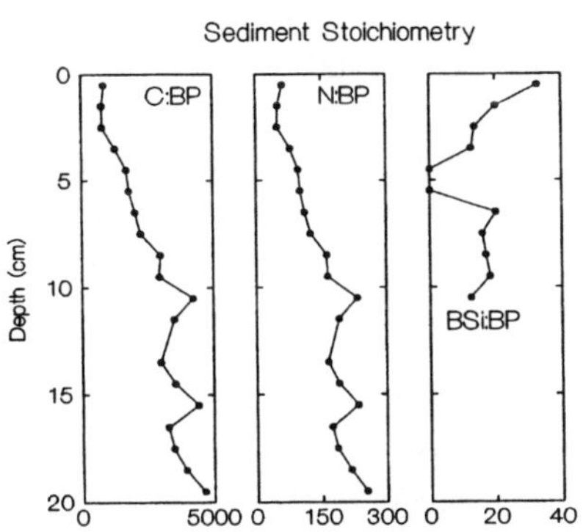

Figure 5. Stoichiometry, by moles, of C, N, and biogenic-Si to biologically available P (BP).

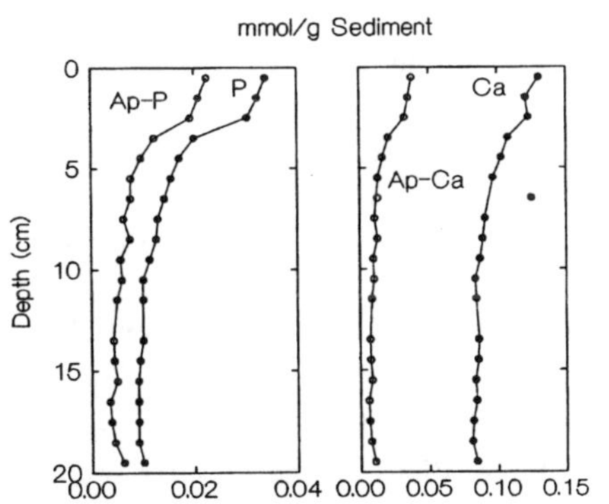

Figure 7. Stratigraphy of total P and total Ca, and putative fractions of each element contributed by apatite minerals.

5. Discussion

5.1. PHOSPHORUS AND CARBONATITE

Lehman and Branstrator (1994) stated that one major remaining problem presented by the eutrophication of Lake Victoria was the source of elevated P concentrations in offshore waters. It now appears that the lake has received massive inputs of terrigenous minerals in recent years, the result of erosion from deforested land. The terrigenous materials are rich in P, predominantly as apatite derived from the weathering of carbonatite formations.

Carbonatites are intrusive carbonates that are common to regions of rift faulting, such as the East African Rift Valley (Reedman 1984; Rutley 1988). These formations often include magnetite, biotite, and apatite, among other minerals. The carbonatite formations of Uganda and Kenya vary in their concentrations of apatite, but the phosphate mineral is always present and may be very abundant in places. Apatite is present as slightly corroded ovoid grains which are readily removed from the carbonatite rocks and concentrated in the residual soils (Reedman 1984). Reedman (1984) reported that the mean P_2O_5 content of residual soils in southeastern Uganda averages nearly 10%, but that apatite can account locally for up to 50% of the soil. This represents a considerable soil enrichment in apatite compared to the parent carbonatite. Physical weathering and transport of only a fraction of these apatite grains could account for the increased P in the lake sediments. Chemical weathering of the P-rich mineral soil in the region could account for the elevated lake P. Recent commercial apatite mining activities in the Tororo and Sukulu complexes of southeastern Uganda and in western Kenya within the Victoria drainage may also elevate the transport of particulate and dissolved P to the lake.

If we use Reedman's (1984) figure of 10% P_2O_5 content (= 4.4% P) in carbonatite-derived soils, it is possible to estimate the magnitude of chemical weathering necessary to establish lakewide concentrations of roughly 2 μM Total P (Lehman and Branstrator 1993, 1994). The volume of Lake Victoria is 2760 km^3, or 2.76×10^{12} m^3, corresponding to a lake bulk P content of 5.5×10^9 moles, or 170000 metric tons. That mass of P could have been weathered from 4×10^6 metric tons of soil, or 1100 tons per km of the lake's 3440 km shoreline, or alternatively about 30 tons per km^2 of the lake's 126200 km^2 terrestrial catchment area. Assuming a mean soil specific gravity of 2 g cm^{-3}, this mass corresponds to a mean overall weathering depth of less than 0.02 mm. Clearly, not all of the regional soils may be derived from weathered carbonatite, but there is no question of the potential of these minerals to produce high concentrations of P in Lake Victoria. The weathering rates required to maintain lake P levels can be estimated from the residence time of P in the lake water column.

Lake Victoria has a water retention time of over a century (Hecky and Bugenyi 1992), which suggests that residence time for lake P is controlled almost entirely by net sediment accumulation. Total P constitutes about 0.1% by weight of the recent sediments. Johnson et al. (1998) report a mean linear sediment accumulation rate of 0.05 cm yr^{-1} for upper Holocene sediments of Lake Victoria, mean sediment porosity of 0.85, and dry sediment density of 2.3 g cm^{-3}. These figures yield P accumulation of 172.5 mg P m^{-2} yr^{-1}. Only about 60% of the Lake Victoria basin is presently accumulating sediment (Scholz et al. 1998). So, averaged across the full surface area of Lake Victoria, sedimentary P accumulates at roughly 100 mg P m^{-2} yr^{-1}.

At 2 μM TP, the bulk reservoir of P in Lake Victoria's water column (40 m mean depth) is about 2.5 g P m^{-2}. Estimating residence time as water column inventory divided by flux, the residence time for P in Lake Victoria is 25 yr. In other words, only about 4% of the Total P in the Lake Victoria water column is permanently lost to sediments each year. Interestingly, this residence time for P (25 yr) is almost identical to the mean residence time for Si (24 yr) estimated by Johnson et al. (1998). The similarity of these residence times for P and Si may be a consequence of a common removal mechanism over long time periods for both elements. The common mechanism is probably the sinking and burial of large diatoms.

Lake Victoria experienced a climate-related elevation of its lake level by 2 m during the early 1960s (Bugenyi and Magumba 1996). Resulting inundation of previously terrestrial soils along the lake's perimeter and on its numerous islands raises the prospect of increased weathering rates. Combined with deforestation for fuel, agriculture, wetland destruction, and livestock range, the potential for accelerated erosion and weathering subsequent to 1960 is considerable. The lake level change of the early 1960s cannot, however, account for changes earlier in the century reported by Lipiatou et al. (1996). The composition of the mineral fraction of the sediments start to change at 10 cm, which corresponds to some time in the late 19th Century, by reference to Hecky's Core 103 and cores by Johnson et al. (1998). Percentages of K, Al, and Mg begin to drop, while composition of Si increases. This may indicate a shift in terrigenous source sediments from clays to silicate sands and silts concurrent with increasing erosive forces with the changing watershed. Changes in weather conditions, including more rain and more erosive force might lead to export and deposition of material of altered composition, such as we observed.

5.2. ORGANIC MATTER COMPOSITION

The upper 2 cm of the core exhibit an increasing organic content, which is a reversal of the trend in the rest of the short core. This modern condition may indicate either an increase in the rate of biogenic sedimentation or reduced terrigenous inputs. Data for biogenic Si indicate a sharp rise in organic content in the upper 2 cm while the percent composition of Ti and mineral Si fall. This apparent dilution of terrigenous source material by organic matter is consistent with increased sedimentation of autochthonous biogenic particles produced during the recent eutrophication of Lake Victoria.

The organic matter content of the core (Fig. 2) ranged from about 20% near the top to 40% at depth. These levels of organic content are similar to the ranges reported for offshore cores by Johnson et al. (1998). The organic C content of the organic matter ranged from 45% to 55% (mean = 50.7%, SD = 2.7%, n = 19), which is higher than the organic C levels reported by Mothersill (1976), but very close to those measured by Verschuren et al. (1998) at offshore sites. We measured H:C stoichiometries ranging from 2.0 by moles near surface to 1.8 downcore (mean = 1.83, SD = 0.10, n = 20 depths), and mean C:N ratios of 17.8 (SD = 1.35, n = 20 depths).

From our measured ratios of H:C, C:N, and C:total O.M., we can calculate the stoichiometry of the sedimentary organic material in Lake Victoria. Assuming that organic matter is mainly C, H, O, and N, the O:N stoichiometry must be 10:1 by moles to satisfy the simultaneous conditions of element ratios and proportion C. The average elemental composition of the organic material is therefore: $C_{18}H_{32}O_{10}N$. This computed stoichiometry is consistent with expectation that the accumulating C is in a very reduced state.

Our sediment stoichiometries, based on non-mineral P rather than total P, are considerably different from those reported by Hecky (1993), but the decline in C:P and N:P toward the surface is consistent with his data. We find that surface sediments have biogenic C:P of about 800 by atoms and N:P of about 50. The ratio of BP:TP in surface sediments is 0.27, however, so the bulk ratios of C:P and N:P are about 200 and

14, respectively. These latter ratios look a lot like the stoichiometries of water column particulate matter (Hecky 1993), but the conformity is only superficial, and is an artifact caused by including mineral P in the calculation. The ratios based on non-mineral P indicate that P is differentially remineralized from biological material during diagenesis. It may also be that some of the material in the core is of terrestrial origin, potentially derived from nearby Bugaia Island. If so, it would have to be highly saturated (H:C = 2), and strongly reduced (O:C < 0.6) organic matter, which is not typical of terrestrial detritus.

5.3. LAKE VICTORIA P DYNAMICS

The model we propose for P dynamics in Lake Victoria has three major elements:

1. The native state of Lake Victoria is one of elevated P concentrations derived from apatite-rich carbonatite deposits.
2. Phosphate is permanently removed from lake water principally through biological uptake and sedimentation of large diatoms.
3. All weathered and dissolved P that enters Lake Victoria is converted to particulate organic form before it leaves the lake either by outflow or sedimentation.

Chemical weathering of eroded particulates appears to be the source of high offshore P concentrations. Modern increases in rainfall and lake level have interacted with land use practices to accelerate erosion of P-rich particulate material into the lake. Phosphate weathered from this material has enriched a lake that does not currently support populations of fast-sinking diatoms. In the absence of rapid sedimentation with large diatoms, P is efficiently recycled from biogenic particles. Offshore, the reservoir of TP is mainly in inorganic phosphate form because the shallow mixed layer and rapid light attenuation constrains the trophogenic zone to near surface. High rates of P regeneration help account for concentrations of orthophosphate that consistently equal or exceed 1 μM in the offshore lake water.

Nearshore, shoaling depths and complete mixing each night contribute to increased development of algal biomass, and phosphate is transformed biologically to particulate organic P. There is little permanent sediment accumulation nearshore, and outflow is the only removal mechanism. Otherwise, P is remineralized and retained efficiently nearshore. On balance and over time, the mean residence time for P in Lake Victoria is 25 yr. More lake P is lost to permanent sediment burial than to outflow, and its sedimentation dynamics parallel those of biogenic Si.

Long residence time for P is additional evidence that the element is not a limiting nutrient for algal production in Lake Victoria. The abundance of the nutrient guarantees that high levels of biomass can be developed whenever mixing conditions provide a satisfactory light climate. It seems that the lake levels of P have indeed risen in Lake Victoria over the past 35 yr. The cause for the increase, however, is not just because inputs accelerated, but also because the major vectors of P removal, the large, heavy diatoms, have declined.

6. Acknowledgments

We thank M. Mugumba and staff of the Fisheries Research Institute, Jinja, Uganda, for help with sediment coring, and B. Betz and C. Carter for help with CHN analyses. M. Julius digested the sediment for diatom remains, and confirmed identification of the diatoms. This research was supported in part by grants from the U.S. National Science Foundation, and from the NOAA National Undersea Research Program.

7. References

Bengtsson, L. and M. Magnus. 1986. Chemical analysis, p. 423-451. *In* B. E. Berglund [ed.], Handbook of Holocene Palaeoecology and Palaeohydrology. Wiley.

Bugenyi, F. W. B., and K. M. Magumba. 1996. The present physicochemical ecology of Lake Victoria, Uganda, p. 141-154. *In* T. C. Johnson and E. Odata [eds.], The limnology, climatology, and paleoclimatology of the East African lakes. Gordon and Breach.

Hecky, R. E. 1993. The eutrophication of Lake Victoria. Verhandlungen Internationale Vereinigung für Limnologie **25**: 39-48.

Hecky, R. E. and F. W. B. Bugenyi. 1992. Hydrology and chemistry of the African Great Lakes and water quality issues: Problems and solutions. Mitteilungen Internationale Vereinigung für Limnologie **23**: 45-54.

Johnson, T. C., Y. Chan, K. Beuning, K. Kelts, G. Ngobi, and D. Verschuren. 1998. Biogenic silica profiles in Holocene cores from Lake Victoria: Implications for lake level history and initiation of the Victoria Nile, p. 75-88. *In* J. T. Lehman [ed.], Environmental change and response in East African Lakes. Kluwer.

Kite, G. W. 1981. Recent changes in level of Lake Victoria. Hydrological Sciences Bulletin 26: 3: 233-243.

Lehman, J. T. and D. K. Branstrator. 1993. Effects of nutrients and grazing on the phytoplankton of Lake Victoria. Verhandlungen Internationale Vereinigung für Limnologie **25**: 850-855.

Lehman, J. T. and D. K. Branstrator. 1994. Nutrient dynamics and turnover rates of phosphate and sulfate in Lake Victoria, East Africa. Limnology and Oceanography **39**: 227-233.

Lehman, J. T. 1996. Pelagic food webs of the East African Great Lakes, p. 281-302. *In* T. C. Johnson and E. Odata [eds.], The limnology, climatology, and paleoclimatology of the East African lakes. Gordon and Breach.

Lipiatou, E., R. E. Hecky, S.J. Eisenreich, L. Lockhart, D. Muir, and P. Wilkinson. 1996. Recent ecosystem changes in Lake Victoria reflected in sedimentary natural and anthropogenic organic compounds, p. 523-541. *In* The limnology, climatology and paleoclimatology of the East African lakes. Gordon and Breach.

Mothersill, J. S. 1976. The mineralogy and geochemistry of the sediments of northwestern Lake Victoria. Sedimentology **23**: 553-565.

Ochumba, P. B. O., and D. I. Kibara. 1989. Observations on blue-green algal blooms in the open waters of Lake Victoria, Kenya. African Journal of Ecology **27**: 23-34.

Ochumba, P. B. O. 1996. Measurement of water currents, temperature, dissolved oxygen and winds on the Kenyan Lake Victoria, p. 155-167. *In* T. C. Johnson and E. Odata [eds.], The limnology, climatology, and paleoclimatology of the East African lakes. Gordon and Breach.

Piper, B. S., D. T. Plinston, and J. V. Sutcliffe. 1986. The water balance of Lake Victoria. Hydrological Sciences Journal **31**: 25-37.

Reedman, J. H. 1984. Resources of phosphate, niobium, iron, and other elements in residual soils over the Sukulu Carbonatite Complex, Southeastern Uganda. Economic Geology **79**: 716-724.

Rutley, F. 1988. Rutley's elements of mineralogy, 27th edition. Unwin Hyman Publishers.

Scholz, C.A., T. C. Johnson, P. Cattaneo, H. Malinga, and S. Shana. 1998. Initial results of 1995 IDEAL seismic reflection survey of Lake Victoria, Uganda and Tanzania, p. 47-57. *In* J. T. Lehman [ed.], Environmental change and response in East African Lakes. Kluwer.

Verschuren, D., D. N. Edgington, H. J. Kling, and T. C. Johnson. 1998. The eutrophication of Lake Victoria: sedimentary signals at offshore stations. Journal of Great Lakes Research (in press).

LAKE VICTORIA PLANKTON ECOLOGY: MIXING DEPTH AND CLIMATE-DRIVEN CONTROL OF LAKE CONDITION

JOHN T. LEHMAN
Department of Biology, University of Michigan
Natural Science Building, Ann Arbor, MI 48109, USA

ROSE MUGIDDE
Fisheries Research Institute
P.O. Box 343, Jinja, Uganda

DONNA A. LEHMAN
Department of Biology, University of Michigan
Natural Science Building, Ann Arbor, MI 48109, USA

1. Abstract

We investigated growth responses of Lake Victoria phytoplankton to nutrient enrichment at different light levels throughout a year. Experiments were conducted at reference sites both nearshore (z_{max} < 15 m) and offshore (z_{max} > 50 m), and were accompanied by water column observations. Nearshore plankton increased growth in response to nitrogen enrichments in all experiments, but episodically deep mixing offshore in August 1995 produced conditions of excess nitrate when light alone appeared to limit biomass development within the lake. Neither phosphate nor silicate enhance growth of existing algal communities nearshore or offshore, but affinities of Lake Victoria plankton for phosphate uptake are greater nearshore than offshore. Ambient concentrations of phosphate are also much lower nearshore. Stoichiometry of algal chlorophyll content to P content varies with light climate, and may be a useful diagnostic of light history. Algae responded to experimental conditions of altered light levels by changing their pigment content over 48 h. Measured relationships among biomass and light attenuation, photosynthetic parameters, and respiration rates are used to predict maximum biomass levels sustainable at different lake mixing depths. We conclude that the master variable controlling the condition of Lake Victoria is mixing depth, and explain its interactions with biomass development, N-limitation, algal species composition, oxygen, and denitrification. Mixing depth, in turn, is controlled by climate, which has changed during the past 35 years toward conditions that promote mainly shallow mixing. Analysis of climate data demonstrates changes in several meteorological variables that affect the heat balance and mixing of the surface layer.

J.T. Lehman (ed.), Environmental Change and Response in East African Lakes, 99–116.

2. Introduction

The modern condition of Lake Victoria has attracted scientific attention for many reasons. Victoria is the second largest freshwater lake on the planet, and as one of the lake sources of the White Nile is an important hydrologic feature of East Africa. Lake Victoria was intensely studied during the late 1950s and early 1960s, and conditions at that time were recorded for food web components, biological productivity, and limnological properties including temperature and dissolved nutrients (Talling 1965, 1966). Potential fishery yield of the lake is high, but the composition of the fish fauna has changed recently. The limnological condition of the lake is different now than 35 years ago. The changes are dramatic and undisputed in their magnitude. Only the cause of the changes has been in dispute.

Numerous changes in lake condition have been summarized from several sources (Hecky 1993; Mugidde 1993; Lehman and Branstrator 1994; Bugenyi and Magumba 1996; Lehman 1996). Algal biomass and particulate chlorophyll have increased. Species composition of algae has changed; there are now proportionally more cyanobacteria and fewer diatoms and chlorophyte algae. The heavily silicified diatom *Aulacoseira,* common earlier in the century, is now absent, and concentrations of dissolved silicate have declined drastically in surface waters. Rates of primary production have accelerated near the surface, but rates of decomposition in deep water have depressed dissolved oxygen to near zero over as much as one-third of the water column. The lake is approximately 0.5 °C warmer than in 1960, and it exhibits reduced water transparency, stronger stability of thermal stratification, and decreased photic zone depth. Concentrations of P have increased offshore, and concentrations of sulfate have decreased. The species richness of its endemic cichlid fish has been drastically reduced, and the abundance of an introduced predatory fish species, Nile perch, has increased so that it now dominates the fishery.

Other changes have probably occurred as well, but they may be difficult to prove because historical data are lacking. For instance, it is likely that rates of both N-fixation and denitrification have increased (Hecky 1993), that algae not capable of N-fixation have become increasingly N-limited (Lehman and Branstrator 1994), and that light limitation of phytoplankton has changed (Mugidde 1993).

Three alternative hypotheses have been offered for observed changes in the lake, particularly its altered phytoplankton composition, increased biomass, and increased hypolimnetic oxygen deficits:

H1: Eutrophication resulting from land clearing, growth of human and cattle populations, and increased P income to the lake, has caused Lake Victoria to resemble Temperate Zone lakes similarly subjected to increased P loading. In particular, this hypothesized nutrient enrichment led to the loss of large diatoms, increases in bluegreen algae as well as biomass generally, increased loss of particulate organic matter to the hypolimnion, and deficits of hypolimnetic oxygen.

H2: Trophic alterations caused by a top-down cascade of predator-prey interactions triggered by the introduction of Nile perch has eliminated endemic herbivores and released the algae from control by grazing, thereby increasing biomass and causing attendant changes to the lake.

H3: Climate variations reduced lake mixing depths, elevated temperatures, and caused a suite of chemical and biological consequences of altered mixing state.

Hypothesis 2 initially attracted most attention, probably because it was linked to the loss of biodiversity in the endemic fish fauna (Baskin 1992). Subsequently this trophic cascade model became discounted, because stratigraphic evidence in recent sediments indicated that biotic changes in the flora preceded introduction of Nile perch (Lipiatou et al. 1996). Thus, if cause must precede effect, the predator did not cause the major lake system changes, although there is little doubt that the introduced piscivore decimated the endemic haplochromine fauna of the lake.

Hypothesis 1 initially seemed plausible as well, because the modern changes were believed to be unique (Hecky 1993). Now, this hypothesis, too, has come under skeptical review because detailed fossil stratigraphy of diatoms during the last 10,000 years indicates that the modern events may not be unique (Stager 1998). Instead, the dominance of heavily silicified diatoms may have waned repeatedly and episodically over millennia, potentially in response to climate variations.

Hypothesis 3 remains as a sufficient explanation for the observed changes to the lower food web and water chemistry of Lake Victoria. Climate variation is linked to lake condition through changes to the heat budget and mixing regime. Biological and chemical changes in Lake Victoria are consistent with diminished vertical mixing and thus with increased physical stability of stratification (Lehman 1996). Recently Diaz and Graham (1996) have demonstrated that above average SST in the tropical Pacific from the late 1950s to 1990 seem to cause increased surface air temperatures at tropical sites greater than 1000 m in elevation. With its surface elevation at more than 1100 m, Lake Victoria is within the zone susceptible to this identified climate change.

Whether caused by nutrient enrichment or climate change, there must be proximate physiological mechanisms that link water column conditions to productivity and biomass levels. This study tries to identify some of these mechanistic links and to develop a predictive model for algal biomass at different mixing depths. We wanted to evaluate interactions between light and nutrient effects, and to assess the seasonality of nutrient limitation in Lake Victoria. Previous studies (Lehman and Branstrator 1993, 1994) documented N-limitation by Lake Victoria phytoplankton. Addition of inorganic N to lake water experimentally increased algal biomass over control conditions, and over effects of other nutrients.

Diazotrophic cyanobacteria are present in Lake Victoria and N-fixation rates can be considerable (Hecky 1993). N-fixation competes with C-fixation for chemical reductant derived from light energy. Thus, N limited algae with diazotrophic capabilities could meet some of their requirements by fixing N. High light conditions should supply enough energy for fixation of both N and C, because at high light C-fixation becomes saturated at a maximum rate (P_{max}). At high light levels, energy from photons could be diverted into N-fixation without compromise of photosynthetic photosystem processes. Under low light levels, however, C-fixation increases proportionally with available light energy. Photosynthetic processes are light-limited at low light, and hence competition among metabolic pathways for chemical reducing compounds should intensify. This means that rates of N-fixation could compromise rates of C-fixation, and vice versa. We consequently made the *a priori* hypothesis that evidence for N-limitation would be stronger at low light than at high light.

3. Methods

Experiments were conducted at nearshore and offshore stations in Ugandan waters of Lake Victoria. Nearshore sites in Napoleon Gulf and Pilkington Bay, and the offshore site south of Bugaia Island, have been described by Branstrator et al. (1996).

3.1. GROWTH RESPONSE EXPERIMENTS

Growth Response Experiments followed methods described by Lehman and Branstrator (1994), with the following additions. Epilimnetic water collected from 1 m depth was filtered through 100 μm aperture Nitex before being dispensed to incubation containers. Water baths were maintained within 2 °C of ambient lake surface temperature, and containers were exposed at two light levels for 48 h incubations. Layers of neutral density light filter (plastic windowscreen) were used to produce incubation light intensities at 58% (High Light) and 12% (Low Light) of surface incident light (I_0) as measured by a LiCor PAR 4-π sensor. Nutrient treatments (4 replicates) were Control, +P (+1 μM as Na_2HPO_4), +N (+7.5 μM NH_4NO_3), +Si (+10 μM as Na_2SiO_3), and +NPSi (three nutrients in combination). SRP was measured immediately on filtrate (GF/C) samples. Unfiltered and filtered (GF/C) lake water samples, as well as desiccated filters, were shipped by international courier service to Ann Arbor for analyses described by Lehman and Branstrator (1994). Chlorophyll samples were measured in Ann Arbor by extraction in 90% acetone and fluorometric assay (Turner Designs) for experiments conducted in Sept 1994 and Aug 1995. For experiments in Nov 1994 and May 1995, Chl was extracted in 90% methanol and the pigment was measured spectrophotometrically in Jinja. Water column limnological properties were established from measurements and samples at the time of experiment setup.

3.2. PHOSPHATE UPTAKE

Kinetics of phosphate uptake were determined with water collected from a nearshore station in Napoleon Gulf during March 1995. Lake water aliquots were sieved through 100 μm Nitex and were dispensed to experimental polycarbonate flasks where they received phosphate additions ranging from 0 to 2 μM Na_2HPO_4. Concentration changes were measured as SRP over 2 h at ambient temperature (ca. 25 °C) in subdued light inside the FIRI laboratory building in Jinja. Uptake rates were calculated as changes in concentration with respect to time. Additional lake water samples were filtered (GF/C), dessicated, and returned to Ann Arbor for pigment analysis and particulate chemistry. Maximum rates of uptake and half-saturation constants were determined by non-linear regression (SYSTAT 5.03).

3.3. PHOTOSYNTHESIS AND MIXING DEPTH

Parameters for the Photosynthetic Model were obtained as follows. An empirical relationship between light attenuation and algal biomass measured as chlorophyll was determined from historical data reported by Talling (1965) together with modern values for downwelling PAR attenuation measured with a Li-Cor 2-π sensor. Daily and seasonal variation in incident PAR was calculated according to Fee (1990). The maximum rate of light-saturated photosynthesis, P_{max} = 25 mg O_2 (mg Chl)$^{-1}$ h^{-1} was according to Talling (1965). I_K= 11 mE m^{-2} min^{-1} was according to Mugidde (1993). Respiration was taken as 25% of P_{max} (Talling 1965, and unpublished data).

Integral daily photosynthesis was computed with a spreadsheet-implemented version of Fee's (1990) numerical photosynthesis model (Lehman 1997). Computations were made for cloud-free conditions and for a location directly on the equator, corresponding to northern Lake Victoria. In computations, the P vs I relationship was represented as a rectilinear model with a light-limited region of slope I_K and a light saturated region where rate of photosynthesis = P_{max}. As with the original Fee model, no provision was incorporated for inhibition at high light intensities. In addition to integral daily photosynthesis, the model calculations also reported values for the depth of 1% I_O and the critical mixing depth, i.e., the uniform mixing depth at which integral daily (24 h) net photosynthesis equals zero.

3.4. CLIMATE VARIATION

The case for climate control of lake conditions was investigated by comparing mean monthly meteorological observations from National Climate Data Center station 637050 at Entebbe, Uganda, on the northwest shore of Lake Victoria. Earlier studies have demonstrated that weather conditions vary systematically from south to north across the lake (Newell 1960), but that the same temporal and seasonal trends occur across the entire basin. Properties investigated were temperature, dew point, station-level barometric pressure, visibility, average wind speed, and maximum wind speed. Relative humidity was calculated from temperature and dew point. Time periods and sources of data are as follows: 1959 to 1961, and 1993: U.S. Air Force Combat Climatology summary of day records (National Climate Data Center archives); 1994 to 1996: NCDC Global summary of day records.

The 36 months from Jan 1959 to Dec 1961 were defined as the reference period. Reference values for each property by month were computed as the mean for each month in each of the 3 years. Month-specific anomalies were then computed for each property by subtracting the reference means from each monthly value individually.

Analysis of anomalies was treated in two ways. First, anomalies were subjected to a two sample t-test (modern versus reference periods), under *a priori* assumptions (one-tail test) that deviations should be in the expected direction from zero. The *a priori* tests were for deviations that would contribute to observed changes in the lake: increased temperature, increased dew point, increased relative humidity, decreased atmospheric pressure, decreased visibility, and decreased wind speed. Second, the total numbers of positive and negative anomalies were tabulated and compared to a null hypothesis of 1:1 ratio.

4. Results

4.1. PHOTOSYNTHESIS MODEL

Talling (1965) reported attenuation of the most penetrating spectral components in Lake Victoria (green light) at chlorophyll concentrations ranging from 1 to 5 mg m^{-3}. That was the effective range of epilimnetic biomass at the time. We supplemented the historical values with measurements at modern concentrations ranging from 4 to 16 mg m^{-3} (Fig. 1) and fit the augmented data set to a linear regression model (r^2 = 0.87, n = 17):

$$\text{Light Attenuation } (m^{-1}) = 0.036[\text{Chl}] + 0.15 \tag{1}$$

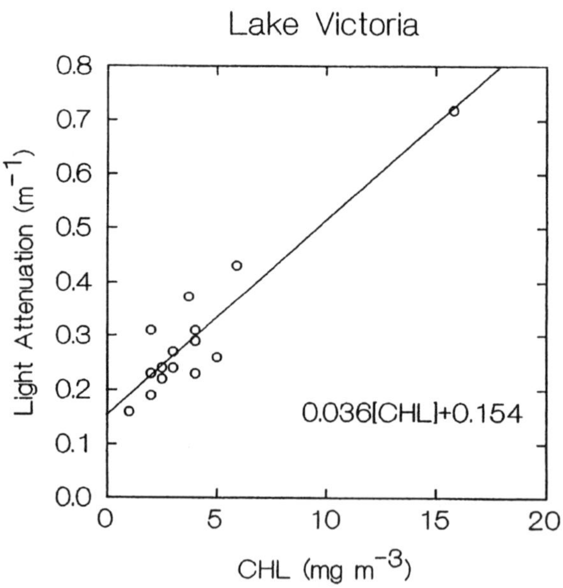

Figure 1. Relationship between epilimnetic algal biomass measured as Chl *a* and vertical attenuation of PAR.

This relationship defines a deterministic link between algal biomass and the vertical extent of primary production. As biomass measured by Chl increases, attenuation increases proportionally, and integral photosynthesis is constrained to an asymptotic upper limit (Fig. 2). The isolume of 1% I_0 varies reciprocally with Chl biomass, as does the critical mixing depth (Fig. 2). The curve identified as Z_{crit} in Fig. 2 defines the constraint relationship between mixing depth and Chl concentrations. If mixing depth is less than Z_{crit}, photosynthetic energy balance is positive and vertically mixed algal populations can exist in a state of cellular growth and division. If actual mixing depth is greater than Z_{crit}, algae spend too little time in the light to compensate for dark respiration over 24 h, and biomass levels are unsustainable. Deep mixing depths are compatible only with very low concentrations of Chl. The high Chl concentrations of the modern lake require mixing depth of about 10 m or less.

One additional parameter that can be derived from the photosynthetic model is compensation depth (Z_{comp}), or the static depth at which photosynthesis balances respiration on average over 24 h, and thus where net photosynthesis is zero on average. This derived property is relevant to Growth Response Experiment conditions, which simulate in situ depths according to light intensity. Table 1 summarizes Growth Response Experiment conditions that are relevant to the Photosynthesis Model. Epilimnetic concentrations of Chl are reported, as well as model-derived light attenuation (Fig. 1) and results of the photosynthesis model. In particular, it should be noted that the Low Light treatment (12% I_0) was still well above the expected compensation light level. Z_{comp} was computed to occur at 1.5% to 3.8% of I_0.

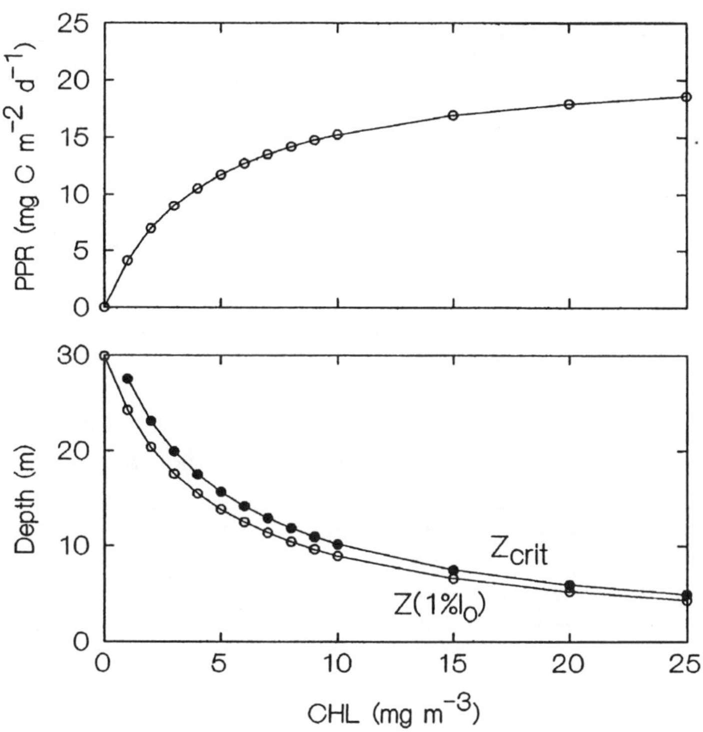

Figure 2. Model relationship between algal biomass measured as Chl *a* and integral photosynthesis (upper) or mixing depth (lower panel).

TABLE 1. Experiment conditions at offshore station: k is light attenuation, Z_{crit} is critical mixing depth, Z_{comp} is compensation depth by Photosynthesis Model.

Date	Chl mg m^{-3}	k m^{-1}	Z_{crit} m	Z at 58% I_o	Z at 12% I_o	Z_{comp}	%I_o at Z_{comp}
SEP 94	6.5	0.39	13.5	1.4	5.5	9.0	3.0
NOV 94	22.3	0.96	5.5	0.6	2.2	4.4	1.5
MAY 95	16.6	0.75	7.0	0.7	2.8	5.4	1.8
AUG 95	3.8	0.29	17.9	1.9	7.3	11.2	3.8

4.2. GROWTH RESPONSE EXPERIMENTS

Epilimnetic conditions of temperature and of chemical properties at the times of offshore experiments are reported in Table 2. Nearshore conditions are identified in Table 3. Offshore regions exhibited SRP concentrations of nearly 1 µM or higher at all times. Nitrate was low or undetectable near surface at all times other than August 1995. August conditions were indicative of recent deep mixing; chemical profiles were relatively homogeneous with depth, and deep waters were enriched with oxygen. August was also the time of lowest Chl concentrations, and elevated nitrate near surface.

TABLE 2. Conditions at Offshore station at experiment setup.

Date	T °C	Chl mg m^{-3}	SRP µM	NO_3 µM	SO_4 µM	Si µM
SEP 94	25.27	4.9	1.7	<0.1	4.8	7.6
NOV 94	25.01	19.1	1.3	<0.1	4.5	15.0
MAY 95	25.79	12.2	0.8	0.5	4.7	4.3
AUG 95	25.19	3.5	1.8	3.3	4.2	45.9

TABLE 3. Conditions at Nearshore stations at experiment setup. na = not available.

Date	Chl mg m^{-3}	SRP µM	NO_3 µM	Si µM
SEP 94	22.6	0.09	<0.1	16.7
NOV 94	56.3	na	na	na
MAY 95	34.8	0.34	0.4	11.6
AUG 95	26.9	0.29	0.2	na

Results of the Growth Response Experiments offshore and nearshore were similar. Addition of N either singly or in combination with P and Si produced growth enhancements with respect to control treatments at all times except the deep mixing event of August 1995 at the offshore station. August was also when nitrate concentrations were elevated in surface water.

Results of Growth Response Experiments were subjected to statistical comparisons by Analysis of Variance (SYSTAT 5.03). Statistical analyses were guided by *a priori* expectations based on previous experience (Lehman and Branstrator 1993, 1994) with nutrient responses by Lake Victoria plankton, and by our hypothesis about the interactive effects of light and nitrogen addition. These statistical tests are summarized in Tables 4 to 6.

TABLE 4. Summary results of Two-Way Analysis of Variance performed on Growth Response Experiments. This table reports the effects of nutrient treatment (CTL, +P, +N, +Si, and +PNSi) and Light (High and Low). ns = not significant ($\alpha = 0.05$).

Site	Date	Source	F-Ratio	P
Offshore	Sept 1994	Nutrient	6.9	<0.0005
		Light	41.2	<0.0005
		Interaction	1.1	ns
	Nov 1994	Nutrient	3.4	0.020
		Light	18.1	<0.0005
		Interaction	2.0	ns
	May 1995	Nutrient	10.3	<0.0005
		Light	3.2	ns
		Interaction	0.2	ns
	Aug 1995	Nutrient	1.7	ns
		Light	6.4	0.017
		Interaction	2.3	ns
Nearshore	Sept 1994	Nutrient	13.1	<0.0005
		Light	0.0	ns
		Interaction	1.3	ns
	Nov 1994	Nutrient	14.5	<0.0005
		Light	9.0	0.005
		Interaction	1.5	ns
	May 1995	Nutrient	18.0	<0.0005
		Light	2.7	ns
		Interaction	4.3	0.007
	Aug 1995	Nutrient	13.9	<0.0005
		Light	26.6	<0.0005
		Interaction	0.5	ns

Summary analysis of the data set (Table 4) is consistent with *a priori* expectations of strong effects of nutrients (7 of 8 trials) and light (5 of 8 trials), but almost no evidence of nutrient × light interaction. Also according to expectation, a subset of the experiment results including only Control, +P and +Si treatments was typically homogeneous with respect to nutrient treatment (Table 5), but effects of light level were present. We found that +N and +NPSi treatments were indistinguishable offshore in all experiments, but that the two treatments could be discriminated nearshore in September 1994 ($P = 0.024$) and August 1995 ($P = 0.002$). As a result of these occasional differences between chlorophyll responses to the two N treatments, we tested the *a priori* hypothesis of increased N-limitation at Low Light by contrasting CTL, +N, and +NPSi at both High and Low Light levels using a Two-Way ANOVA model (Table 6). Despite strong effects of N treatments in all experiments except August 1995 Offshore, significant interaction between N and light could be demonstrated in only 2 cases of 8 possible (August 1995 Offshore and May 1995 Nearshore).

TABLE 5. As Table 4, but for *a priori* contrast of CTL, +P, and +Si treatments (Nutrient) at High and Low Light treatments, testing null hypothesis of zero nutrient effect. No interaction terms were statistically significant ($\alpha = 0.05$).

Site	Date	Source	F-Ratio	P
Offshore	Sept 1994	Nutrient	0.4	ns
		Light	35.7	<0.0005
	Nov 1994	Nutrient	2.6	ns
		Light	2.4	ns
	May 1995	Nutrient	1.5	ns
		Light	2.0	ns
	Aug 1995	Nutrient	1.9	ns
		Light	0.5	ns
Nearshore	Sept 1994	Nutrient	1.5	ns
		Light	0.0	ns
	Nov 1994	Nutrient	1.0	ns
		Light	12.2	0.003
	May 1995	Nutrient	5.8	0.011
		Light	10.2	0.005
	Aug 1995	Nutrient	0.6	ns
		Light	14.1	0.001

TABLE 6. Summary results of Two-Way Analysis of Variance performed on Growth Response Experiments, contrasting CTL, +N, and +PNSi (Nitrogen treatments) and Light (High and Low). ns = not significant ($\alpha = 0.05$).

Site	Date	Source	F-Ratio	P
Offshore	Sept 1994	Nitrogen	6.2	0.009
		Light	20.9	<0.0005
		Interaction	1.3	ns
	Nov 1994	Nitrogen	13.8	<0.0005
		Light	50.8	<0.0005
		Interaction	0.9	ns
	May 1995	Nitrogen	8.2	0.003
		Light	1.5	ns
		Interaction	0.2	ns
	Aug 1995	Nitrogen	0.1	ns
		Light	4.5	0.048
		Interaction	6.4	0.008
Nearshore	Sept 1994	Nitrogen	16.3	<0.0005
		Light	1.4	ns
		Interaction	0.9	ns
	Nov 1994	Nitrogen	12.8	<0.0005
		Light	4.7	0.044
		Interaction	2.4	ns
	May 1995	Nitrogen	36.4	<0.0005
		Light	0.0	ns
		Interaction	6.4	0.008
	Aug 1995	Nitrogen	33.7	<0.0005
		Light	23.6	<0.0005
		Interaction	1.7	ns

In order to make sure we had thoroughly tested our hypothesis about N and light effects, we conducted explicit contrasts between CTL and +N treatments at both high and low light levels for each experiment. We expected the hypothesized interaction would manifest itself as a larger effect of N addition at low light than high light. We reasoned that N-fixation could more readily compensate for N-limitation at high light, where C-fixation is already saturated at P_{max}, than at low light when light levels approach photosynthetic compensation, and the additional energy costs of N-fixation would become a disadvantage. By this logic, addition of inorganic N at low light could encourage development of algal biomass by providing N without the necessary metabolic costs of N-fixation. Standardized differences between +N and CTL treatments (i.e., t-statistic values for +N versus CTL treatments) should thus be greater at Low Light than at corresponding High Light treatments. Out of 8 such possible contrasts, the expected result occurred in only 5 cases (3 of 4 offshore and 2 of 4 nearshore).

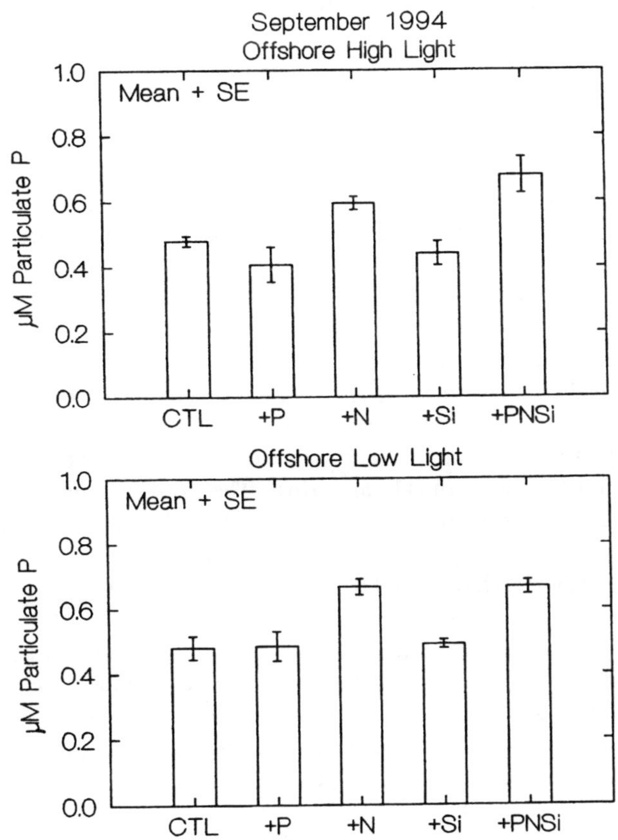

Figure 3. Growth Response Experiment results for September 1994 at the Offshore station, measured as Particulate P.

Further evidence that ambient concentrations of phosphate in Lake Victoria are greatly in excess of ambient biomass requirements by the algae can be seen in results from the Sept 1994 offshore experiment (Fig. 3). Particulate P increased in treatments which received N addition. The source of additional Particulate P is the large pool of dissolved phosphate present in the offshore lake water (Table 2). Increases in algal biomass were expressed not only as increased Chl, but also as increased Particulate P.

The data summarized in Fig. 3 were subjected to Analysis of Variance, treating light level as a categorical variable and Chl as a covariant. Effects of both light and Chl were statistically significant ($P < 0.03$ and $P < 0.0005$, respectively). The stoichiometry of Chl:P in new algal biomass thus differed with light treatment. From simultaneous measurements of Particulate P (μM) and Chl (mg m^{-3}), the stoichiometry of Chl:P in new algal biomass was found by linear regression:

$$\text{Chl:P (High Light)} = 1.65 \quad (\text{SE} = 0.41, n = 21) \quad \text{mg mmol}^{-1} \qquad (2)$$

$$\text{Chl:P (Low Light)} = 2.96 \quad (\text{SE} = 0.77, n = 21) \quad \text{mg mmol}^{-1} \qquad (3)$$

In short, nearly twice as much Chl was produced per unit assimilated P at low light compared with high light levels.

Effects of the light treatments were consistent across all experiments. Low Light treatments typically yielded higher Chl levels than did parallel High Light treatments, even though both sets were initiated with the same source water. Contrasts between light levels by nutrient treatment demonstrated that mean Chl yields at Low Light exceeded High Light Chl by numerical trend in 18 of 20 comparisons offshore and 15 of 20 comparisons nearshore. The differences were large enough to be statistically different by t-test in 12 cases offshore and 7 cases nearshore. These differences were manifest in all nutrient treatments, including control. Thus, incubation of lake water at different light levels, independent of any nutrient amendments, resulted in differential changes. Across the 8 experiments, ratios between the average Chl measured in Low Light and in High Light averaged 1.18 (range = 0.70 to 1.60) for treatments that did not include N addition (i.e., control, +P, and +Si). The same average ratio (1.18: range = 0.79 to 1.67) occurred for treatments that included N addition (+N and +PNSi).

4.3. PHOSPHATE UPTAKE KINETICS

Results of phosphate uptake experiments at the nearshore station are shown in Figure 4. A rectangular hyperbola representing Michaelis-Menten kinetics was fit to the data by nonlinear regression, and estimated parameters for maximum uptake rate (V_m) and half saturation constant (k_s, μM) are reported in Table 7. Also shown in Table 7 are summary kinetic data from the offshore station as reported by Lehman and Branstrator (1994). From these original and published parameters, the initial slopes, or affinity coefficients for P-uptake are calculated as V_m/k_s (Table 7).

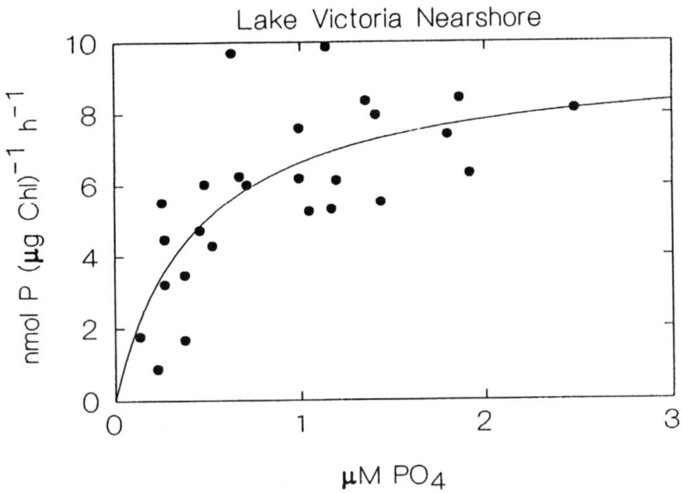

Figure 4. Kinetics of P uptake, Nearshore Lake Victoria, March 1995.

TABLE 7. Phosphate uptake parameters for nearshore Lake Victoria (Napoleon Gulf, NPG), March 1995, and offshore station BG (Bugaia). Data for BG are from Lehman and Branstrator (1994).

Lake	Site	Chl a mg m^{-3}	SRP µM	V_m nmol P L^{-1} h^{-1}	V_m nmol P (µg Chl)$^{-1}$ h^{-1}	k_s nM P	V_m/k_s h^{-1}
Victoria	NPG	18.8	0.04	180 (23)	9.6 (1.2)	447 (168)	0.403
Victoria	BG	11.3	1.65	23 (4)	2.0 (0.4)	1200 (600)	0.019

TABLE 8. Comparison of climate data anomalies for 36 months (April 1993 to March 1996), with reference to means by month from Jan 1959 to Dec 1961 (36 month reference period). Entebbe, Uganda. (Station 637050). T = values of t-statistic testing the difference from zero for mean anomalies.

Property	Mean anomaly	SE	t	positive anomalies	negative anomalies	χ^2
temperature (C)	0.418	0.090	3.88	29	7	7.4
dew point (C)	0.717	0.039	11.77	36	0	24.0
relative humidity (%)	1.585	0.456	2.67	26	10	3.7
pressure (mb)	-0.697	0.099	-6.26	4	31	12.4
visibility (km)	-20.00	0.589	-26.73	0	36	24.0
wind speed (m s^{-1})	-0.187	0.066	-1.76	13	23	1.4
maximum wind speed	-2.006	0.686	-2.52	8	28	6.0

4.4. CLIMATE VARIATION

Deviations have occurred in all climate data categories compared between recent years and 30 years ago (Table 8). In each case the deviations are in the direction that could contribute to elevated lake water temperature and reduced vertical mixing. In particular, increased atmospheric humidity and decreased wind speeds probably contribute to a favorable water balance for the lake which cause its level to remain about 1 m higher than the base level that was maintained before the intense rains of late 1961, 1963, and again in the late 1970s (Bugenyi and Magumba 1996).

5. Discussion

Experiments and model calculations demonstrate that algal biomass in Lake Victoria is limited by available N except during deep vertical mixing offshore. Excess phosphate concentrations offshore can readily be assimilated into new algal biomass if inorganic N is added to the lake water. N-fixation is not sufficient to eliminate the existence of N limitation in short term experiments. The only time that N-limitation is ameliorated is when deep mixing places the plankton under light limitation, thereby causing biomass to decline and nutrients to be released from biological control.

Mixing depth appears to be a master variable for Lake Victoria. Changes in light attenuation are mainly regulated by algal biomass changes (Fig. 1), and maximum biomass levels are constrained by mixing depth (Fig. 2) through interaction between photosynthesis and vertical light climate. When mixing depths are shallow, high algal biomass can develop. The algae are nourished by persistent, high concentrations of phosphate ultimately derived from P-rich carbonatite deposits in the catchment (Holtzman and Lehman 1998).

The ratios at which N and P are supplied to surface water by vertical mixing processes are strongly influenced by mixing depth and the stability of thermal stratification. Table 9 summarizes the stoichiometries of concentration gradients across the main thermocline and oxycline of the lake on three experiment dates (Sept 1994, Nov 1994, and Aug 1995) for which deep profiles exist, and from two previous occasions (Lehman and Branstrator 1993, 1994) when Growth Response Experiments demonstrated N-limitation. The molar gradient ratios of N to P represent the ratios at which these new nutrients would be supplied to the epilimnetic plankton by entrainment or vertical eddy diffusion. It must be recognized that the very low ratios of N to P correspond to periods of sustained thermal stratification. Denitrification removes nitrate from hypolimnetic water at low oxygen concentrations (Hecky 1993), and thus drives the lake toward flux gradients indicative of N-limitation.

TABLE 9. Concentration gradients across the main thermocline in offshore Lake Victoria on different dates, and ratios among gradients referenced to SRP.

	Concentration Gradients ($mmol\ m^{-4}$)				Gradient Ratios (by moles)			
Date	SRP	NO_3	Si	$-O_2$	SRP	NO_3	Si	$-O_2$
APR 92	0.11	0	9.1	15.2	1	0	83	138
OCT 92	0.049	0.08	1.65	6.2	1	3	34	4
SEP 94	0.069	1.04	0.99	3.4	1	15	14	49
NOV 94	0.063	0.58	1.80	7.8	1	9	29	124
AUG 95	0.014	0.29	0.21	2.6	1	21	15	183

Deep mixing produces two effects that act together to relax N-limitation. First, mixing supplies oxygen to the deep water, thus reducing denitrification in the water column and increasing the availability of nitrate for upward flux. Second, deep mixing produces a condition of light limitation. To the extent that mixing depths equal or exceed the critical mixing depth, algal growth ceases and biomass levels decline. The Photosynthesis Model predicts that lower total biomass of algae (mg Chl m^{-2}) can be sustained in a deeply mixed water column than in a shallow mixed layer. Nitrate is thus released from biological control and accumulates to measurable concentrations at the surface (Table 2: August 1995) when deep mixing occurs. The algae at these times are not limited by nutrients. But they cannot increase their biomass because of the feedback between biomass and light attenuation, with consequences for net rates of photosynthesis integrated across the depth of mixing.

Previous workers have called attention to the fact of light limitation in Lake Victoria (e.g., Hecky 1993; Mugidde 1993), and have described it as a consistent feature of the lake plankton. It is true that light climate always plays a critical role in production processes of Lake Victoria, but light limitation is not a complete explanation for the present condition of the lake.

The condition of Lake Victoria is principally determined by its mixing depth. Under climatic circumstances that promote deep mixing, dissolved nutrients are abundant and competition among algae is based on photosynthetic efficiency at low average light. Diatoms are the typical winners of that contest, and deep mixing keeps the cells in suspension. When climate conditions promote shallow mixed layers well stabilized by stratification, epilimnetic biomass increases (Fig. 2), some surface nutrients decrease, diatoms are lost, denitrification contributes to N-limitation, and diazotrophic cyanobacteria appear in the plankton. This basic model is in accord with all recorded changes in the lower food web and nutrients of the lake. The condition of Lake Victoria is set by its mixing state.

Phosphate is present at perennial excess offshore because diurnal mixing depths are still too deep to permit development of algal biomass levels that could consume all of the P at the surface. Simultaneously there is insufficient inorganic N offshore to permit such biomass development. The constraints on algal biomass become relaxed nearshore, where shallow lake morphometry permits mixing to the bottom, keeps cells in suspension, inhibits water column denitrification, and thus allows the abundant inorganic P to be converted into algal particulates. As a consequence, nearshore concentrations of algae are much higher than offshore, SRP concentrations are low or negligible, and the uptake affinities of the algae for phosphate are much greater nearshore (Table 7). Nevertheless, the nearshore algae remain primarily limited by N. The ultimate source of new nitrogen for the elevated algal biomass nearshore must be N fixation, because N income from the catchment is insufficient to support the production rates (Lehman and Branstrator 1992).

We were not able to demonstrate amelioration of N-limitation by N-fixation at high light compared with low light treatments in our experiments. The results do not discount the role of N-fixation in plankton dynamics over long time periods, but instead they demonstrate that inputs of inorganic N release the algae from N limitation most efficiently and rapidly.

There were statistically significant effects of light level in most of the experiments. Light effects were typically manifested as higher Chl concentrations in Low Light treatments than in corresponding High Light treatments. We ascribe these experimental differences to physiological response to light level by the algae. The mechanism is most likely a change in the accessory pigment of the photosynthetic unit, akin to sun-adaptation and shade-adaptation. It is possible that the ubiquitous effects of pigment adaptation at high and low light levels masked any underlying effects of N-fixation. As we noted, both high and low light treatments were above the estimated compensation light level for photosynthesis, and so positive effects of N addition were evident. Differences in the Chl:P stoichiometry of algae did occur as a function of light level (Eqs. 2 and 3). The differential response represents a possible way to diagnose the effective light climate experienced by algae in situ.

5.1. MIXING DEPTH AND CLIMATE CHANGE

For Lake Victoria, surface water temperatures exceed mean air temperatures all year round (Newell 1960; Ochumba 1996). Thus heat gained from irradiance is lost to the atmosphere at rates that maintain surface water temperatures near or above 25 °C. Heat loss occurs mainly by black body radiation, conduction, and evaporation; outflow effects are negligible (Newell 1960; Talling 1966).

Net heat loss by black body radiation depends on the absolute temperature difference (K) between surface water and sky. Variation in the aerosol or particulate content of the atmosphere could reduce this heat loss, depending on the specific absorption characteristics of the aerosols. Heat loss by conduction is commonly formulated as a linear function of wind speed and the temperature gradient from lake surface to air. Either decreased wind speed or elevated air temperatures can decrease this flux. Heat loss by evaporation is linked with conduction loss and is typically formulated as a linear function of wind speed and humidity differences in the air compared with a saturated vapor atmosphere in equilibrium with the lake surface. Decreased wind speed or increased relative humidity decrease this flux. Heat balance of surface waters is fundamentally self-correcting; a temporary decrease in heat loss will be compensated by elevated water temperature, and resulting increased heat flux.

Diaz and Graham (1996) report reduced ice cover in the tropical Andes, which parallel similar reductions in African tropical highlands (Hastenrath and Kruss 1992). It appears that a mechanism for the changes exists in increased tropical SST. The authors also observed that the changes in temperature at high elevation should be linked to increased moisture contents in the lower troposphere. They hypothesized that the combination of increased temperatures and humidity was likely to have an impact on hydrological and ecological balances of high altitude ecosystems in the tropics. Virtually every recent change in Lake Victoria from limnological conditions recorded by Talling (1966) can be ascribed by cause and effect to consequences of altered mixing state (Lehman 1996). It thus appears that Lake Victoria presents an example of ecosystem response to these climatic forcing functions, and thus that Hypothesis H3, the climate change hypothesis, is a valid explanation for the modern condition of the lake.

Climate variations intensify and accelerate the effects of nutrient income. Shallow mixing permits higher algal biomass to develop in nutrient rich surface waters, promotes the loss of heavy diatoms, and reduces the ventilation of deeper waters with oxygen. Essentially all inorganic P is converted to organic form in the nearshore before the water exits the lake (Lehman and Branstrator 1994). Consequently, phosphate levels place a constraint on maximum algal biomass levels of nearshore regions. The changes in the condition of the lake can be seen in the composition of its recent sediments. Verschuren et al. (in press) identify changes attributable to landscape alteration during this century that affected the export of both biogenic P and Si to the sediments of the lake. Ngobi et al. (1998) demonstrate that much of the recent sediment has come from the watershed. And Holtzman and Lehman (1998) show that the land-derived material is rich in apatite phosphates. It seems most accurate to regard climate change and eutrophication not as alternative explanations for lake condition, but as elements of a single explanatory theory about Lake Victoria.

6. Acknowledgments

This research was aided in part by grants from the National Geographic Society, the U.S. National Science Foundation, NOAA National Undersea Research Center, and the University of Michigan Global Change Program. We thank the government and people of Uganda for their assistance in facilitating this scientific study.

7. References

Baskin, Y. 1992. Africa's troubled waters. Bioscience **42**: 476-481.

Branstrator, D. K., L. M. Ndawula, and J. T. Lehman. 1996. Zooplankton dynamics in Lake Victoria, p. 337-356. *In* T. C. Johnson and E. Odata [eds.], Limnology, Climatology and Paleolimnology of the East African Lakes. Gordon and Breach.

Bugenyi, F. W. B., and K. M. Magumba. 1996. The present physicochemical ecology of Lake Victoria, Uganda, p. 141-154. *In* The limnology, climatology and paleoclimatology of the East African lakes. Gordon and Breach.

Diaz, H. F., and N. E. Graham. 1996. Recent changes in tropical freezing heights and the role of sea surface temperature. Nature **383**: 152-155.

Fee, E. J. 1990. Computer programmes for calculating in situ phytoplankton photosynthesis. Canadian Technical Report of Fisheries and Aquatic Science.

Hastenrath, S., and P. D. Kruss. 1992. Greenhouse indicators in Kenya. Nature **355**: 503.

Hecky, R. E. 1993. The eutrophication of Lake Victoria. Verhandlungen der Internationale Vereinigung für Limnologie **25**: 39-48.

Holtzman, J., and J. T. Lehman. 1998. Role of apatite weathering in the eutrophication of Lake Victoria, p. 89-98. *In* J. T. Lehman {ed.], Environmental Change and Response in East African Lakes. Kluwer.

Lehman, J. T., and D. K. Branstrator. 1993. Effects of nutrients and grazing on the phytoplankton of Lake Victoria. Verhandlungen der Internationale Vereiniging für Limmnologie **25**: 850-855.

Lehman, J. T., and D. K. Branstrator. 1994. Nutrient dynamics and turnover rates of phosphate and sulfate in Lake Victoria, East Africa. Limnology and Oceanography **39:** 227-233.

Lehman, J. T. 1996. Pelagic food webs of the East African Great Lakes, p. 281-301. *In* T. C. Johnson and E. Odata [eds.], The limnology, climatology and paleoclimatology of the East African lakes. Gordon and Breach.

Lehman, J. T. 1997. A modification of Fee's numerical primary production model for spreadsheets. Technical report.

Lipiatou, E., R. E. Hecky, S. J. Eisenreich, L. Lockhart, D. Muir, and P. Wilkinson. 1996. Recent ecosystem changes in Lake Victoria reflected in sedimentary natural and anthropogenic organic compounds, p. 523-541. *In* T. C. Johnson and E. Odata [eds.], The limnology, climatology and paleoclimatology of the East African lakes. Gordon and Breach.

Mugidde, R. 1993. The increase in phytoplankton primary productivity and biomass in Lake Victoria (Uganda). Verhandlungen der Internationale Vereinigung für Limnologie **25**: 846-849.

Newell, B. S. 1960. The hydrology of Lake Victoria. Hydrobiologia **15**: 363-383.

Ngobi, G. N., K. Kelts, T. C. Johnson, and P. A. Solheid. 1998. Environmental magnetism of the Late Pleistocene Holocene sequences from Lake Victoria, East Africa, p. 59-73. *In* J. T. Lehman [ed.], Environmental Change and Response in East African Lakes. Kluwer.

Ochumba, P. B. O. 1996. Measurement of water currents, temperature, dissolved oxygen and winds on the Kenyan Lake Victoria, p. 155-167. *In* The limnology, climatology and paleoclimatology of the East African lakes. Gordon and Breach.

Stager, J. C., B. Cumming, and L. Meecher. 1997. A high-resolution 11,400-year diatom record from Lake Victoria, East Africa. Quaternary Research **47**: 81-89.

Talling, J. F. 1965. The photosynthetic activity of phytoplankton in East African lakes. Internationale Revue der gesamten Hydrobiologie **50**: 1-32.

Talling, J. F. 1966. The annual cycle of stratification and phytoplankton growth in Lake Victoria, East Africa. Internationale Revue der gesamten Hydrobiologie **51**: 545-621.

Verschuren, D., Edgington, D. N., Kling, H. J., and Johnson, T. C. in press. The eutrophication of Lake Victoria: sedimentary signals at offshore stations. Journal of Great Lakes Research.

CARIDINA NILOTICA: SPATIAL DISTRIBUTION AND EGG PRODUCTION IN LAKE VICTORIA, UGANDA

GODFREY B. MBAHINZIREKI
Department of Molecular Genetics, Ohio State University
Columbus, OH 43210-1292 USA

JOHN T. LEHMAN
Department of Biology, University of Michigan
Ann Arbor, MI 48109-1048 USA

HANNINGTON OCHIENG
Uganda Fisheries Research Institute
P.O. Box 343, Jinja, Uganda

1. Abstract

The freshwater prawn *Caridina nilotica*, a major prey for juvenile Nile perch, *Lates niloticus*, is abundant in both shallow and deep regions of Lake Victoria. The shrimp exploits the full water column of the lake, with most individuals close to the bottom during day time and most distributed throughout all lake depths during the night. Diel vertical migration by *Caridina* occurs at all lake regions, but it is most pronounced at deep, offshore locations. In offshore regions, *Caridina* populations are dominated by juveniles, and few individuals attain lengths exceeding 18 mm. Larger *Caridina* occur mainly nearshore, among littoral rooted macrophytes and expanses of water hyacinth (*Eichhornia crassipes*). Minimum length of gravid females was 12 mm, and clutch sizes vary allometrically with female size. We combine empirical measurements of fecundity with embryo development rates to develop a model that predicts population birth rates from size frequency data. The model provides a convenient estimation of rate parameters for population growth rates from simple measurements for an important invertebrate forage item. Average birth rate of offshore *Caridina* populations is about 2% d^{-1}, corresponding to a potential population doubling time of about 33 d.

2. Introduction

Caridina nilotica (Roux), a freshwater atyid prawn, is one of the most abundant and ecologically important macroinvertebrates in Lake Victoria. The decapod can attain a length of 25 mm (Fryer 1960), and has been described as a microphagous chelate

J.T. Lehman (ed.), Environmental Change and Response in East African Lakes, 117–124.

raptatory feeder (Fryer 1960). Pelagic specimens are often transparent, mainly colorless, though sometimes the prothorax is tinged yellow-brown. Littoral specimens found among the weed beds are darkly pigmented, by contrast. Worthington (1931) first recorded the presence of *Caridina* in Lake Victoria, as well as in other lakes of East Africa. Some authors have suggested anecdotally that abundances of the shrimp may have increased in recent years, coincident with decline of native haplochromine cichlid populations. Decline of haplochromines has been ascribed to predation by Nile perch. In turn, *Caridina* is supposed to have replaced the extirpated fish species, and coincidentally to have become a major forage item of juvenile Nile perch (Ogutu-Ohwayo 1990). Hughes (1992) reported that *Caridina* constitute important rations for Nile perch from 5 cm to 70 cm, but not for larger fish, which are mainly piscivorous. Before the rise of Nile perch in Lake Victoria, the major consumers of *Caridina* were *Bagrus dogmac* (Forsk.) and *Clarias mossambicus* (Peters), both catfishes (Corbet 1961).

Lake Victoria has undergone considerable environmental changes during the last three decades. Algal biomass has increased 3 to 5 fold since the 1960s (Talling 1965, Mugidde 1993). Algal communities have changed dramatically; chlorophytes have been replaced by cyanophytes, which dominate the flora at present. Hypolimnetic oxygen depletion results in anoxia that now extends from sediments upward to 40 or 50 m depths. The nutrient chemistry of the lake has also changed (Talling and Talling 1965, Hecky 1993). The harvestable fish fauna has changed from a multispecies fishery to one that targets three species: *Lates niloticus*, *Oreochromis niloticus*, and *Rastrineobola argentae*. The role of *Caridina* as a forage item in the diets of all of these fish makes its status in Lake Victoria a topic of considerable interest.

The objective of our research was to characterize the modern distribution of *Caridina* within the Ugandan waters of Lake Victoria. We recorded abundances, size at maturity, and clutch sizes, to permit calculations of biomass and secondary production.

3. Materials and Methods

3.1. SITES AND SAMPLING

Inshore stations within northern Lake Victoria, Napoleon Gulf (15 m depth), Buvuma Channel (20 to 23 m depth) and Pilkington Bay (10 m depth), were sampled monthly, as was a reference offshore station south of Bugaia Island (Branstrator et al. 1996). Other stations throughout Ugandan territorial waters were sampled irregularly (Table 1) over a period of two years. Most samples were collected during daylight hours (0700 to 1900 h). Triplicate collections from sediments to the lake surface were obtained with a 1 m^2 frame net fitted with 300 μm Nitex mesh. The net was lowered to the sediment surface while the ship lay at anchor and was retrieved after 5 to 10 minutes in order to allow time for *Caridina* to redistribute after initial disturbance. Retrieval was by manual hauling. In some cases the vertical distribution of the animals was determined by lowering the net to specific depths in the water column, above the sediments, and then drawing it to the surface. Collections were preserved in 10% formalin.

In the laboratory, samples were split quantitatively and the subsamples were examined. *Caridina* were counted and measured for body length, from midpoint of the eyes to the base of the telson. Numbers of ovigerous females were recorded and clutch sizes were counted to assess fecundity variations. Samples were subsequently archived in 5% formalin. Statistical analyses were performed with SYSTAT version 5.03.

TABLE 1. *Caridina nilotica* (individuals m^{-2}) in day collections at various nearshore stations in northern Lake Victoria. Nets were lowered to the bottom and towed to surface.

Station	Longitude	Latitude	Depth (m)	Date	*Caridina* m^{-2}
Napoleon Gulf	33° 17' E	0° 24.5' N	15	Aug 94	2200
				Sep 94	3231
				Oct 94	2
Pilkington Bay	33° 19.6' E	0° 15' N	8	Aug 94	2300
				Sep 94	0
				Oct 94	0
Macdonald Bay	33° 41.7' E	0° 12.3' N	18	Mar 94	1454
S. of Buvuma Is.	33° 21' E	0° 5.5' N	12	Jun 94	225
SE of Buvuma Is.	33° 23.5' E	0° 9.5' N	23	Jun 94	589
S. of Kasuri Is.	32° 17' E	0° 35.4' N	8	Oct 94	4
Kisima-Nsadzi Is.	32° 38' E	0° 12.3' N	24	Oct 94	101
Zinga Bay	32° 10' E	0° 8.2' N	8	Oct 94	78
NW of Sigulu Is.	33° 43.7' E	0° 6.8' N	13	Nov 94	12
Itome Bay	33° 14.3' E	0° 13' N	16	Dec 94	239

4. Results

Abundances of *Caridina* at a series of shallow stations (Table 1) reveal a high degree of spatial and temporal variability in population sizes. Abundances encountered at the offshore reference station (Bugaia) are reported in Table 2. Vertical distributions of temperature and oxygen at this station are shown in Fig. 1. Vertical distribution patterns were assessed by raising the net through successively greater depth intervals during either day or night. The observed patterns are consistent with previous reports of diel vertical migration, because the animals are rare near the surface during daytime, but not at night. Similar avoidance of surface waters during daytime is evident at a more shallow, nearshore station as well (Buvuma Channel, Table 3).

The smallest gravid female encountered in this study was a 12 mm individual, but the smallest size encountered with any regularity was 13 mm. Gravid females typically constitute only a few percent of the population at any collection site. The frequency of occurrence of fecund females by size class was assessed by examining 49 samples from the deep offshore reference station and 5 sample collections from shallow nearshore stations (Table 4). Reproductive females were found during every month except January and July. The largest crustaceans encountered were exclusively females.

TABLE 2. Abundances of *Caridina nilotica* (individuals m^{-2}) by depth strata during day or night in collections at the offshore reference station (depth = 60 m). Nets were lowered to specified depths and were towed to the lake surface. Missing values are denoted by "-".

	Depth of Day Collections (m)						Depth of Night Collections (m)					
	10	20	30	40	50	60	10	20	30	40	50	60
Jan 93	0	2	-	49	-	54	85	201	-	244	306	342
Feb 93	0	7	-	120	-	230	-	75	69	120	-	127
Mar 93	0	10	-	64	-	102	-	-	-	-	-	-
Jun 93	0	-	-	-	-	904	-	446	-	629	-	983
Dec 93	0	52	141	617	619	405	46	80	283	277	542	423
Jan 94	0	0	3	23	43	33	-	-	-	-	-	-
Feb 94	0	2	142	67	119	48	-	-	-	-	-	-
Mar 94	0	0	93	675	876	1519	-	-	-	-	-	-
Apr 94	0	13	6	222	1077	2206	-	-	-	-	-	-
May 94	0	0	117	1720	1165	1025	-	-	-	-	-	-
Jun 94	0	4	225	153	202	1106	-	-	-	-	-	-
Jul 94	0	3	11	150	285	1101	-	-	-	-	-	-
Aug 94	-	0	13	0	288	-	-	-	-	-	-	-
Sep 94	-	13	-	99	-	107	376	340	176	632	572	1144
Oct 94	0	0	12	32	42	216	-	-	-	-	-	-
Nov 94	0	2	9	21	46	64	30	88	63	78	57	98

Clutch size increases significantly with body length at the offshore station (Fig. 3; $P < 0.0005$). Analysis of variance on log-transformed data detected no differences in the clutch size vs length relationship among 8 different dates (June 92, Aug 92, June 93, Dec 93, Feb 94, May 94, June 94, Sept 94). The empirical allometric relationship, determined by nonlinear regression (Fig. 2), between body length (L, mm) and clutch size (CS, eggs/female) for offshore Lake Victoria is

$$CS = a \times L^{b} \tag{1}$$

where a = 0.00086 (SE = 0.00126), and b = 4.31 (SE = 0.54).

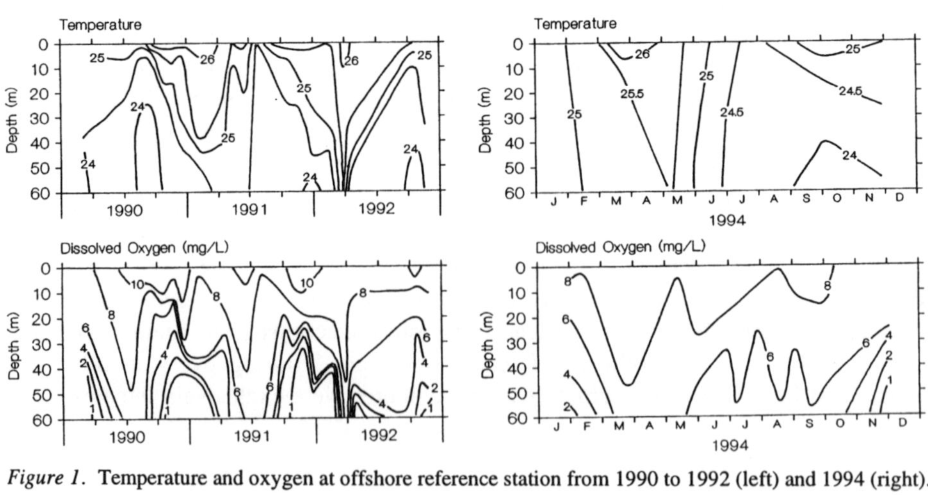

Figure 1. Temperature and oxygen at offshore reference station from 1990 to 1992 (left) and 1994 (right).

TABLE 3. Abundances of *Caridina nilotica* (individuals m^{-2}) by depth strata during day collections at a nearshore reference station in Buvuma Channel (depth = ca. 20 m). Nets were lowered to specified depths and were towed to the lake surface.

	Depth of Collections (m)			
	5	10	15	20
Aug 93	1	1	42	1186
Sep 93	5	2	52	125
Oct 93	0	0	38	117

TABLE 4. Frequency distribution of gravid females among 49 collections at offshore reference station, and from 5 collections at shallow, nearshore sites.

Length	Offshore		Nearshore	
(mm)	N	% fecund	N	% fecund
13	310	2.6	304	0
14	375	3.2	370	0
15	662	32.3	206	19.9
16	65	52.3	0	-
17	2	100	0	-
18	1	100	0	-

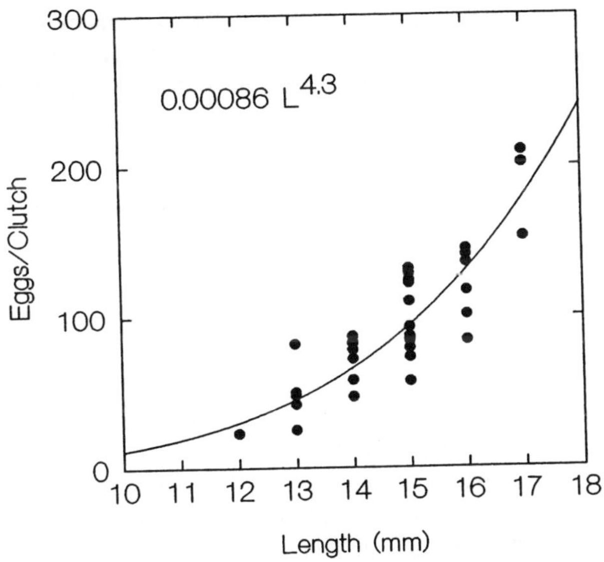

Figure 2. Clutch size versus body length for *Caridina* collected at the offshore reference station.

5. Discussion

Patterns of diel vertical migration exhibited by *Caridina* at our offshore reference station (Bugaia) have been reported by Lehman et al. (1996) based on vertical net collections and direct video observations. More limited data about night distributions at shallow stations indicate that migration patterns are similar at those sites to that recorded at Bugaia. During the day, most *Caridina* remain in the lower half of the water column, but they migrate toward the surface at night. Diel migration is a characteristic behavior to avoid predation by visual planktivores. Previous model studies based on physiological data have established that assimilation efficiencies of detritus-fed animals are low (Ignatow et al. 1996). It is conceivable, therefore, that the vertical migration represents an effort by the shrimp to gain access to food with increased assimilation potential at shallow depths.

Caridina is not restricted from entering deep water by the low concentrations of oxygen that occur there seasonally. Branstrator and Ndawula (1997) in this volume have demonstrated that *Caridina* can survive and grow at low oxygen levels for more than a week at a time. The population size structure offshore is always dominated by juveniles (Ignatow et al. 1996), which indicates the likelihood of intense predation on the adults. In fact, the large size categories of *Caridina* reported by Fryer (1960) are no longer present in limnetic samples, probably owing to predation by juvenile Nile perch. Among the thousands of *Caridina* measured for this study, no individual exceeded 19 mm length, and specimens larger than 15 mm were rare. Most reproduction, therefore, appears to come from primaparous females, whose life expectancy may be short. At present, large *Caridina* can be found only in littoral regions, among thick mats of water hyacinth (*Eichhornia crassipes*). The weed beds may provide a structural refuge from fish predation.

Caridina eggs are roughly 0.5 mm in diameter and are closely packed and glued ventrally among the five pairs of pleopods of adult females. The greatest numbers of gravid females were found during June 1993 and October 1994, when they comprised 7% and 9%, respectively, of the total populations.

Our new information about the fecundities and clutch size variation of *Caridina* complements and extends the size-structured secondary production model developed by Ignatow et al. (1996). Now, in addition to estimates of somatic growth, size frequency data can be used to estimate proportions of the population that are fecund, and the expected clutch sizes of the individuals. Combining these statistics with egg development rates measured by Hart (1980), population birth rates can be estimated by Edmondson's Egg Ratio method (Edmondson 1974, Paloheimo 1974). For example, Table 5 shows a representative estimation of birth rate calculated for the same sample used by Ignatow et al. (1996) to illustrate their secondary production model (their table 2). Fecund proportions of each size class were taken from Table 4, and duration of embryo development (D_e, days) was calculated from Hart (1980) according to the equation:

$$D_e = 20240\ T^{-2.172} \tag{2}$$

At the ambient lake temperature of 25 ^{0}C, D_e = 18.6 days. As demonstrated in Table 5, the instantaneous birth rate (b, day^{-1}) for the population is estimated according to equation 10 of Paloheimo (1974), modified for multiple size categories:

$$b = \Sigma \ln(f_i \times p_i \times CS_i + 1)/D_e \tag{3}$$

where f_i is the empirical relative frequency of *Caridina* in size class i, and p_i is the expected proportion of fecund animals in that size class, from Table 4. Clutch size for each size class (CS_i) is calculated by Eq. 1. The calculation reveals that birth rate is roughly 6 percent per day for the population, and that the larger individuals, although rare, contribute significantly owing to their large clutch sizes.

Mean birth rates determined for *Caridina* at our offshore reference station between October 1992 and June 1994 (N = 27) were 0.021 (SE = 0.005) d^{-1}. This birth rate represents a potential population doubling time of 33 days. Birth rate was not significantly correlated with either population abundance (R = 0.355, n.s.) or month of year (R = 0.332, n.s.) within our data set. Future, more comprehensive investigation of the temporal variation in birth rate would be desirable, however, particularly with attention to any detectable differences between times of stratification and vertical mixing, and among regions of the lake.

Model development for growth and reproduction of *Caridina nilotica* now makes it feasible to estimate production, biomass, and population dynamics from spatial surveys of population size structure. Because the animals exhibit considerable heterogeneity, our effort was directed at developing methods which could be applied to size-structured data collected by automated instruments or by nets. Our model approach is appropriate for Lake Victoria and other large tropical lakes because temperature variations are slight both seasonally and vertically, thus simplifying physiological rate calculations which involve temperature dependences. We believe that the approaches used by Ignatow et al. (1996) and by us in this report have considerable generality and can be applied to the estimation of potential fishery production in the Lake Victoria ecosystem, as well as in other lakes where *Caridina* is an important forage resource.

TABLE 5. Representative birth rate calculations for Lake Victoria *Caridina* sampled 27 Oct 1992 at 0930 h from 60 m to surface. Size frequency data from Ignatow et al. (1996, table 2). E_i is the egg ratio ($f_i \times p_i \times CS_i$).

L (mm)	Relative Frequency (f_i)	Proportion Fecund (p_i)	Mean Clutch Size (CS_i)	$\ln(E_i+1)/D_e$ (d^{-1})
2	0.005	0	0	-
3	0.016	0	0	-
4	0.033	0	0	-
5	0.096	0	0	-
6	0.146	0	0	-
7	0.112	0	0	-
8	0.225	0	0	-
9	0.101	0	0	-
10	0.050	0	0	-
11	0.022	0	0	-
12	0.096	0	0	-
13	0.022	0.026	54	0.0016
14	0.039	0.032	75	0.0048
15	0.016	0.323	101	0.0226
16	0.011	0.523	133	0.0306
17	0	1.0	173	0

Population Birth Rate b (d^{-1}) = $\Sigma \ln(E_i + 1)/D_e = 0.060$

6. Acknowledgments

We thank the limnology team of the Fisheries Research Institute, Jinja, Uganda, as well as the captains and crew of the *MV Ibis* and *MV Imputa* for their cooperation and field assistance. This research was supported by NSF grant DEB-9318035 as well as by funds from the International Development Research Center, Ottawa, Canada, and from the Government of Uganda.

7. References

Branstrator, D. K., L. M. Ndawula, and J. T. Lehman. 1996. Zooplankton dynamics in Lake Victoria, p. 337-355. *In* T. C. Johnson and E. O. Odada [eds.], The limnology, climatology, and paleoclimatology of the East African lakes. Gordon and Breach.

Branstrator, D. K., and L. M. Ndawula. 1998. Oxygen tolerance of the atyid prawn, *Caridina nilotica*, in Lake Victoria (East Africa) and implications for habitat refuge from Nile perch, p. 125-133. *In* J. T. Lehman [ed.], Environmental change and response in East African lakes. Kluwer.

Corbet, P. 1961. The food of non-cichlid fishes in the Lake Victoria basin, with remarks on their evolution and adaptation to lacustrine conditions. Proceedings of the Zoological Society of London **186**: 1-101.

Edmondson, W. T. 1974. Secondary production. Mitteilungen Vereinigung Internationale Limnologie **20**: 229-272.

Fryer, G. 1960. The feeding mechanism of some atyid prawns of the genus *Caridina*. Transactions of the Royal Society of Edinborough **64**: 217-244.

Hart, R. C. 1980. Embryonic duration, and post-embryonic growth rates of the tropical freshwater shrimp *Caridina nilotica* (Decapoda: Atyidae) under laboratory and experimental field conditions. Freshwater Biology **10**: 297-315.

Hecky, R. E. 1993. The eutrophication of Lake Victoria. Verhanglungen der Internationale Vereinigung für Limnologie **25**: 39-48.

Hughes, N. F. 1992. Nile perch (*Lates niloticus*) predation on the freshwater prawn, *Caridina nilotica* in the Nyanza Gulf, Lake Victoria, E. Africa. Environmental Biology of Fish **33**: 307-309.

Ignatow, M., G. Mbahinzireki, and J. T. Lehman. 1996. Secondary production and energetics of the shrimp *Caridina nilotica* in Lake Victoria, East Africa: model development and application. Hydrobiologia **332**: 175-181.

Lehman, J. T., G. B. Mbahinzireki, and L. Mwebaza-Ndawula. 1996. *Caridina nilotica* in Lake Victoria: Abundance, biomass, and diel vertical migration. Hydrobiologia **317**: 177-182.

Mugidde, R. 1993. The increase in phytoplankton primary productivity and biomass in Lake Victoria (Uganda). Verhanglungen der Internationale Vereinigung für Limnologie **25**: 846-849.

Ogutu-Ohwayo, R. 1990. Changes in the prey ingested and the variations in the Nile perch and other fish stocks of Lake Kyoga and the northern waters of Lake Victoria (Uganda). Journal of Fish Biology **37**: 55-63.

Paloheimo, J.E. 1974. Calculation of instantaneous birth rate. Limnology and Oceanography **19**: 692-694.

Talling, J. F. 1965. The photosynthetic activity of phytoplankton in East African lakes. Internationale Revue der gesamten Hydrobiologie **50**: 1-32.

Talling, J. F., and I. B. Talling. 1965. The chemical composition of African lake waters. Internationale Revue der gesamten Hydrobiologie **50**: 421-463.

Worthington, E. B. 1931. Vertical movements of freshwater zooplankton. Internationale Revue der gesamten Hydrobiologie und Hydrographie **25**: 394-436.

LOW-OXYGEN TOLERANCE OF THE ATYID PRAWN, *CARIDINA NILOTICA*, IN LAKE VICTORIA (EAST AFRICA): IMPLICATIONS FOR REFUGE FROM NILE PERCH PREDATION

DONN K. BRANSTRATOR
Department of Biological Sciences
DePaul University
Chicago, IL 60614, USA

LUCAS MWEBAZA-NDAWULA
Fisheries Research Institute
P. O. Box 343, Jinja, Uganda

1. Abstract

Eutrophication of Lake Victoria has led to substantial depletion of hypolimnetic dissolved oxygen concentrations. One major consequence of this altered state is the potential loss of habitat for fishes and invertebrates that are sensitive to hypoxia. Laboratory experiments with the meroplanktonic prawn, *Caridina nilotica*, show that survivorship and body growth are not significantly different between animals reared under low-oxygen conditions (3-4 mg L^{-1} O_2 and those reared in well-oxygenated waters (6-7 mg L^{-1} O_2). Previous, short-duration observations of field populations by remotely operated video camera and by plankton net tows show that *C. nilotica* inhabit poorly-oxygenated regions of Lake Victoria, and our data demonstrate that *C. nilotica* can withstand these conditions for extended periods of more than one week at a time. *C. nilotica* is the major food resource of the introduced Nile perch, a fish which itself demands well-oxygenated waters. Extreme deoxygenation in the hypolimnion of Lake Victoria has become more widespread and prolonged in recent years in response to eutrophication and decomposition of massive algal blooms. If eutrophic conditions persist, spatial expansion of hypoxic waters may reduce habitat for Nile perch and consequently reduce their access to *C. nilotica* prey that occupy low-oxygen environments. This mechanism could negatively impact Nile perch production, but at the same time it could reduce top-down effects by Nile perch on the lower food chain and thus help stabilize the fishery.

2. Introduction

The limnological conditions of Lake Victoria, East Africa, have changed profoundly since pioneering studies of the lake in the 1960s (Talling 1966; Kudhongania and

J.T. Lehman (ed.), Environmental Change and Response in East African Lakes, 125–133.

Cordone 1974). The lake is now more eutrophic with attendant increases in algal productivity and biomass, and accelerated rates of sedimentation (Hecky 1993). Deoxygenation (O_2 < 1.0 mg L^{-1}) of the hypolimnion is frequent during the stratified period of October through March (Hecky et al. 1994; Ochumba 1996). This modern condition contrasts sharply with infrequent historical deoxygenation that occurred at depths usually greater than 60 m (Talling 1966).

As early as the 1980s, hypoxic conditions in Lake Victoria severely threatened habitat boundaries for fishes and altered food-web processes and fish production in the lake. Hecky (1993) and Kaufman and Ochumba (1993) summarize evidence that suggests that episodes of hypolimnetic deoxygenation during that period displaced the once abundant haplochromine cichlids to marginal, shallow-water habitats where Nile perch were easily able to exploit them. The opposing constraints of hypoxic conditions in deep water and severe predation risk in shallow water left haplochromines no viable refuge and their populations collapsed rapidly during the early 1980s (Hecky 1993).

The oxycline in Lake Victoria now demarcates water deeper than 45 m and extends horizontally over half of the lake's sediments during stratification (Hecky et al. 1994). The low-oxygen boundary limits year round habitation of the offshore, benthic zone to species with high tolerance to deoxygenation. The meroplanktonic prawn, *Caridina nilotica*, is one species that appears to tolerate such conditions. Estimates by net collection (Lehman et al. 1996) and direct observation by underwater video at several offshore locations around Bugaia Island, Uganda, (DKB, personal observation) confirm that *C. nilotica* are present in low-oxygen zones (O_2 < 2.0 mg L^{-1}). Whether use by *C. nilotica* of these zones is periodic or chronic is unknown. The question is relevant in Lake Victoria where use of oxygen-scarce waters by the prawn could impact Nile perch production. Nile perch is the cornerstone species of both commercial and subsistence fisheries in the lake and in recent years its production has been based energetically almost entirely on *C. nilotica* (Ogari and Dadzie 1988; Ogutu-Ohwayo 1990). Physiological studies suggest that Nile perch are relatively intolerant of hypoxia (Fish 1956), and this raises the possibility that *C. nilotica*, as forage to Nile perch, may be limited seasonally by consequence of the prawn using areas of the lake that are acutely hypoxic to Nile perch.

In this chapter we assess by laboratory experiments the survival and growth of *C. nilotica* under different oxygen regimes. We then discuss implications of the results to future production and sustainability of the Nile perch fishery in Lake Victoria.

3. Materials and Methods

C. nilotica were collected in November 1995 from a reference sampling station (24 m depth: 00° 23.92' N, 33° 15.09' E) in Napoleon Gulf, Uganda, using a 1-m^2 frame net. The net was deployed to the sediments and rested there for one minute to allow prawns to redistribute across the net mouth before it was retrieved. Animals were transferred to the laboratory at the Fisheries Research Institute, Jinja, Uganda in 20-L carboys that were kept in darkness by placing them inside black plastic containers. In the laboratory, *C. nilotica* were held in plastic basins with well-oxygenated water and were acclimated and observed for 1 to 2 days before experimentation so that damaged and weak animals could be excluded. Less than 5% of collected individuals were thus excluded from experiments.

Experiments were conducted in round, plastic basins filled with 10 L of filtered (50 μm) lake water. Basins were kept on laboratory bench tops and covered with two layers of neutral density window screen to reduce indirect sunlight. Water temperatures and dissolved oxygen (DO) concentrations were measured by a portable thermister and DO meter (ATI Orion model 835). Temperature and DO were recorded twice daily, between 7 A.M. and 9 A.M. local time, and again between 6 P.M. and 8 P.M. Oxygen measurements are presented here as daily averages of these two values. Water temperature fluctuated with room temperature and ranged from 21.9 °C to 25.2 °C across all basins during both experiments. Oxygen concentration was regulated by combined use of pumped air and natural bacterial respiration. Air was delivered to half of the basins (high oxygen basins) by steady, gentle flow through air stones submerged about two cm below the water surface. Bacterial respiration was stimulated in all basins by addition of a *C. nilotica* powder, prepared by drying and crushing one kilogram of *C. nilotica* from Lake Victoria. About 1 g of powder was dispensed almost daily to the surface of every basin. Non-aerated basins responded rapidly to the addition, and we were able to regulate dissolved oxygen in the high-oxygen treatments (6-7 mg L^{-1} O_2) and low-oxygen treatments (3-4 mg L^{-1} O_2) to within about 1 mg L^{-1} of the target concentrations (Fig. 1).

In 1992, we recorded video observations of numerous *C. nilotica* resting on the sediment surface at 62 m depth when oxygen in the overlying water at 60 m depth was 1.3 mg L^{-1} (Lehman and Branstrator 1994). We thus suspect that *C. nilotica* can withstand hypoxic conditions more extreme than we tested here. However, we were unable to maintain stable experimental dissolved oxygen concentrations in the basins for longer than a few days at levels below 2 mg L^{-1} and thus could not test the lower limit of tolerance.

We were able to maintain *C. nilotica* in the laboratory on the powder made from dried *C. nilotica* and we used this as the food source during both experiments. Once hydrated, the powder formed a detrital layer on the bottom of the experimental basin that was in texture similar to natural organic sediment from the lake. Fryer (1960) reports that species of the genus *Caridina* are primarily detritivorous, and this seems likely for *C. nilotica* from Lake Victoria as well. Stable nitrogen isotope signatures of *C. nilotica* from Lake Victoria indicate that these prawns feed low in the food chain and are not carnivores (DKB, unpublished data). The dietary contribution of living phytoplankton versus suspended and settled detritus, however, is yet unknown.

Experiment A measured the number of *C. nilotica* surviving one week at high-oxygen (6-7 mg L^{-1} O_2) versus low-oxygen (3-4 mg L^{-1} O_2). Three replicate basins were used at each oxygen level and each basin was stocked with 30 *C. nilotica* to begin the experiment. The maximum density in each basin was about 400 *C. nilotica* m^{-2} which is similar to estimated natural densities in Lake Victoria (Lehman et al. 1996). We scored the total number of *C. nilotica* alive at the end of each day, but we report here only the final numbers that survived one week. All *C. nilotica* that survived the experiment were preserved in Formalin and later measured for carapace length by ocular micrometer and dissecting microscope. Carapace length was based on Hart's (1980) standard as the "maximum distance laterally from the anterior margin of the carapace behind the insertion of the eyestalk to the most posterior margin of the carapace".

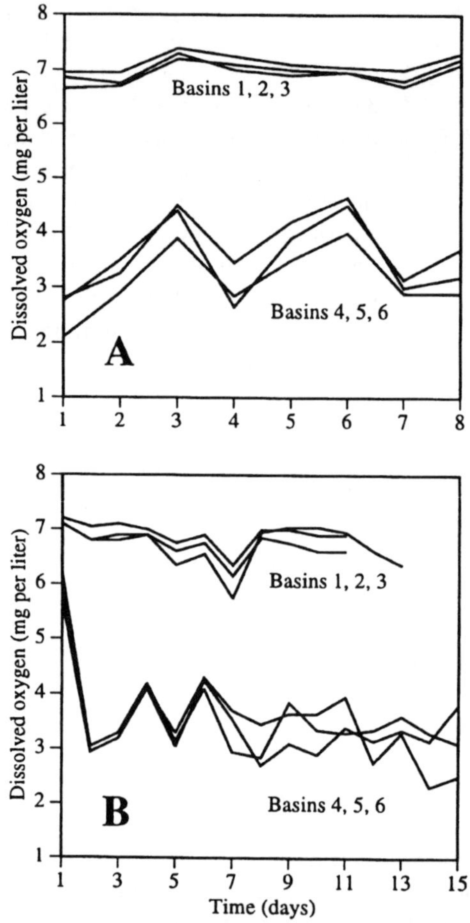

Figure 1. Dissolved oxygen concentrations (mg L^{-1} O_2) in experiments A and B. Low-oxygen treatments in experiment B initiated at about 6 mg L^{-1} O_2 and reduced to target concentration during the first 24 h.

Experiment B measured net somatic growth of *C. nilotica* during 11 to15 days at high-oxygen versus low-oxygen. As in experiment A, each basin was stocked with 30 *C. nilotica*. Carapace length of each specimen was measured at the start of the experiment by ocular micrometer and dissecting microscope. Surviving individuals at the end of incubation were preserved in Formalin and measured later for carapace length. Individual carapace length was converted to dry weight (Eq. 1) and net growth in each basin was estimated from the difference in mean dry weight of pre- and post-measured animals. Hart (1980), working with *C. nilotica* from Lake Sibaya, South Africa, reported that the smallest breeding individuals had a carapace length of 5 mm.

We thus intentionally began our experiment with non-reproductive *C. nilotica* that were strictly 3.7 to 4.1 mm carapace length, in order that their growth during the experiment would be as somatic tissue.

Midway through the experiment on Day 7, the water in each basin was exchanged with fresh, filtered (50 μm) lake water and the accumulated debris in each basin (exuviae, fecal pellets, and uneaten food) was removed.

For construction of a live-length to dry-weight relationship, *C. nilotica* were collected from the sampling station. Live specimens were measured for carapace length and then dried at 60 °C for subsequent weighing (Sartorius Basic Analytical Balance).

4. Results

The following empirical relationship between dry weight (DW) and carapace length (CL) for *C. nilotica* was established by nonlinear regression (r^2 = 0.945, n= 104, SYSTAT 5.0):

$$DW = 0.081(SE = 0.010)CL^{3.213(SE = 0.085)}. \quad (1)$$

Eq. 1 compares favorably with Hart (1980) who reports a similar equation for the relationship between dry weight and carapace length in *C. nilotica* from South Africa.

TABLE 1. Results of experiment A including mean dissolved oxygen level (D.O., mg L^{-1}), final number of *C. nilotica* surviving, and mean individual carapace length (CL, mm) of surviving *C. nilotica* in each replicate.

Treatment and replicate	D.O.	Number surviving	CL
High Oxygen			
1	7.1	24	4.1
2	6.9	27	4.0
3	7.0	26	4.0
Low Oxygen			
1	3.6	27	3.9
2	3.1	25	4.0
3	3.6	26	3.9

Results of experiment A indicate that survivorship was uniformly high among the low-oxygen and high-oxygen basins. Survivorship ranged from 80% to 90% for all basins and did not differ between populations reared at the two different oxygen levels (Mann-Whitney U-Test compared final numbers of survivors in low-oxygen versus high-oxygen basins, P= 0.82; Table 1). The mean lengths of post-measured *C. nilotica* were not different between the low-oxygen and high-oxygen basins either (Mann-Whitney U-Test, P= 0.10).

In experiment B (Table 2), all three high-oxygen basins were terminated early due to substantial mortality on Day 11 (basins 1 and 2) and Day 13 (basin 3). Low-oxygen treatments experienced minimal mortality and they ran for 15 days. Analysis by Independent T-test compared population mean pre-weights versus post-weights in each basin and detected statistically significant growth in 5 of 6 basins (P-values were all <0.05 except for basin 5 where $P = 0.07$). The estimated monthly growth increments for *C. nilotica* were not significantly different between the low-oxygen and high-oxygen treatments (Mann-Whitney U-Test, $P = 0.38$; Table 2).

TABLE 2. Results of experiment B. Experiment duration (d), mean dissolved oxygen (D.O., mg L^{-1}), mean initial and final dry weights (DW_0 and DW_f, mg ± SD, n), and estimated mean monthly net growth increment G (mg DW).

Treatment and replicate	Duration (d)	D.O.	DW_0	DW_f	G
High					
1	11	6.7	6.34 ± 0.60 (30)	6.91 ± 0.52 (21)	1.56
2	11	6.9	6.32 ± 0.51 (30)	6.64 ± 0.66 (28)	0.87
3	13	6.7	6.33 ± 0.53 (30)	6.96 ± 0.61 (21)	1.72
Low					
4	15	3.5	6.41 ± 0.56 (30)	7.19 ± 1.00 (27)	1.56
5	15	3.5	6.52 ± 0.64 (30)	6.94 ± 0.95 (27)	0.84
6	15	3.6	6.61 ± 0.61 (30)	7.16 ± 0.78 (26)	1.10

5. Discussion

Experiments demonstrate that *C. nilotica* can survive and grow for at least one to two weeks in sustained, low-oxygen conditions. Average monthly growth rates per individual ranged from 0.84 to 1.72 mg DW across six experimental units, and survivorship was from 80% to 90% after one week in both high oxygen and low oxygen treatments. These results are relevant to Lake Victoria where periodic use of low-oxygen regions could be an important refuge from Nile perch predation.

Growth rates achieved by *C. nilotica* in our experiment (Table 2) are smaller than values reported by Hart (1981) who worked with *C. nilotica* from Lake Sibaya, South Africa. Hart (1981) found that animals of 3.3 to 4.2 mm carapace length increased 3 to 8 mg dry weight during one month in the laboratory, which is several fold greater than growth rates measured here. Hart used algae and detritus foods from the lake, and acclimated animals to experimental conditions for one month prior to start of his experiment. We acclimated animals to experimental conditions for only two days. *C. nilotica* may be capable of achieving variable growth rates under different conditions.

Members of the genus *Caridina*, and the Atyidae in general, are successful in tropical systems where year-round rates of biological metabolism and oxygen demand at sediment surfaces are high in comparison to temperate lakes (Beadle 1981). In Lake Victoria, adaptations that permit *C. nilotica* to adjust physiologically to hypoxic conditions, and the ultimate limits of their hypoxic exposure, remain unknown. Other aquatic organisms adapted to hypoxia enhance their respiration through increased relative gill surface area and gill ventilation rate, both of which are potentially relevant in *C. nilotica* (Eckert et al. 1988). Respiratory pigments in the hemolymph of *C. nilotica*, that probably include hemocyanin and possibly hemoglobin, may also be important. During our three months of weekly collecting, we consistently noted a contrast in body color between adult-sized *C. nilotica* collected from sediments (24 m depth) at the sampling station and adult-sized *C. nilotica* collected from the roots of water hyacinth (*Eichhornia crassipes*) floating nearby. *C. nilotica* associated with water hyacinth had a characteristic blue hue, which is the color of hemocyanin in oxygenated form (Eckert et al. 1988). In contrast, *C. nilotica* collected from sediments at the sampling station were often deep red, and never blue, which is characteristic of hemoglobin in its oxygenated form. This variation in body pigmentation between individuals which were collected in close spatial proximity suggests that different selective pressures may be operating in these two environments. We suspect that different local characteristics of dissolved oxygen level, visual predation risk, and food type between the two environments may underlie variation in body color that we observe.

C. nilotica migrate vertically in offshore Lake Victoria; the majority of the population is below the thermocline at day and above the thermocline at night (Lehman et al. 1996). Worthington (1931) initially documented this diel pattern of migration during September 1927, prior to massive changes in the lake's fish community and seasonal depletion of deep-water oxygen. His observations, coupled with our laboratory results on survivorship and growth, suggest that other constraints, in addition to oxygen availability, may control vertical migration of *C. nilotica* in Lake Victoria. Lehman et al. (1996) identified fish planktivory as an important influence on vertical migration by *C. nilotica*, and we suspect that variation in food quality and quantity between the sediments and water column also contribute to migration behavior. Ecological tradeoffs among food, predation risk, and oxygen are well-recognized factors structuring vertical migratory patterns of aquatic insect larvae in Lake Victoria and similar tradeoffs are probably also relevant to *C. nilotica*'s success in the lake (MacDonald 1956; LaRow 1970).

Following collapse of haplochromine cichlids in Lake Victoria, *C. nilotica* became the dominant food for juvenile Nile perch, and adult Nile perch became cannibalistic on their own juveniles (Ogutu-Ohwayo 1990). Net productivity of Nile perch is currently contributed by the food of juveniles which makes *C. nilotica* a focus of models that aim to predict future fish production in the lake.

The ability of *C. nilotica* to tolerate low-oxygen environments could have negative consequences for Nile perch because the fish is relatively intolerant of hypoxia. Fish (1956) found that Nile perch require water with high dissolved oxygen (5 mg L^{-1}). Though they can withstand short-term exposure to extreme hypoxia (4.9 minutes without access to the surface at $O_2 < 0.02$ mg L^{-1}), they have a higher threshold for aquatic surface respiration than do several cichlids from the region (Schofield and Chapman, person communication). Low tolerance to hypoxia by Nile perch may account for their preference for relatively shallow waters in Lake Victoria (Kudhongania and Cordone 1974). One likely scenario for the distribution and abundance of *C. nilotica* is tied to the lake's physical mixing regime and suggests that *C. nilotica* builds up high abundances in offshore, deoxygenated areas of the lake during stratification. Seasonal presence of a low-oxygen environment in deep offshore waters, which now occurs annually, should lessen the extent to which Nile perch can exploit *C. nilotica*. This scenario exists in contrast to recent historical conditions under which Nile perch were able to overexploit the haplochromine cichlids that themselves could not tolerate the low-oxygen environment. A refuge for *C. nilotica*, even if only seasonal and offshore, may reduce consumer control of the food chain and help stabilize Nile perch production in the future. Changes in phytoplankton productivity, mean lake temperature, and water column stability that expand the spatial and temporal extent of deoxygenation are likely to further limit the habitat of Nile perch and reduce their spatial overlap with *C. nilotica*. However, *C. nilotica* will likely remain the dominant food resource of juvenile Nile perch in the vast shallow bays and gulfs fringing the main lake where effects of stratification and hypolimnetic anoxia are mitigated by wind-induced mixing. It is unlikely that *C. nilotica* in these areas would be protected by a low-oxygen refuge to any great extent, even if eutrophic conditions worsen.

Despite a dearth of data about historical and modern abundances of *C. nilotica*, it seems reasonable to infer that stocks increased following collapse of haplochromine cichlids. Many haplochromine species that are now either extinct or scarce were once competitors of *C. nilotica* for algae and detritus; others were specialized predators on the prawns (Goldschmidt 1996). Their loss likely favored *C. nilotica*. A low-oxygen refuge from fish predation, and increased production of algae and detrital foods should continue to encourage growth of *C. nilotica*. This invertebrate is a critical link in the Nile perch food chain and cannot be overlooked in trophic dynamic models. Understanding its production and seasonal distribution, both vertical and horizontal, is imperative to development of predictive models and informed management of the Nile perch fishery. In particular, more data are needed to characterize the distribution of oxygen and *C. nilotica* simultaneously, in order to sharpen definitions of the ability of *C. nilotica* to exploit low-oxygen habitats. Energy flow in the modern food chain is remarkably direct, from algae and detritus to *C. nilotica* to Nile perch. Combined application of bioenergetics models (Ignatow et al. 1996) and distribution models for *C. nilotica* and Nile perch, particularly as they relate to oxygen concentration, offer a predictive framework for biologists and lake managers concerned with understanding the sustainable nature of the Nile perch food chain and how eutrophication might interact with fish production in this system.

6. Acknowledgments

This research was financially supported in part by the U.S. National Science Foundation International Programs Postdoctoral Fellowship 9403733 to D. K. Branstrator. Additional support was from U.S. National Science Foundation grants to J. T. Lehman. We gratefully acknowledge S. Wasige and staff at the Fisheries Research Institute, Uganda, for field and laboratory assistance. Comments from L. Chapman and J. T. Lehman substantially improved the manuscript.

7. References

Beadle, L. C. 1981. The Inland Waters of Tropical Africa. Longman.

Eckert, R., D. Randall, and G. Augustine. 1988. Animal Physiology. W. H. Freedman.

Fish, G. R. 1956. Some aspects of respiration of six species of fish from Uganda. Journal of Experimental Biology **33**: 186-195.

Fryer, G. 1960. The feeding mechanism of some atyid prawns of the genus *Caridina*. Transactions of the Royal Society of Edinburgh **64**: 217-244.

Goldschmidt, T. 1996. Darwin's Dreampond. MIT Press.

Hart, R. C. 1980. Embryonic duration and post-embryonic growth rates of the tropical freshwater shrimp *Caridina nilotica* (Decapoda: Atyidae) under laboratory and experimental field conditions. Freshwater Biology **10**: 297-315.

Hart, R. C. 1981. Population dynamics and production of the tropical freshwater shrimp *Caridina nilotica* (Decapoda: Atyidae) in the littoral of Lake Sibaya. Freshwater Biology **11**: 531-547.

Hecky, R. E. 1993. The eutrophication of Lake Victoria. Verhandlungen der Internationale Vereinigung für Limnologie **25**: 39-48.

Hecky, R. E., F. W. B. Bugenyi, P. Ochumba, J. F. Talling, R. Mugidde, M. Gophen, and L. Kaufman. 1994. Deoxygenation of the deep water of Lake Victoria, East Africa. Limnology and Oceanography **39**: 1476-1481.

Ignatow, M., G. Mbahinzireki, and J. T. Lehman. 1996. Secondary production and energetics of the shrimp *Caridina nilotica* in Lake Victoria, East Africa: model development and application. Hydrobiologia **332**: 175-181.

Kaufman, L., and P. Ochumba. 1993. Evolutionary and conservation biology of cichlid fishes as revealed by faunal remnants in northern Lake Victoria. Conservation Biology **7**: 719-730.

Kudhongania, A. W., and A. J. Cordone. 1974. Batho-spatial distribution pattern and biomass estimate of the major demersal fishes in Lake Victoria. African Journal of Tropical Hydrobiology and Fisheries **3**: 15-31.

LaRow, E. J. 1970. The effect of oxygen tension on the vertical migration of *Chaoborus* larvae. Limnology and Oceanography **15**: 357-362.

Lehman, J. T., and D. K. Branstrator. 1994. Nutrient dynamics and turnover rates of phosphate and sulfate in Lake Victoria, East Africa. Limnology and Oceanography **39**: 227-233.

Lehman, J. T., G. Mbahinzireki, and L. Mwebaza-Ndawula. 1996. *Caridina nilotica* in Lake Victoria: abundance, biomass, and diel vertical migration. Hydrobiologia **317**: 177-182.

MacDonald, W. W. 1956. Observations on the biology of chaoborids and chironomids in Lake Victoria and on the feeding habits of the 'Elephant-Snout Fish' (*Mormyrus kannume* Forsk). Journal of Animal Ecology **25**: 36-53.

Ochumba, P. B. O. 1996. Measurement of water currents, temperature, dissolved oxygen and winds on the Kenyan Lake Victoria, p. 155-167. *In* T. C. Johnson and E. O. Odada [eds.], The Limnology, Climatology, and Paleoclimatology of the East African Lakes. Gordon and Breach.

Ogari, J., and S. Dadzie. 1988. The food of the Nile perch, *Lates niloticus* (L.), after the disappearance of the haplochromine cichlids in the Nyanza Gulf of Lake Victoria (Kenya). Journal of Fish Biology **32**: 571-577.

Ogutu-Ohwayo, R. 1990. Changes in the prey ingested and the variation in the Nile perch and other fish stocks of Lake Kyoga and the northern waters of Lake Victoria (Uganda). Journal of Fish Biology **37**: 55-63.

Talling, J. F. 1966. The annual cycle of stratification and phytoplankton growth in Lake Victoria (East Africa). Internationale Revue der gesamten Hydrobiologie **51**: 545-621.

Worthington, E. B. 1931. Vertical movements of freshwater zooplankton. Internationale Revue der gesamten Hydrobiologie und Hydrographie **25**: 394-436.

SECONDARY PRODUCTION BY THE LAKE FLY *CHAOBORUS* IN LAKE VICTORIA, EAST AFRICA: IMPLICATIONS FOR TROPHIC DYNAMICS OF THE MODERN LAKE

JOHN T. LEHMAN
KATHLEEN HALAT
BRYAN BETZ
Department of Biology, University of Michigan
Natural Science Building, Ann Arbor, MI 48109 USA.

LUCAS M. NDAWULA
VINCENT KIGGUNDU
Fisheries Research Institute
P.O. Box 343, Jinja, Uganda

1. Abstract

We modified a general secondary production model for tropical *Chaoborus* to make it more specific and applicable to Lake Victoria, where *Chaoborus* is an important invertebrate predator and forage item for planktivorous fish. We incorporated in situ development times and length-weight relationships of larvae and pupae collected from offshore waters throughout 1994. The model predicted *Chaoborus* biomass to be 55 mg C m^{-2} during 1994 with a production to biomass ratio of 0.11 d^{-1}. Field data show that nearshore populations of *Chaoborus* exhibit a different length-weight relationship and different growth strategy from offshore populations; nearshore animals weigh less than offshore animals per unit length until pupation. The model permits abundance and size frequency data of *Chaoborus* from lake surveys or annual time series to be used to estimate rates of biomass and production on an areal basis. Model calculations also permit estimation of ingestion rates for *Chaoborus* referenced to their own body carbon contents, which average about 0.78 d^{-1} for Lake Victoria populations. Error propagation demonstrates that uncertainties of abundance estimates and of length-weight conversions both contribute to the calculated uncertainties of in situ biomass estimates with about the same magnitudes of error. Current estimates are subject to propagated errors of about 40%. These biomass and production estimates can be improved substantially only by reducing both sources of uncertainty, possibly through the use of automated measurements.

Secondary production and estimated consumption by *Chaoborus* in Lake Victoria is consistent with interpretations that major proportions of lake primary production do not enter the pelagic food web leading to higher consumers. Even though *Chaoborus*

J.T. Lehman (ed.), Environmental Change and Response in East African Lakes, 135–145.

secondary production is about ten times greater than reported fishery yields, it is 1000-fold less than primary production. Inventory and flux calculations suggest that most of the primary production is consumed in detrital respiration directly rather than participating in higher trophic level processes.

2. Introduction

Lake flies, including Chironomidae and *Chaoborus*, are conspicuous members of tropical lake communities. In lakes such as Victoria in Africa, the flies exhibit spectacular synchronized emergences which testify to their biomass accrual and potential food web significance. The insects are planktivorous as aquatic larvae, but do not feed as winged adults. Generation times are typically two months in duration, and it is typical for two cohorts to be present, resulting in a broad range of size classes present in the lake water (MacDonald 1956).

This study develops a secondary production model for *Chaoborus* in Lake Victoria, applies it to size-frequency data, evaluates the precision of resulting estimates, and interprets the results in the context of trophic level interactions and pathways for biological production. Little quantitative data previously existed for biomass and production by offshore *Chaoborus* in relation to other ecosystem components, even though the animal is a forage resource for planktivorous fish. The quantitative importance of the lakefly as both predator and prey needed to be evaluated. Specifically, estimates for biomass and production were needed so that the measurements might be compared with estimates of lake primary production, fishery yield, and the biomass of associated plankton populations. Initial expectation was that *Chaoborus* could be an important reservoir of biomass, and an even more important conduit for biological production.

Crustacean zooplankton, *Chaoborus*, and the freshwater shrimp *Caridina* together constitute the main invertebrate elements of the offshore food web of the modern lake. All three of these groups may function to varying degrees as forage resources to the fish community.

Because *Chaoborus* is a predator on small zooplankton, its *a priori* expected contribution to secondary production was at least two orders of magnitude below rates of primary production. These expectations are based on very general ecological efficiencies of ca. 10% per trophic link applied to the fraction of primary production that enters conventional pelagic food chains rather than detrital ones directly. The expected consumptive demands of *Chaoborus* was expected *a priori* to be a relatively large fraction of zooplankton secondary production. The rationale for the thinking, as expressed by Halat and Lehman (1996), was that conspicuous presence of *Chaoborus* means that planktivory by fish is less a force in the community, compared to lakes like Tanganyika where chaoborids are absent. The reason for relative success by *Chaoborus* in lakes like Victoria may be the presence of refuge conditions in deep water of low oxygen content, which larvae exploit on a diel basis. Fish are excluded from the low oxygen habitats, thereby offering invertebrates a refuge from predation in the hypoxic zone. Branstrator and Ndawula (1998) invoke similar reasoning to explain continued coexistence of *Caridina nilotica* with fish in the lake.

3. Methods

We adapted the secondary production model of Halat and Lehman (1996) to the unique circumstances of Lake Victoria. The model applies weight and temperature dependent functions to estimate respiration, development and growth rates for tropical *Chaoborus*. In the case of Lake Victoria, the lake temperature is almost constant throughout the year at 25 °C, and generation time is known to be two lunar months, based on the work of MacDonald (1956). These facts were used to develop a specific rate function for respiration (R, $\mu g\ C\ d^{-1}$):

$$R = \alpha C^{0.75} \tag{1}$$

where $\alpha = 0.971$ and $SE(\alpha) = 0.194$. We fit a growth function to the length and age data of MacDonald (1956) combined with our empirical length-weight relationship for Lake Victoria *Chaoborus*. Following the approach of Ignatow et al. (1996), we modeled carbon content (C) versus age with the function:

$$C = C_0 \exp[u\ Age/(Age+v)] \tag{2}$$

where C_0 is the weight of the egg. Growth rate (dC/dt, $\mu g\ C\ d^{-1}$) is the first derivative of the growth function:

$$dC/dt = C\ u/(Age+v) \times [1 - Age/(Age+v)] \tag{3}$$

where u = 14.59 (SE = 2.46) and v = 67.9 (SE = 20.4).

We investigated the variation in length-weight relationships for *Chaoborus* through an annual time series. Quantitative archival samples had been collected monthly from an offshore reference station in northern Lake Victoria, south of Bugaia Island (60 m depth) and from a nearshore station in Pilkington Bay (9 m depth). Station locations and sampling methods were reported by Branstrator et al. (1996). All measurements were obtained from quantitative sample splits archived in Ann Arbor. Standard length of larval *Chaoborus* was measured as the distance between midpoints of the segments that contain the gas bladders. Size frequency distributions and abundances were estimated from at least 100 individuals per sample, except in two cases where less than 100 larvae were present in the entire sample split. Weights of larvae and pupae were measured by electrobalance (Cahn 29).

Application of the growth and production model at the population level requires accurate estimates of *Chaoborus* abundance in situ. Expected levels of inter-replicate sampling variability were established from 16 sets of replicate vertical net collections obtained in 1991 and 1992. We used stepwise linear regression to predict the natural logarithm of the inter-replicate standard deviation from the natural logarithms of abundance (*Chaoborus* m^{-2}) and nominal sampled volume (m^3), assuming 100% net collection efficiency. These calculations were used to define the precision of the field abundance estimates.

Statistical calculations were conducted with SYSTAT 5.03. When regression equations from logarithmically transformed data were back-transformed to original units, the corrections prescribed by Bird and Prairie (1985) were applied. Error propagation was by first-order approximation (Meyer 1975).

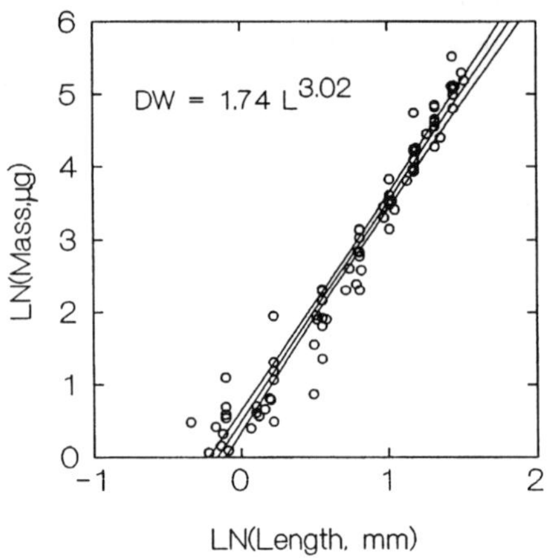

Figure 1. Natural logarithms of *Chaoborus* dry mass (µg) plotted against standard length (mm) for data pooled by size class (0.5 mm intervals) from 10 dates in 1994. Offshore animals.

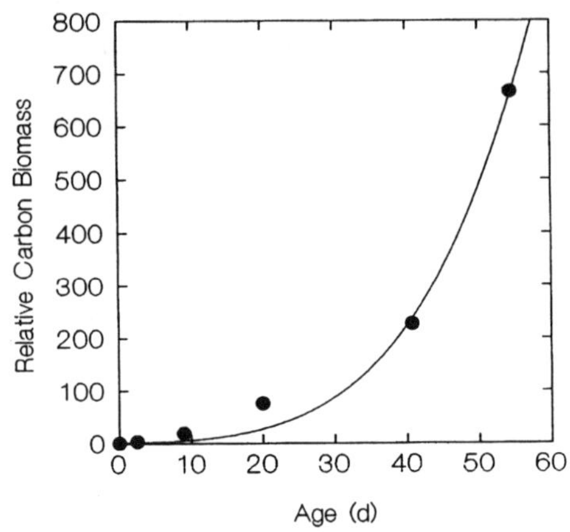

Figure 2. Dry mass versus age for offshore Lake Victoria *Chaoborus*. Fitted curve is Eq. 2.

4. Results

4.1. BIOMASS AND PRODUCTION

The length-weight relationship for offshore Lake Victoria *Chaoborus* is shown in Figure 1. Analysis of variance did not detect any seasonal effect among dates and consequently the data were pooled to produce a single function applicable to larvae throughout the year.

$$\mathrm{DW}\ (\mu g) = a\,L^{b} \tag{4}$$

where L = standard length (mm), a = 1.743 (SE = 0.097), and b = 3.019 (SE = 0.066). We used our empirical estimate for mean pupal weight of 371.2 µg (SE= 17.1, n = 16), and C:DW of 0.45 (McGowan 1974). Individual eggs contain 0.251 µg C (McGowan 1974). The growth function and its fit to data are shown in Figure 2.

Chaoborus collected from a nearshore station in Pilkington Bay of Buvuma Island, Uganda exhibited a different length-weight dependency than did offshore animals.

There were no statistically significant differences detected by ANOVA among nearshore animals from 4 different dates (20 June 1992, 1 March 1993, 23 March 1993, 13 May 1993), but application of Eq. 4 to pooled nearshore data resulted in parameter values (a = 0.805, SE(a)= 0.103; b = 3.515, SE(b)= 0.087) for nearshore populations that were different from offshore animals. Nonetheless, weights of nearshore pupae were indistinguishable from offshore specimens (mean DW = 383.4 μg). The results imply differences in clutch size and reproductive strategy between *Chaoborus* in nearshore and offshore environments, because nearshore animals appear to begin life with smaller mass, but they attain the same pupal weight within a two month life cycle. MacDonald reported the presence of two *Chaoborus* species in Lake Victoria, and our observations are consistent with that report.

TABLE 1. Representative calculations for Lake Victoria *Chaoborus* sampled 30 May 1994. Column entries (X_i) are weighted by relative frequencies of size class (f_i), summed, and totals calculated by multiplying the column sums by measured areal abundance of *Chaoborus* (N, 8330 individuals m^{-2}).

L mm	f_i	C_i μg C	G_i μg C d^{-1}	R_i μg C d^{-1}
0.5-1.0	0.433	0.40	0.08	0.48
1.0-1.5	0.298	0.79	0.14	0.82
1.5-2.0	0.144	3.18	0.47	2.31
2.0-2.5	0.010	4.50	0.62	3.00
2.5-3.0	0.029	15.33	1.70	7.52
3.0-3.5	0.029	23.59	2.40	10.4
3.5-4.0	0.019	38.25	3.53	14.9
4.0-4.5	0.039	111.9	8.14	33.4
$\Sigma f_i \cdot X_i$		7.1	0.65	2.9
	N Indiv m^{-2}	C_{tot} μg C m^{-2}	G_{tot} μg C m^{-2} d^{-1}	R_{tot} μg C m^{-2} d^{-1}
	8330	58.9	5.4	24.2

The size-specific model developed for production and metabolism by offshore *Chaoborus* is illustrated in Table 1. Individual rates by size class are scaled by relative proportions of the population in each class and by total larval abundance. Model calculations produce estimates for biomass, growth, and respiration:

$$X_{tot} = N \cdot \Sigma f_i \cdot X_i \tag{5}$$

where N is *Chaoborus* m^{-2}, X_i represents individual carbon content, growth rate, or respiration in size class i, and f_i is the proportion of the population in size class i. Production to biomass ratio (P:B, d^{-1}) is calculated as the ratio of growth rate (G_{tot}) to biomass (C_{tot}). Prey consumption rate (I_{tot}, d^{-1}) is calculated as:

$$I_{tot} = (\text{Growth} + \text{Respiration})/\text{Biomass}/\text{AE} \tag{6}$$

where AE = assimilation efficiency = 0.8 (Halat and Lehman 1996). Annual variation in biomass, P:B, and I_{tot} is shown in Figure 3.

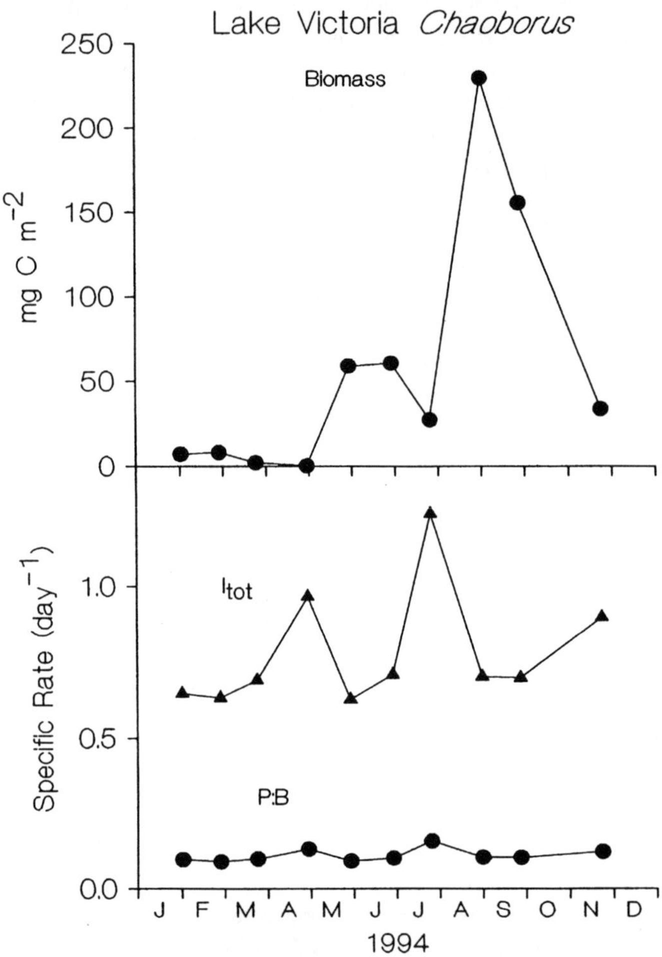

Figure 3. Estimated biomass (C_{tot}), production to biomass ratio (P:B), and per capita ingestion rate (I_{tot}) for offshore *Chaoborus*.

Inter-replicate sampling variability among 16 sets of replicate vertical net collections in 1991 and 1992 is reported in Table 2. Abundance alone proved to be a statistically significant predictor of the variation (Fig. 4). The results indicate that the standard deviation (SD) among replicate samples can be calculated as:

$$SD = 1.078\ N^{0.850} \tag{7}$$

where N is *Chaoborus* m^{-2}. On Lake Victoria, we routinely used four replicate collections at each site. Thus, our expected coefficient of variation for mean estimates (CV_X) is

$$CV_X = SE/Mean = 0.539\ N^{-0.15} \tag{8}$$

TABLE 2. Variability of *Chaoborus* vertical net collections from Lake Victoria. BG= offshore station, Bugaia; PB= nearshore station, Pilkington Bay. N= mean *Chaoborus* m^{-2}; SD = standard deviation; V= nominal volume sampled.

Date	Site	Z (m)	N	SD	V (m^3)
30 Apr 91	BG	50-0	7006	1457	9.8
30 Apr 91	BG	50-0	5823	1898	9.8
10 Apr 92	PB	8-0	520	250	1.6
10 Apr 92	PB	8-0	1290	681	1.6
11 Apr 92	PB	8-0	122	117	1.6
11 Apr 92	PB	8-0	107	53	1.6
12 Apr 92	PB	8-0	167	50	1.6
12 Apr 92	PB	8-0	164	63	1.6
24 Oct 92	BG	50-0	1001	203	9.8
24 Oct 92	BG	10-0	446	92	2.0
27 Oct 92	BG	20-0	242	167	3.9
27 Oct 92	BG	40-0	978	35	7.8
28 Oct 92	BG	20-0	75	36	3.9
28 Oct 92	BG	40-0	757	199	7.8

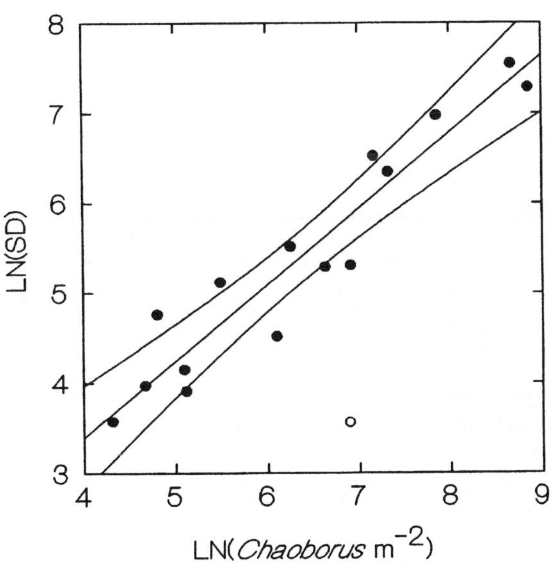

Figure 4. Relationship between interreplicate variation, expressed as standard deviation (SD), and mean larval abundance m^{-2} from Table 3. Fitted line is Eq. 6. One point was excluded from regression.

The prescribed number of replicate vertical net collections (n) required to achieve standard errors within 10% of the mean (where SE = SD/√n) are shown in Table 3 for the range of *Chaoborus* abundances encountered in Lake Victoria.

TABLE 3. Number of replicate samples calculated from inter-replicate SD (Eq. 6) that permit estimation of *Chaoborus* abundance to a precision of 10% (SE/Mean < 0.1), at different abundances of larvae.

Chaoborus m^{-2}	Replicates Required
100	29
300	21
1,000	15
3,000	11
10,000	8
30,000	6

4.2. PRECISION OF MODEL ESTIMATES

At low population abundance, the uncertainty associated with estimates of *Chaoborus* biomass is caused almost equally by uncertainties in abundance and by uncertainties in the conversion from lengths to weights (Table 4). At higher population abundances, the estimation errors for biomass are dominated by the length-weight uncertainties. On average, over the 11 sampling dates our estimates of biomass are subject to uncertainties (SE/Mean) of 44%.

TABLE 4. Error analysis for Lake Victoria *Chaoborus* biomass. Biomass estimates (C_{tot}, mg C m^{-2}) are calculated as the products of abundances (N) and size-specific mass (W). Proportions of estimation error from each source are listed as percentages.

Date	C_{tot}	$SE(C_{tot})$	Var(N)	Var(W)
31 Jan 94	7.2	3.4	44.7	55.3
27 Feb 94	8.2	3.9	42.9	57.1
25 Mar 94	2.2	1.1	51.8	48.2
30 Apr 94	0.3	0.2	56.2	43.8
30 May 94	58.9	24.7	29.2	70.8
29 Jun 94	60.6	25.5	29.6	70.4
26 Jul 94	27.4	10.7	16.8	83.2
31 Aug 94	229.6	90.8	19.2	80.8
27 Sep 94	155.6	62.3	21.4	78.6
23 Nov 94	33.8	14.0	29.1	70.9

5. Discussion

Rates of secondary production by *Chaoborus* are small, as expected, compared with rates of total primary production in Lake Victoria. Mugidde (1993) has reported that modern average primary productivity measured offshore is equivalent to about 5.2 g C m^{-2} d^{-1}. The rates recorded by Mugidde are double those measured 30 years earlier by Talling (1965). *Chaoborus* production averaged 5.9 (SE = 2.3, n = 11) mg C m^{-2} d^{-1} during 1994, or 1000-fold less than primary production. No historical estimates of *Chaoborus* production are available for comparison. The dominating components of variability in estimates of secondary production are in situ abundances of the organisms and determination of individual mass from standard length categories. Because *Chaoborus* is a predator on zooplankton, expectations were that its secondary production would be at least 100-fold below primary production.

Ignatow et al. (1996) reported that *Caridina nilotica*, a freshwater prawn that is an important forage item in the diets of Nile perch, exhibits an average rate of secondary production in Lake Victoria of 11.4 mg C m^{-2} d^{-1}, which far exceeds the current annual fishery yield from the lake at 0.2 to 0.8 mg C m^{-2} d^{-1}. *Chaoborus* biomass is less than that of *Caridina*, but the rates of secondary production are similar. The results imply that *Chaoborus* may represent a forage resource nearly equal in potential to that of the prawn. The results also provide an important commentary on the relative importance of pathways to pelagic versus benthic and detrital secondary production in the lake.

Branstrator and Ndawula (1998) cite evidence that *Caridina* express stable isotope ratios indicative of a detrital algal source for their main diet. Even assuming that detrital food is of such low quality to permit only about 1% trophic transfer efficiency, the *Caridina* production can still be supported by only about 20% of primary production. Using related reasoning, at 10% transfer efficiency, *Chaoborus* production can be supported through two trophic links by about 10% of the primary production. These numbers, together with estimated fishery yield 10-fold lower than *Chaoborus* production, imply that primary production is not efficiently channeled to higher trophic levels in the modern lake. This interpretation is consistent with observable excessive accumulation of biomass as plant material and detritus.

Specific rates of prey consumption by *Chaoborus* average 0.78 (SE = 0.06, n = 11) d^{-1}. The quantitative impact of these rates on zooplankton must be assessed at the population level with reference to the abundance of potential prey. Branstrator et al. (1996) found that total microcrustacean zooplankton biomass in offshore Lake Victoria averaged 4544 mg DW m^{-2} during 1992. Copepod and cladoceran biomass reached maximum levels in July that were approximately double the annual mean biomass, and then declined to about one-half the mean. Thus, zooplankton biomass varies only about 4-fold through the year. We found maximum biomass of *Chaoborus* in August 1994 (Fig. 3) and an annual biomass variation much greater than 4-fold.

At peak *Chaoborus* biomass found in this study (ca. 500 mg DW m^{-2}), consumptive demands by the dipteran planktivores would account for loss of somewhat less than 9% of the annual mean crustacean plankton biomass each day. At its mean biomass, *Chaoborus* could account for loss rates of between 2 and 3% of the mean zooplankton standing crop each day. These low figures do not disallow the possibility that *Chaoborus* may exert selective mortality on individual taxa, at rates well above the population average. They imply that variations in *Chaoborus* abundance might at times have a measurable effect on zooplankton population dynamics, but they are not obviously consistent with *a priori* expectation that *Chaoborus* is a dominating force on zooplankton. Resolution of the question as to whether *Chaoborus* is a major loss vector for crustacean microzooplankton in offshore Lake Victoria will require a quantitative assessment of zooplankton secondary production. At this time, however, our interpretation is that planktivory by fish, though not as extreme as in Lakes Tanganyika, Kariba, or Kivu, is nonetheless the main source of mortality to the zooplankton.

6. Conclusions

Chaoborus is a integral member of the offshore plankton of Lake Victoria, and its biomass and rates of production make it a potentially important forage resource for fish. Individuals increase in body mass by several hundred fold during their two month life cycle. *Chaoborus* production amounts to about 6 mg C m^{-2} d^{-1} on average, and the animals consume about 80% of their weight each day. These rates make them potentially important agents of mortality on offshore zooplankton. Estimates of population biomass in the lake would benefit from increased precision in estimation of mean individual biomass. At present, the time-consuming process of picking and weighing each sample is an impediment to comprehensive lake surveys. Improved precision of abundance estimates can be achieved by increasing the sampling replication, but the required sampling effort to achieve such precision is large, particularly at low abundance levels (Table 3). Efforts to streamline and automate the process of biomass and numerical determinations would improve estimations considerably if the level of sampling replication can be substantially increased.

Secondary production by *Chaoborus* is about 1000-fold less than primary production, and is about one-half that of the atyid shrimp *Caridina nilotica*. A quantitative assessment of zooplankton secondary production is now needed to resolve how much of the production gap between algae and chaoborids is controlled by ineffective grazing on the modern algae, thereby limiting zooplankton production, and how much of the gap is because production by herbivorous zooplankton is intercepted into fish production.

The environmental changes which have produced the modern state of Lake Victoria have produced a food web that accumulates biomass at the level of plants and detritus. Based on evidence to date, it is very likely that the fate of more than half of its primary production is direct consumption in detrital pathways without intervening contribution to higher trophic elements.

7. Acknowledgments

This research was supported by grants from the NOAA National Undersea Research Center and the U. S. National Science Foundation, through its Environmental Biology and Climate Dynamics programs. We thank the staff of the Fisheries Research Institute at Jinja, Uganda for field sampling assistance.

8. References

Bird, D. F. and Y. T. Prairie. 1985. Practical guidelines for the use of zooplankton length-dry weight regression equations. Journal of Plankton Research **7**: 955-960.

Branstrator, D. K. and L. M. Ndawula. 1998. Low-oxygen tolerance of the atyid prawn, Caridina nilotica, in Lake Victoria (East Africa): Implications for refuge from Nile perch predation, p. 125-133. *In* J. T. Lehman [ed.], Environmental change and response in East African lakes. Kluwer.

Branstrator, D. K., L. M. Ndawula, and J. T. Lehman. 1996. Zooplankton dynamics in Lake Victoria, p. 337-355. *In* T. C. Johnson and E. O. Odada [eds.], The limnology, climatology and paleoclimatology of the East African lakes. Gordon and Breach Publishers.

Halat, K. and J. T. Lehman. 1996. Temperature dependent energetics of *Chaoborus* populations: hypothesis for anomalous distribution in the great lakes of East Africa. Hydrobiologia **330**: 31-36.

Ignatow, M., G. Mbahinzireki, and J. T. Lehman. 1996. Secondary production and energetics of the shrimp *Caridina nilotica* in Lake Victoria, East Africa: model development and application. Hydrobiologia **332**: 175-181.

MacDonald, W. W. 1956. Observations on the biology and behavior of chaoborids and chironomids in Lake Victoria and on the feeding habits of the 'elephant-snout' fish (*Morymyrus kannume* Forsk). Journal of Animal Ecology **25**: 36-53.

McGowan, L. M. 1974. Ecological studies on *Chaoborus* (Diptera, Chaoboridae) in Lake George, Uganda. Freshwater Biology **4**: 483-505.

Meyer, S. L. 1975. Data analysis for scientists and engineers. Wiley.

Mugidde, R. 1993. The increase in phytoplankton primary productivity and biomass in Lake Victoria (Uganda). Verhandlungen der Internationale Vereinigung für Limnologie **25**: 846-849.

Talling, J. F. 1965. The photosynthetic activity of phytoplankton in East African lakes. Internationale Revue der gesamten Hydrobiologie **50**: 1-32.

ABRUPT CLIMATIC CHANGES ASSOCIATED WITH THE ARID YOUNGER DRYAS INTERVAL IN AFRICA

KRISTINA R. M. BEUNING
Department of Earth and Environmental Sciences
Wesleyan University
Middletown, CT 06459-0139, USA

KERRY KELTS
Limnological Research Center,
University of Minnesota
Minneapolis, MN 55455, USA

J. CURTIS STAGER
Math/Science Division, Paul Smith's College
Paul Smiths, NY 12970, USA

1. Abstract

The European Younger Dryas (YD) event, a cool period that interrupted northern hemisphere deglaciation ca. 11 to 10 × 10^3 ^{14}C yr BP (ka), has an arid counterpart in Africa. We present new high-resolution palynological evidence from Lake Albert, East Africa indicating that tropical aridity began at 11.5 ka and terminated abruptly around 9.8 ka. This ca. 1700 yr arid event was punctuated by a brief period of somewhat moister conditions between 10.9 and 10.4 ka. The multi-phasic nature of this event appears to be widespread in Africa. These observations carry implications for the mechanisms applied to explain North Atlantic Younger Dryas signals.

2. Introduction

Northern hemisphere cooling has commonly been linked to aridity in tropical Africa (Kendall 1969; Street and Grove 1976; Nicholson and Flohn 1980; Gasse et al. 1990; Street-Perrott and Perrott 1990; Karpuz and Jansen 1992; Mayewski et al. 1993; Roberts et al. 1994). However, evidence of an aridity event in Africa coincident with Younger Dryas (YD) cooling in Europe has remained equivocal. Lakes Bosumtwi and Magadi in tropical Africa provide convincing records of abrupt change to drier conditions during the deglacial period (Talbot and Johannessen 1992; Roberts et al. 1994) but temporal and spatial resolution are lacking to grasp the magnitude and extent

J.T. Lehman (ed.), Environmental Change and Response in East African Lakes, 147–156.

of this cool poles/dry tropics correlation. In comparison to the Northern Hemisphere, paleoclimate records from Africa extending beyond the Holocene are relatively sparse and few have adequate radiocarbon chronologies. Here we present a high-resolution pollen sequence from a large lake in East Africa to confirm that abrupt hydrological changes did occur in East Africa that can be correlated with YD cooling in the Northern Hemisphere and to delimit the nature of these changes with high temporal resolution.

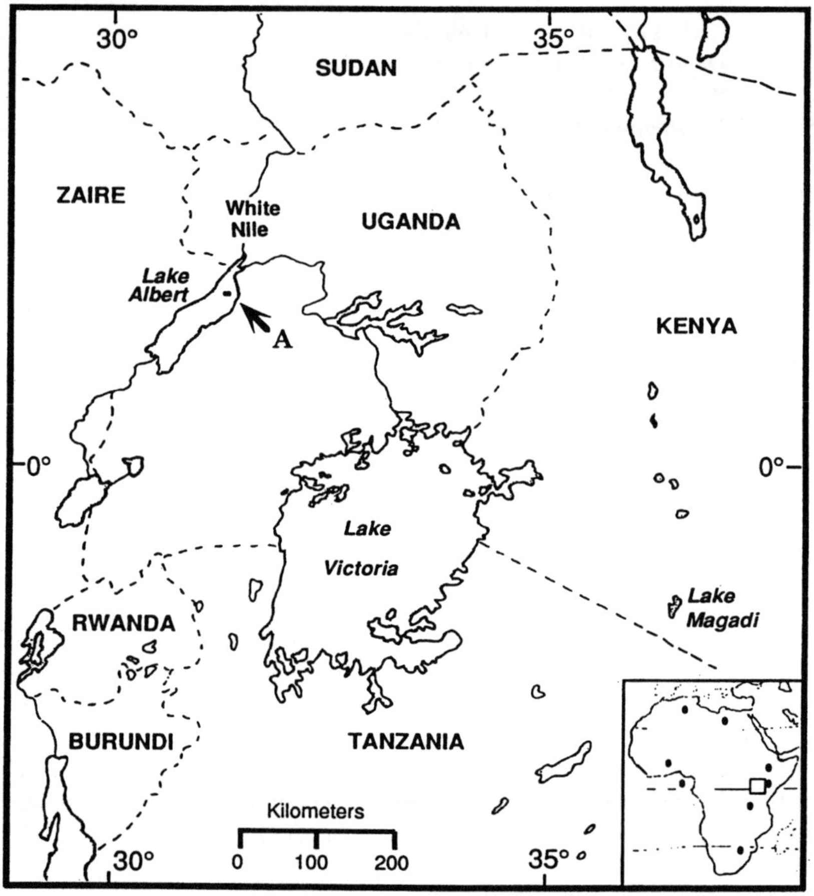

Figure 1. Map of Africa, showing Lake Albert and adjacent area. Symbol "A" points to location of the F core. Inset shows sites of other paleoclimate records cited in the text.

3. Study Site and Core Chronology

Lake Albert, Uganda (1° 3' to 2° 15' N, 30° 22' to 31° 24' E, 620 m above sea level) is the northernmost lake of the Western Rift system of East Africa (Fig. 1). This half-graben lake was formed in the Miocene and is currently about 5,300 km^3, with a maximum depth of 58 m. Most inflow to Lake Albert derives from the Ruwenzori highlands to the south via the Semliki River and from Lakes Victoria and Kyoga to the northeast via the Victoria Nile. All outflow from Lake Albert is northward, forming the start of the White Nile River.

Our palynological analyses from Lake Albert are from a sediment core (core F) taken in 1971 by D. Livingstone and T. Harvey in 46 m of water in the NE corner of the lake. Chronological fidelity across the interval 8.2 to 12.5 ka is based on five conventional bulk radiocarbon dates and one accelerator-mass spectrometer (AMS) date on a pollen concentrate (Table 1; Harvey 1976; Sowumni 1991). The terrestrial AMS date confirms the uniform sedimentation rate across the deglacial and early-Holocene period suggested by the bulk dates from the core. A line connecting the 8.23 ka and 12.5 ka points encompasses all 6 dates and falls within the 1 SD dating error as well as the stratigraphic range of sediment collected for each of the dates (Fig. 2). This linear extrapolation indicates a high sedimentation rate of 130 cm kyr^{-1} for the deglacial/early Holocene interval. Based on this sedimentation rate, pollen samples analyzed for this study represent 50 to 100 yr intervals. Ages are reported as conventional ^{14}C yr BP to facilitate comparisons with older literature and because of still unresolved problems with the YD time range.

TABLE 1. Absolute depths, radiocarbon and calendar ages and lab numbers for dated samples within the deglacial and early Holocene segment of the Lake Albert F core. Samples with Gif and "I" laboratory numbers are dates from bulk sediment samples completed at the Centre des Faibles Radioactivités in France and at Isotopes Inc., New Jersey, U.S.A., respectively (Sowumni 1991; Harvey 1976). The CAMS date was completed on the 30 to 70 μm pollen/lignin/charcoal fraction of the sediment at the Lawrence Livermore Laboratory in California. Calendar dates were calculated using the computer program CALIB 3.0.

Core Depth (cm)	^{14}C yr BP	Calendar yr BP	Uncertainty (±1 SD)	Laboratory Number
85-90	8230 ± 60	9210	9362-9000	CAMS #20671
140-150	8750 ± 400	ca. 9710	10043-9376	Gif 6695
160-200	9005 ± 145	9980	10044-9889	I-6303
360-400	10700 ± 150	12630	12786-12463	I-6142
450-460	10980 ± 260	12898	13193-12645	Gif 6696
640-665	12500 ± 190	14648	14975-14350	I-6141

4. Results and Discussion

Our results confirm an arid interval in Africa coincident with the cool YD event of the northern hemisphere. The key feature of this aridity event is its multi-phase nature. The data imply that aridity in East Africa preceded the start of the European-defined YD interval by ca. 500 yr (Phase P: 11.5 to 10.9 ka) although dating correlations are difficult (Zolitschka 1996). In addition, we suggest that YD-interval aridity in this tropical African system was apparently interrupted by a slight climatic amelioration between 10.9 and 10.4 ka (Phase A) with a return to more arid conditions from 10.4 to 9.8 ka (Phase B, Fig. 3). The radiocarbon age differences of ca. 500 yr may seem suspect, but we believe the linear results are robust to within 150 yr.

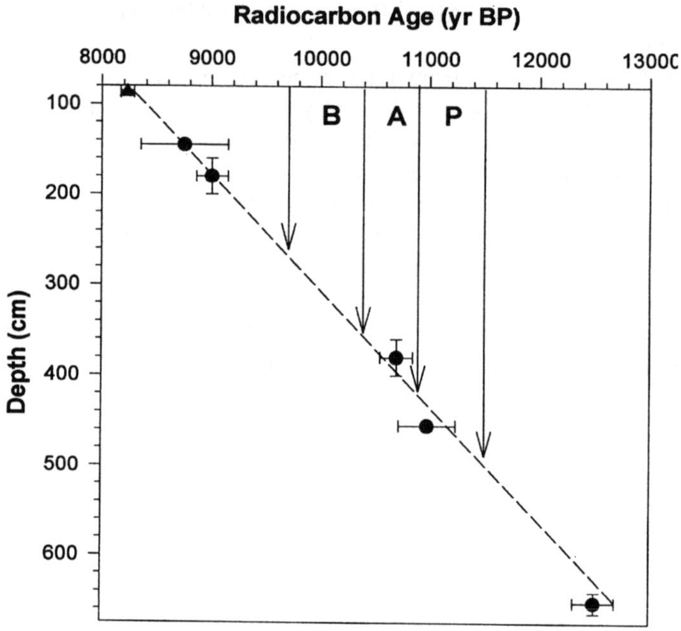

Figure 2. Radiocarbon chronology of the Lake Albert F core. Solid circles are bulk radiocarbon dates and triangle is an AMS date. The dashed line refers to the line used for calculation of dates across the deglacial and early Holocene interval for this core. P, A, and B refer to temporal intervals defined in the text and Fig. 3.

The onset of aridity at ca. 11.5 ka, marking the start of Phase P, is documented in the Lake Albert record by a 3-fold increase in Gramineae pollen percentages (Fig. 3). Independent of changes in grass pollen abundance, percentages of several forest taxa decline across this interval (Fig. 4). From ca. 11.5 to 10.9 ka, vegetation around Lake Albert was dominated by arid wooded grasslands. Savanna expansion and increased fire activity in the Albert region, as evidenced by abundant charcoal and blackened macrophyte shreds, document regional aridity from 11.5 to 10.9 ka (Beuning et al. 1997).

During Phase A (10.9 to 10.4 ka), a rising ratio of precipitation to evaporation is indicated by increasing Moraceae pollen percentages near Lake Albert, independent of shifts in the percentage of Gramineae, and by a drop in Gramineae pollen percentages from 53% to 38% (Figs. 3 and 4). The increase in Moraceae pollen at this time could be attributed to the expansion of forest some distance from Lake Albert. Moraceae pollen is thin-walled, small (less than 15 μm in diameter) and wind-pollinated. This type of pollen is more likely to be transported long distances than other forest taxa that are commonly found with Moraceae trees (i.e., *Alchornea* and *Celtis*), but whose percentages do not increase during this interval.

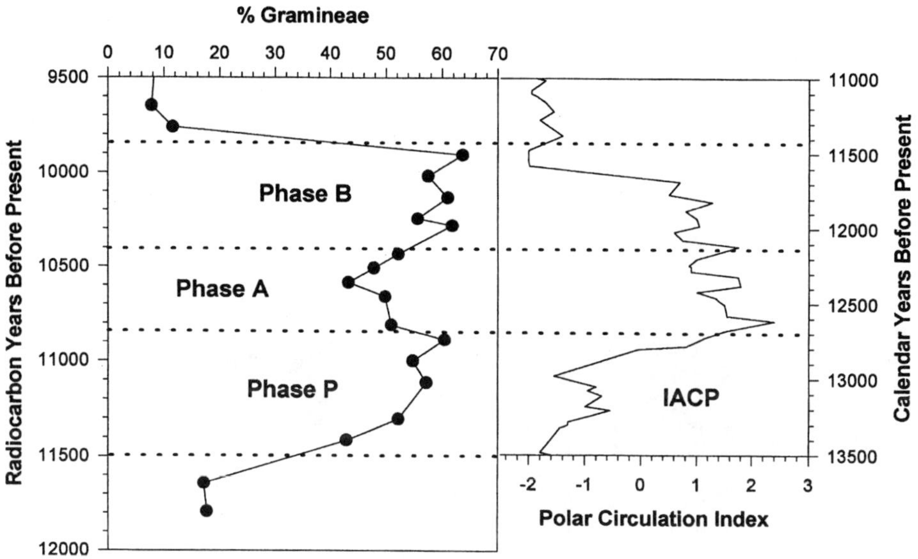

Figure 3. Comparison of East African and Greenland paleoclimate records for the deglacial period. IACP = Intra-Allerød Cold Period, Phase P = 11.5 to 10.9 ka, Phase A = 10.9 to 10.4 ka, Phase B = 10.4 to 9.85 ka. Calibration of ^{14}C years based on Bard et al. (1990). GISP2 Polar Circulation Index reflects atmospheric cooling over Greenland. Gramineae pollen abundance reflects the degree of aridity at Lake Albert. Gramineae pollen percentages fell to 7% by 9.7 ka.

During Phase B (10.4 to ca. 9.85 ka) aridity increased again at Lake Albert, as indicated by Gramineae pollen percentages even higher than in Phase P. Moraceae percentages decline across this interval. The relative abundance of *Olea hochstetteri*-type pollen (including *Olea hochstetteri* and *O. africana*) in both Phases A and B supports the expansion of wooded grassland near Lake Albert. *Olea hochstetteri*-type trees are locally common in the fire-protected depressions within the wooded grassland environments of the rift valley (Beuning, personal observation). At the close of Phase B, forest taxa dominate the Lake Albert pollen spectra as Gramineae pollen percentages decline to one-fifth their value in less than 200 yr (Figs. 3 and 4). A similar abrupt 30% decline of Gramineae pollen in a core from the southern basin of Lake Albert occurs at an extrapolated age consistent with the end of Phase B in the F core data reported here (Ssemmanda and Vincens 1993). These large and abrupt changes in grassland development require major shifts in the water balance of the region.

Although recognition of an arid event in Africa coincident with the Younger Dryas interval is not new, our documentation of the timing and multi-phasic nature suggests this event is more complex than previously reported. We find in tropical East Africa a structured, ca. 1700 yr overall arid interval from 11.5 to 9.8 ka with three characteristic abrupt fluctuations in the severity of the aridity lasting approximately 600 yr each. The early phase of this African arid event (Phase P) from 11.5 to 10.9 ka implies that the onset of tropical aridity may have preceded the northern hemisphere YD cooling, as defined by the GISP2 polar circulation index (Mayewski et al. 1994), by about 500 yr (Fig. 3). Increased aridity at this time may be related to the circum-North Atlantic's

Intra-Allerød Cold Period or to the Antarctic Cold Reversal, which preceded the northern YD signal (Nilsson 1983; Jouzel et al. 1995).

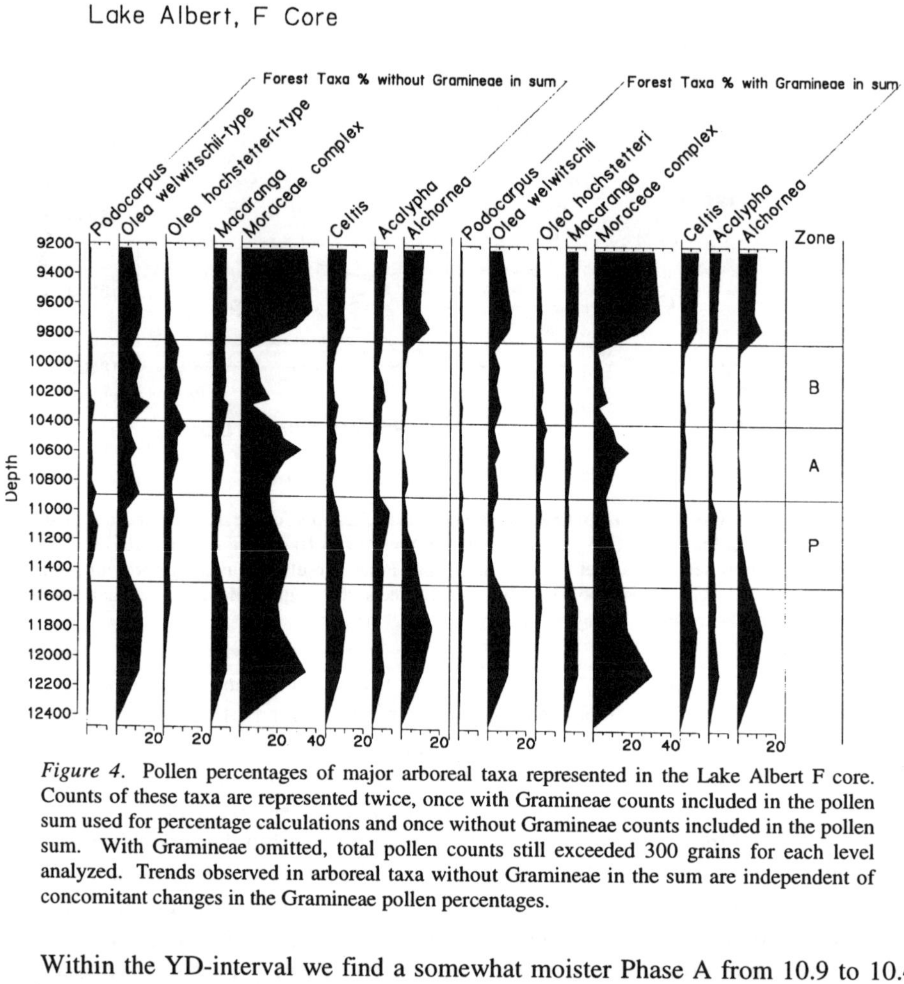

Figure 4. Pollen percentages of major arboreal taxa represented in the Lake Albert F core. Counts of these taxa are represented twice, once with Gramineae counts included in the pollen sum used for percentage calculations and once without Gramineae counts included in the pollen sum. With Gramineae omitted, total pollen counts still exceeded 300 grains for each level analyzed. Trends observed in arboreal taxa without Gramineae in the sum are independent of concomitant changes in the Gramineae pollen percentages.

Within the YD-interval we find a somewhat moister Phase A from 10.9 to 10.4 ka and more arid Phase B from 10.4 to 9.8 ka. This pattern appears to be widespread in African paleoclimate records. An earlier survey of 13 African closed basin lakes found that lake regressions were more widespread during the late YD than during the early YD (Street-Perrott and Roberts 1983); our results from Lake Albert suggest that this was due to greater aridity in Phase B. In the Sahel, we believe a brief moist phase with extrapolated dates of ca. 10.7 to 10.3 ka is linked within the range of dating resolution to the slightly moister conditions of Phase A in Lake Albert (Gasse et al. 1990). In addition, eastern Mediterranean sediments registered heavy Nile floods ca. 11 ka (Rossignol-Strick 1983). Records from the northern Saharan margin (Gasse et al. 1990), North Africa's Tibesti Mountains (Flohn and Nicholson 1980), the Ziway-Shala Basin (Gasse 1980), Kenya's Bogoria and Mt. Satima (Street-Perrott and Perrott 1990; Tiercelin et al. 1981), Lake Tanganyika (Gasse et al. 1989a), South Africa's Transvaal

(Scott 1982), Lake Bosumtwi (Talbot and Johannessen 1992) and Congo deep-sea sites (Gasse et al. 1989b) all present evidence of a similar two phase pattern during this interval (Fig. 1). Despite relatively low-resolution dating at most of these sites, the data appear to be consistent with that from Lake Albert and support a multi-phase, continent-wide aridity event at this time.

A few reconstructions of paleohydrology during the YD interval differ from that at Lake Albert. A peat core from the Burundi Highlands suggests maximum aridity during the early YD but lacks dating resolution across this zone (Jolly et al. 1994). At Lake Magadi, Kenya, a well-dated laminated sequence with multi-proxy evidence suggests that maximum deglacial aridity occurred around 10.7 ka during the relatively moister Phase A at Lake Albert (Roberts et al. 1994).

Evidence for climatic structure within the northern European YD time frame is also postulated for recent high-resolution records from Europe, Asia, and South America (NASP Members 1994; Figge and White 1995; Goslar et al. 1995; Zolitschka 1996). An apparent global climate instability at this time is evidenced by multiple short-term climatic fluctuations in all of these records. Short-term fluctuations in sea-surface temperature (SST) provide a possible mechanism for the fluctuating tropical moisture patterns reported here. Long-term precipitation trends in tropical Africa reflect slow, orbitally driven insolation changes, but these gradual adjustments are an unlikely cause of century-scale climatic fluctuations (Rossignol-Strick 1983; Lehman and Keigwin 1992; Van Campo and Gasse 1993). However, changes in global thermohaline circulation resulting from abrupt ice-rafting or meltwater events (Lehman and Keigwin 1992; Edwards et al. 1993) during Phase A might have briefly increased tropical ocean SST and landward transport of moisture in tropical Africa, interrupting an overall cool, dry period in the region (Street-Perrott and Perrott 1990). Alternatively, continental warming alone might have increased Phase A precipitation by intensifying land-sea thermal gradients. Although temperature and moisture are often difficult to distinguish in proxy records, evidence from glacier and sediment records suggest somewhat warmer conditions in Africa during the slightly moister conditions of Phase A found in Lakes Albert (Hastenrath 1984; Rosqvist 1990; Taylor 1990).

5. Conclusions

The Lake Albert region of East Africa appears to have experienced a rapid-onset overall aridity event from 11.5 to 9.8 ka punctuated by slightly moister conditions from 10.9 to 10.4 ka. This aridity event is coincident with, but could also lead, northern hemisphere cooling during the YD interval. Most other records in Africa do not have the temporal resolution to confirm the timing of this event or a potential tropical lead, but many paleoclimate reconstructions throughout the continent show similar multi-phase, short-term fluctuations consistent with this interval. The ice-core polar index (Fig. 3) is essentially one of temperature, whereas the Albert pollen record is essentially tropical moisture balance and may thus not be expected to exhibit the same detail in structure as the polar data. If the timing and three-phase nature of this aridity event found in the Lake Albert region can be validated with the recovery of additional records in Africa, we need to look further to high-resolution sequences from the tropical Indian and Atlantic oceans to confirm a tropical lead. The energy of water vapor makes the tropics a tantalizing, although yet unproven, trigger of short-term climatic fluctuations.

6. Acknowledgments

This work was made possible by generous financial, technical, and advisory support from D. A. Livingstone, the National Science Foundation, M. Talbot, and D. Trinh. An anonymous reviewer's comments greatly improved the quality of this manuscript, although we remain convinced of the pollen evidence for tri-phase behavior.

7. References

Bard, E., B. Hamelin, R. G. Fairbanks, and A. Zindler. 1990. Calibration of the ^{14}C timescale over the past 30,000 years using mass spectrometric U-Th ages from Barbados corals. Nature **345**: 405-410.

Beuning, K. R. M., M. R. Talbot, K. Kelts, and D. A. Livingstone. 1997. A revised 30,000 paleohydrologic and paleoclimatic history of Lake Albert, East Africa. Palaeogeography, Palaeoclimatology, Palaeoecology **136**: 259-279.

Edwards, R. L., J. W. Beck, G. S. Burr, D. J. Donahue, J. Chappell, A Bloom, E. Druffel, and F. Taylor. 1993. A large drop in atmospheric $^{14}C/^{12}C$ and reduced melting in the Younger Dryas, documented with ^{230}Th ages of corals. Science **260**: 962-968.

Figge, R. A., and W. C. White. 1995. High-resolution Holocene and late glacial atmospheric CO_2 record: Variability tied to changes in thermohaline circulation. Global Biogeochemical Cycle **9**: 391-403.

Flohn, H., and S. Nicholson. 1980. Climatic fluctuations in the arid belt of the Old World since the last glacial maximum; possible causes and future implications. Paleoecology of Africa and the Surrounding Islands **12**: 3-21.

Gasse, F., R. Téhet, A. Durand, E. Gibert, and J. Fontes. 1990. The arid-humid transition in the Sahara and the Sahel during the last deglaciation. Nature **346**: 141-146.

Gasse, F., V. Lédée, M. Massault, and J. Fontes. 1989a. Water-level fluctuations of Lake Tanganyika in phase with oceanic changes during the last glaciation and deglaciation. Nature **342**: 57-59.

Gasse, F., B. Stabell, E. Fourtanier, and Y. van Iperen. 1989b. Freshwater diatom influx in intertropical Atlantic: Relationships with continental records from Africa. Quaternary Research **32**: 229-243.

Gasse, F. 1980. Late Quaternary changes in lake-levels and diatom assemblages on the south-eastern margin of the Sahara. Paleoecology of Africa and the Surrounding Islands **12**: 333-350.

Goslar, T., M. Arnold, E. Bard, T. Kuc, M. Pazdur, R. Ralska-Jasiewiczowa, K. Rózanski, N. Tisnerat, A. Walanus, B. Wicik, and K. Wieckowski. 1995. High concentration of atmospheric ^{14}C during the Younger Dryas cold episode. Nature **377**: 414-417.

Harvey, T. 1976. The paleolimnology of Lake Mobutu Sese Seko, Uganda-Zaire: the last 28,000 years. Unpublished Ph.D. thesis, Duke University.

Hastenrath, S. 1984. The Glaciers of Equatorial East Africa. D. Reidl Publishing.

Jolly, D., R. Bonnefille, and M. Roux. 1994. Numerical interpretation of a high resolution Holocene pollen record from Burundi. Palaeogeography, Palaeoclimatology, Palaeoecology **109**: 357-370.

Jouzel, J., R. Vaikmae, J. R. Petit, M. Martin, Y. Duclos, M. Stievenard, C. Lorius, M. Toots, M. A. Melieres, L. H. Burckle, N. I. Barbov, and V. M. Kotlyakov. 1995. The two-step shape and iming of the last deglaciation in Antarctica. Climate Dynamics **11**: 151-161.

Karpuz, K. and E. Jansen. 1992. A high-resolution diatom record of the last deglaciation from the SE Norwegian sea: Documentation of rapid climatic changes. Paleoceanography **7**: 499-520.

Kendall, R. L. 1969. An ecological history of the Lake Victoria basin. Ecological Monographs **39**: 121-176.

Lehman, S. J., and L. D. Keigwin. 1992. Sudden changes in North Atlantic circulation during the last deglaciation. Nature **356**: 757-762.

Mayewski, P. A., L. D. Meeker, S. Whitlow, M. S. Twickler, M. C. Morrison, P. Bloomfield, G. C. Bond, R. B. Alley, A. J. Gow, P. M. Grootes, D. A. Meese, K. C. Taylor, and W. Wumkes. 1994. Changes in atmospheric circulation and ocean ice cover over the North Atlantic during the last 41,000 years. Science **263**: 1747-1751.

Mayewski, P. A., L. D. Meeker, S. Whitlow, M. S. Twickler, M. C. Morrison, R. B. Alley, P. Bloomfield, and K. Taylor. 1993. The atmosphere during the Younger Dryas. Science **261**: 195-197.

NASP Members. 1994. Climatic changes in areas adjacent to the North Atlantic during the last glacial-interglacial transition (14-9 ka BP): a contribution to IGCP-253. Journal of Quaternary Science **9**: 185-198

Nicholson, S., and H. Flohn. 1980. African environmental and climatic changes and the general at,ospheric circulation in Late Pleistocene and Holocene. Climatic Change **2**: 313-348.

Nilsson, T. 1983. The Pleistocene Geology and Life in the Quaternary Ice Age. Reidl Publishing.

Roberts, N., M. Taleb, P. Barker, B. Damnati, M. Icole, and D. Williamson. 1994. Timing of the Younger Dryas event in East Africa from lake-level changes. Nature **366**: 146-148.

Rosqvist, G. 1990. Quaternary glaciations in Africa. Quaternary Science Review **9**: 281-287.

Rossignol-Strick, M. 1983. African monsoons, an immediate climate response to orbital insolation. Nature **304**: 46-49.

Scott, L. 1982. A late Quaternary pollen record from the Transvaal Bushveld, South Africa. Quaternary Research **17**: 339-370.

Sowunmi, M. A. 1991. Late Quaternary environments in equatorial Africa: palynological evidence. Palaeoecology of Africa **22**: 213-238.

Ssemmanda, I., and A. Vincens. 1993. Végétation et climat dans le Bassin du lac Albert (Ouganda, Zaïre) depuis 13,000 ans B..P.: Apport de la Palynologie. C. R. Academy of Science **316**: 561-567.

Street, F. A., and A. T. Grove. 1976. Global maps of lake-level fluctuations since 30,000 yr. BP. Nature **261:** 385-389.

Street-Perrott, F. A., and R. A. Perrott. 1990. Abrupt climate fluctuations in the tropics: the influence of Atlantic Ocean circulation. Nature **343**: 607-612.

Street-Perrott, F. A., and N. Roberts. 1983. Fluctuations in llcosed-basin lakes as an indicator of past atmospheric circulation patterns, p. 331-345. *In* A. Street-Perrott et al. [eds.], Variations in the Global Water Budget. D. Reidl.

Talbot, M. R., and T. Johannessen. 1992. A high resolution paleoclimatic record for the last 27,500 years in tropical West Africa from the carbon and nitrogen isotopic composition of lacustrine organic matter. Earth and Planetary Science Letters **110**: 23-37.

Taylor, D. M. 1990. Late Quaternary pollen records from two Ugandan mires: evidence for environmental change in the Rukiga Highlands of southwest Uganda. Palaeogeography, Palaeoclimatology, Palaeoecology **80**: 283-300.

Tiercelin, J. J., R. W. Renaut, G. Delibrias, J. LeFournier, and S. Bieda. 1981. Late Pleistocene and Holocene lake level fluctuations in the Lake Bogoria basin, Northern Kenya rift valley. Paleoecology of Africa and the Surrounding Islands **13**: 105-119.

Van Campo, E., and F. Gasse. 1993. Pollen-and-diatom-inferred climatic and hydrological changes in Sumxi Co. Basin (Western Tibet) since 13,000 yr B.P. Quaternary Research **39**: 300-313.

Zolitschka, B. 1996. Paläoklimatische bedeutung laminierter sedimente. Ph. D. thesis, Universität Potsdam.

NUTRIENTS AND PLANKTON BIOMASS IN THE RIFT LAKE SOURCES OF THE WHITE NILE: LAKES ALBERT AND EDWARD

JOHN T. LEHMAN
Department of Biology, University of Michigan
Ann Arbor, MI 48109 USA

ARNI H. LITT
Department of Zoology, University of Washington
Seattle, WA 98195 USA

ROSE MUGIDDE
Fisheries Research Institute
P.O. Box 343, Jinja, Uganda

DONNA A. LEHMAN
Department of Biology, University of Michigan
Ann Arbor, MI 48109 USA

1. Abstract

The rift lakes Albert and Edward, as well as Lake George, Uganda, were sampled at nearshore and offshore sites during March 1995 with attention to water column chemistry, plankton biomass, and production rates. Biomass of phytoplankton exceeds that of zooplankton in both rift lakes, and the predictive relationship between light attenuation and particulate chlorophyll is consistent with that of Lake Victoria. Cyclopoid copepods dominate Albert and Edward; calanoids are rare (<1 m^{-3}) and Cladocera represent less than ten percent of zooplankton biomass. Although previously reported to be present in other lakes in the Nile system, *Thermodiaptomus galebi* was not confirmed in Lake Albert before this study. Adult males and ovigerous females of the species were collected in Lake Albert and immature females were sampled in Lake Edward. *Mesocyclops ogunnus* was collected from Lake Albert, which is a new record for this species from East African Rift Valley lakes. Present water temperatures of Lake Albert are elevated by about 0.5 °C compared with historical records, a circumstance which parallels changes recorded for Lake Victoria. Both Albert and Edward have lower levels of Total P and nitrate than historical accounts have reported. In shallow Lake George, heterocyst-forming cyanobacteria and centric diatoms were rare compared with historical records, a recent condition that may be the result of metal contamination, especially by copper from mining activities in the Ruwenzoris.

J.T. Lehman (ed.), Environmental Change and Response in East African Lakes, 157–172.

2. Introduction

The Western Rift of East Africa, north of the Virunga volcanoes, contains two ancient lakes that attracted scientific attention since their discovery by Europeans more than a century ago. Underlain by sediments of great antiquity, Lakes Albert and Edward together with Lake Victoria constitute the major lake sources of the White Nile and contribute to the rich limnology of central East Africa. These lakes have not been subject to comparative examination of limnological properties for over two decades, despite changes in land use, fisheries exploitation, population growth, and regional climate. In order to identify modern conditions, we undertook limnological sampling and process rate experiments at these lakes in 1995.

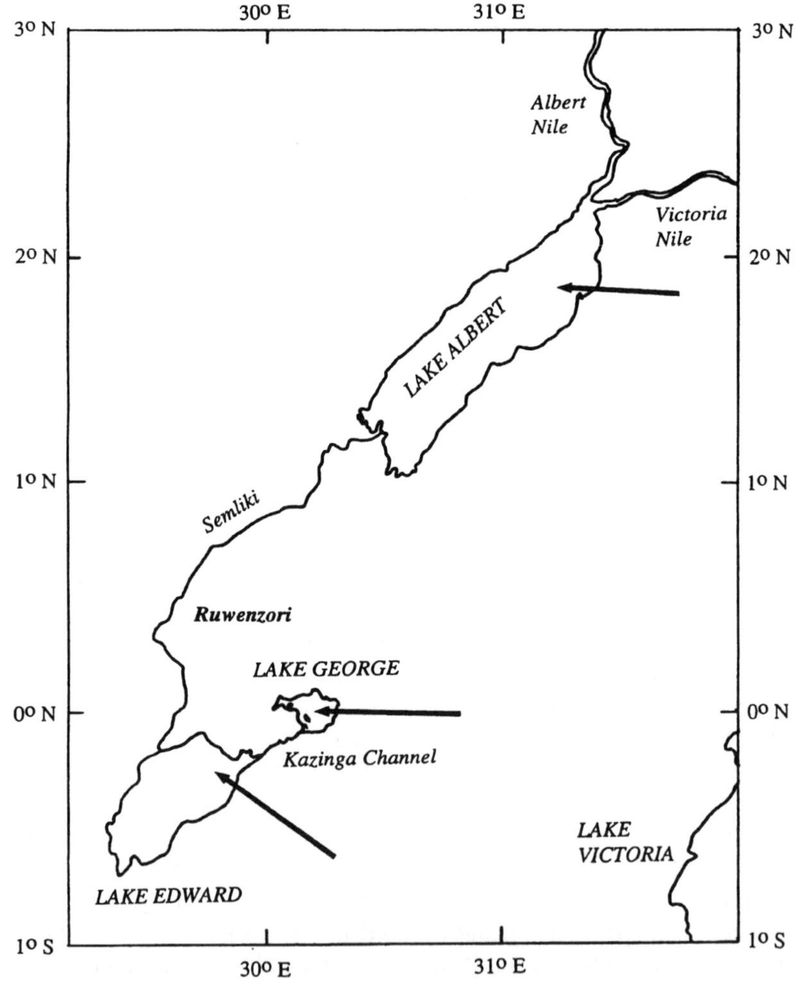

Figure 1. Lakes Albert, Edward, and George, with sampling sites indicated.

Lake Albert is a characteristic half-graben lake of maximum depth 58 m, and 5300 km^2 surface area, lying at surface elevation of 615 m (Fig. 1). The lake receives its main inflow at the south from the Semliki River, which descends from Lake Edward, and which river is both augmented and diluted by tributary streams from the northern slopes of the Ruwenzoris (Viner 1975). The Victoria Nile enters Lake Albert at its northernmost end, where it joins the northward outflow of the Albert Nile.

Lake Edward is a smaller (2325 km^2) and deeper (112 m) tilted basin. The lake reaches its maximum depth within 5 km of the western (Congo) shore, and slopes gradually to the Uganda shore. The lake receives drainage from the southwestern Ruwenzoris, from the Rwanda highlands, and from the Virunga volcanoes to the south. Some inflow occurs through the Kazinga Channel from a shallow (2 m) but optically deep and hypereutrophic northeastern basin, Lake George. Lake George had been well studied during the 1960s as part of a Royal Society component of the International Biological Programme (Burgis et al. 1972). Subsequently the lake has become an area of environmental concern because of heavy metal contamination by copper and cobalt from mining operations in the Ruwenzoris (Denny et al. 1995).

3. Methods

Lake Albert was sampled nearshore (5 m) and offshore (AL1: 40 m) from Butiaba, Uganda in the northeastern half of the lake. Lake Edward was sampled nearshore close to the mouth of the Kazinga Channel (ED1, 4 m) as well as at two offshore locations (ED2, 18 m and ED3, 25 m).

In situ profiles of temperature, oxygen, pH, and conductivity were obtained with an Endeco/YSI 6000 CTD equipped with internal data logger. Oxygen and pH electrodes were calibrated each day of use and in situ profiles of dissolved oxygen were further calibrated against surface water measurements by Winkler titration. Conductivities at in situ temperatures were converted to specific conductance at 25 °C by using a temperature coefficient of 2.3% $°C^{-1}$ (Talling and Talling 1965). Underwater irradiance (μE PAR) was measured with a LiCor cosine-corrected quantum sensor. Discrete water samples were collected by Van Dorn bottle. Samples were filtered on shore within 4 h of collection. SRP and SRSi were measured immediately by spectrophotometric methods (Hach). Filters for chlorophyll and particulate chemistry were dried over silica gel and were shipped to Ann Arbor for analysis. Chlorophyll filters were ground in 90% acetone; pigment was measured by extracted fluorescence, corrected for pheopigment. Both unfiltered water samples and filtrate were also shipped to Ann Arbor. Chloride, nitrite, nitrate, and sulfate were measured by ion chromatography. Alkalinity was measured by Gran titration. Sodium, potassium, magnesium, and calcium were measured by flame atomic absorption spectroscopy. Iron and manganese were measured by graphite furnace atomic absorption. Total dissolved P (DP) and particulate P were measured after persulfate oxidation and Total P (TP) was calculated as the sum of the two.

Zooplankton were sampled with four replicate vertical hauls by nets of both 100 and 60 μm aperture, 0.5 m mouth opening, and 5:1 aspect ratio (Research Nets, Inc.).

Zooplankton dry mass by taxon was estimated by drawing individuals as encountered from plankton collections, rinsing them in deionized water, and then drying them to constant mass at 60 °C. Mass determination was by electrobalance (Cahn 29).

Kinetics of phosphate uptake were measured for nearshore and offshore plankton assemblages. Lake water aliquots received additions ranging from 0 to 2 μM Na_2HPO_4 and concentration changes were measured over ca. 2 h at ambient temperature (ca. 25 °C) in subdued light. Maximum rates of uptake and half-saturation constants were determined by non-linear regression (SYSTAT 5.03).

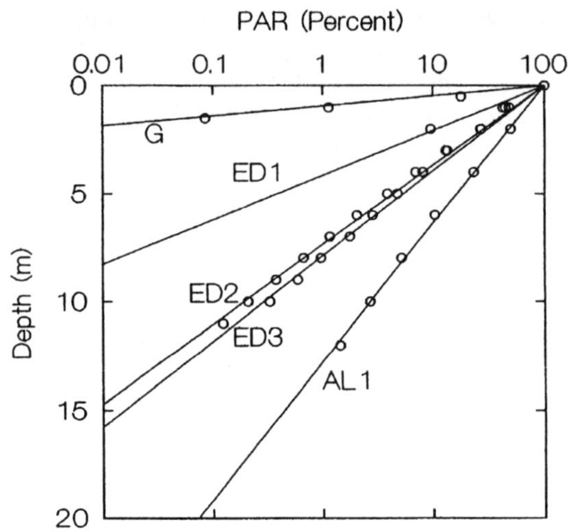

Figure 2. Photosynthetically active irradiance versus depth, as percentage of surface readings, for Lakes Albert, Edward, and George, 1995.

4. Results

4.1. PHYSICAL AND CHEMICAL PROPERTIES

4.1.1. *Light Attenuation*

Vertical attenuation of light varies among lakes, and between nearshore and offshore Lake Edward (Fig. 2). With the exception of Lake George, the range of attenuation coefficients was similar to observations from offshore Lake Victoria (Lehman et al. 1998). The nearshore station ED1 was near the mouth of the Kazinga Channel, linking Edward with its upstream hypereutrophic extension, Lake George. Light attenuation was primarily a function of particulate chlorophyll concentration (Fig. 3) such that a common linear model applies to the three large basins Albert, Edward, and Victoria. Light attenuation could thus be used to predict algal biomass for these lakes when direct measurements are lacking. Light attenuation in Lake George is considerably greater than predicted by the model that fits the deeper lakes. This appears to result from high abiotic turbidity in Lake George. The residual deviation of light attenuation in Lake George (2.54 m^{-1}) from the linear model prediction is similar in magnitude to the high background attenuation value of 2.55 m^{-1} that Ganf (1974) reported for Lake George water, independent of its chlorophyll content.

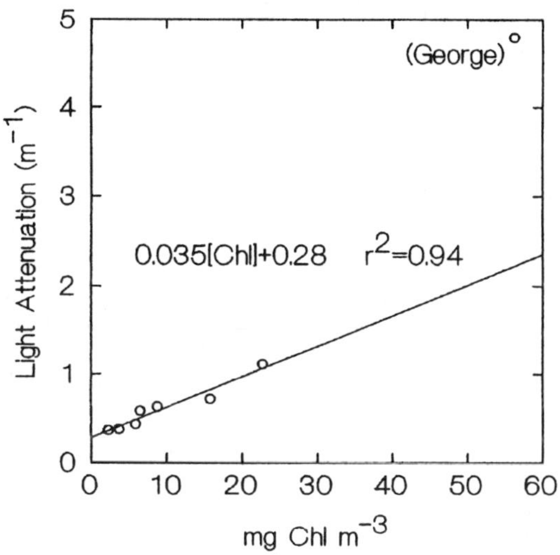

Figure 3. Vertical attenuation of PAR versus chlorophyll for Lakes Albert, Edward, George, and modern Victoria. Modern Victoria data from R. Mugidde (Lehman et al. 1998).

4.1.2. *Lake Albert*

Measurements for Lake Albert at the offshore station AL1 reveal a lake that is about 0.5 °C warmer throughout (Fig. 4) than measured by Talling during March 1961, or during any other of his longitudinal lake surveys (Talling 1963). The lake is well-mixed on a diel basis to a depth of about 25 m. In other respects, the lake is more dilute in terms of specific conductance and concentrations of major ions, and lower in phosphate and nitrate than measured decades ago by Talling and others (Table 1).

Historical values for particulate chlorophyll in Lake Albert are lacking for comparison with our modern profile (Fig. 4), but Talling (1965) reported light attenuation coefficients for green light, the most penetrating spectral component. The values cited by Talling range from 0.35 to 0.72 m^{-1}, including 0.50 m^{-1} during March of 1961. Our estimate for light attenuation (Fig. 2) during the same month of 1995 was 0.36 m^{-1}. As we noted above, the linear model of Fig. 4 can be reformulated to predict chlorophyll from light attenuation. Such a reformulated model predicts that chlorophyll levels in March 1961 were about 6.4 mg m^{-3}, or nearly three times higher than in March 1995. If it is true that historical light attenuation in Lake Albert was controlled by variation in algal abundance, this comparison indicates that chlorophyll levels in Lake Albert have declined.

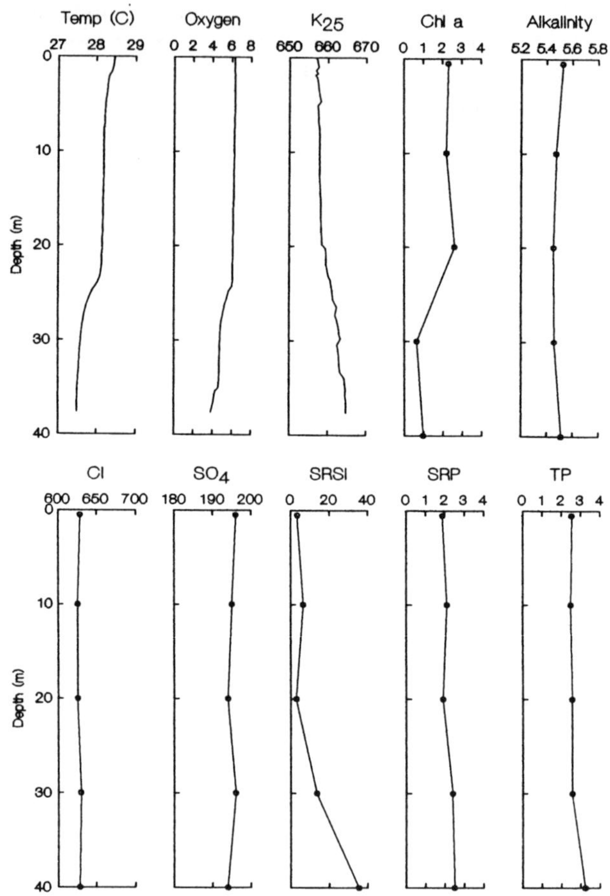

Figure 4. Temperature, oxygen (mg L^{-1}), chlorophyll (mg m^{-3}), and chemical properties: Lake Albert, March 1995. Chemical units are as in Table 1.

4.1.3. *Lake Edward*

Lake Edward also differs from historical record (Fig. 5, Table 2). Talling (1965) reported light attenuation of 1.3 m^{-1} from a 5.5 m northeast station near the Kazinga Channel. Our empirical relationship between attenuation and chlorophyll (Fig. 3) implies that the value corresponds with ca. 29 mg m^{-3} Chl, similar to the 22.7 mg m^{-3} we found at a 2.5 m station immediately offshore of the Channel mouth (at 1.11 m^{-1} light attenuation). Offshore stations at 18 and 25 m had Chl of 8.8 and 6.5 mg m^{-3}, corresponding to measured light attenuation coefficients of 0.63 and 0.58 m^{-1}, respectively. Present P concentrations in Lake Edward are about one half of historical reports and modern sulfate concentrations are lower, as well (Table 2). Lower modern values for sulfate parallel observations made by Hecky and Bugenyi (1992) and by Lehman and Branstrator (1994) for Lake Victoria, where modern sulfate is lower than previously reported. Hecky and Bugenyi speculate that some of the difference may be analytical, owing to modern methods (ion chromatography).

TABLE 1. Chemical composition of Lake Albert surface water. Samples for 29 March 1995 are from offshore station AL1 (40 m depth). Values are μM unless indicated.

	1923[1]	1953[2]	1960-61[3]	1969[4]	1995
K_{25} (μS)		825	803-870	770	660
pH	7.8-9.2	9.12	8.9-9.0		
Na		4220	3960	3567	3430
K		1690	1670	1333	1070
Mg		1310	1300-1340	1150	1010
Ca		230	225-270	260	216
F				44	37
Cl	850-1160	900	930-1080	770	630
SO_4		260	280-470	290	195
Alk (μEq)	8070-8150	7800	7250-7330	6390	5500
SiO_2		57	0.7-18.3		3.4
SRP		4.2	3.9-5.5		1.88
DP					2.17
TP			3.9-6.5		2.54
NO_3		0.6	0.3-2.4		<0.1
Fe			0.8-1.6		0.08
Mn			<0.2		

[1]From Tottenham, cited by Worthington (1930)
[2]From Elskens, cited by Talling (1963)
[3]From Talling (1963), n= 5 sampling dates, and from Talling and Talling (1965)
[4]From Kilham (1972)

TABLE 2. Composition of Lake Edward surface water. Samples for 25 March 1995 are from an offshore station at 0° 12.596' S, 29° 46.799' E (25 m depth). Values and methods as in Table 1.

	1954[1]	1961[2]	1964[3]	1969[4]	1995
K_{25} (μS)	1043	1031	1026	508-580	880
pH	8.89	9.1	9.1	7.9-8.2	8.9
Na	4870	4780	2450	2175-2262	3750
K	2030	2310	2120	818-941	1620
Mg	1850	1990		833-962	1630
Ca	242	310	280	330-340	310
F				32-35	44
Cl	760	1010		440-474	580
SO_4	365	320		350-400	250
Alk (μEq)	10100	9850	9900	4070-4740	7870
SiO_2	33	108		210-270	150
SRP	1.0	0.6			0.6
DP					0.8
TP		4.1			1.4
NO_3	1.6	1.7			<0.1
Fe		4.2			0.06
Mn		<0.2			

[1]From Elskens, cited by Talling (1963)
[2]From Talling and Talling (1965)
[3]From Beadle (1966)
[4]From Kilham (1972)

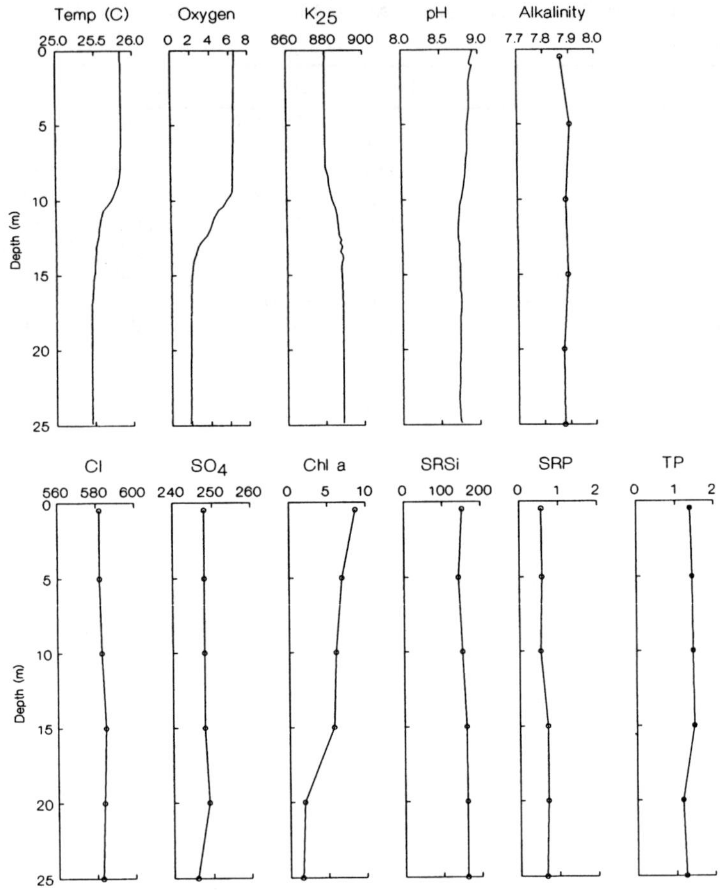

Figure 5. Distribution of temperature, oxygen, chlorophyll, and chemical properties in Lake Edward, March 1995. Units as in Fig. 4.

4.2. PLANKTON COMMUNITIES

Cyclopoid copepods dominate Lakes Albert (Fig. 6) and Edward (Fig. 7). The lakes exhibit similar species richness of crustacean zooplankton, but differ in species assemblages (Table 4). In particular, large unhelmetted *Daphnia lumholtzi* var. *monacha* are present in Lake Albert, consistent with observations by Green (1967, 1971). The species was rare in Lake Edward, and smaller *Daphnia longispina* occurred there instead. *Thermodiaptomus galebi* is very rare in both lakes; only immature specimens were found in Lake Edward despite exhaustive searching. We assume that the immature individuals are *T. galebi* because previous authorities identified the taxon from the lake (Dussart 1989). We found ovigerous *T. galebi* females and adult males in

Lake Albert, despite prior literature assertions that calanoids were absent (Verheye and Dumont 1984).

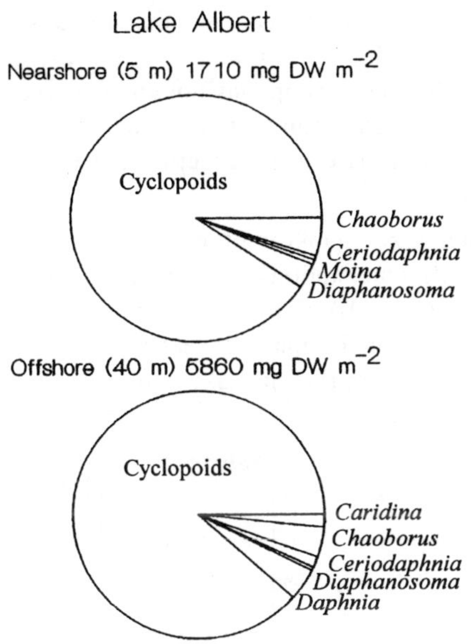

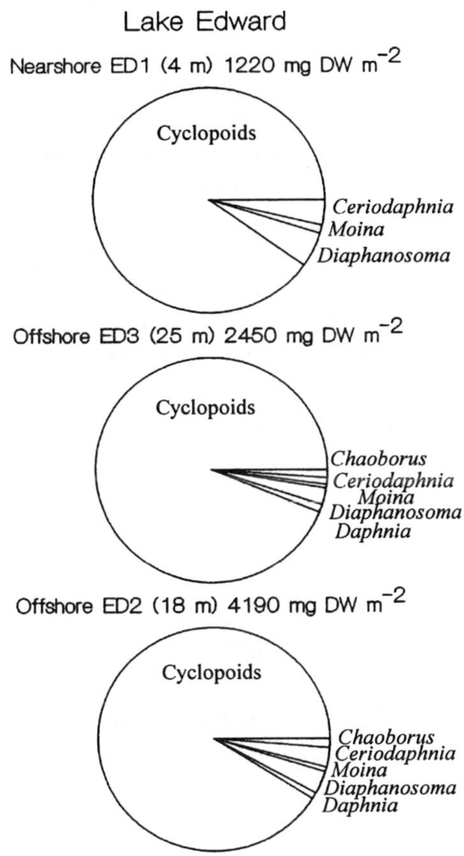

Figure 6. (above) Relative contribution to total water column biomass by major zooplankton taxa in Lake Albert. Sampling sites indicated in Fig. 1.

Figure 7. (right) As Fig. 6, for Lake Edward.

TABLE 4. Crustacean zooplankton taxa in Lakes Albert and Edward, March 1995

Lake Albert	Lake Edward
Mesocyclops aequatorialis aequatorialis	*Mesocyclops aequatorialis similis*
Mesocyclops ogunnus	*Mesocyclops aequatorialis* hybrid
Thermocyclops neglectus	*Thermocyclops oblongatus*
	Thermocyclops consimilis
Thermodiaptomus galebi	*Thermodiaptomus (galebi)*
Daphnia lumholtzi monacha	*Daphnia lumholtzi monacha*
	Daphnia longispina
Diaphanosoma excisum	*Diaphanosoma excisum*
Diaphanosoma mongolianum	*Diaphanosoma mongolianum*
Ceriodaphnia cornuta	*Ceriodaphnia cornuta rigaudi*
Ceriodaphnia dubia	*Ceriodaphnia dubia*
Moina micrura	*Moina micrura*
Caridina nilotica	

Size distributions of the dominant cyclopoid copepod assemblages in the two lakes are similar, but analyses of *Chaoborus* and *Daphnia* populations reveal that animals are larger in Lake Albert than in Lake Edward (Fig. 8). Moreover, *Caridina* was present in offshore plankton collections from Lake Albert, but the shrimp was not encountered in Lake Edward collections. Rotifers made a trivial contribution to zooplankton biomass. They were virtually absent from the plankton of both rift lakes, except for occasional specimens of *Keratella tropica* in Lake Edward.

Water column inventories of phytoplankton and zooplankton biomass are plotted in Figure 8 for both shallow nearshore and deeper offshore stations of both lakes. Algal C was estimated from chlorophyll by calculating C:Chl = 100:1 by mass. Zooplankton C was calculated as 50% of dry mass. Phytoplankton biomass exceeds zooplankton biomass in both basins, by a greater ratio in Lake Edward than in Lake Albert.

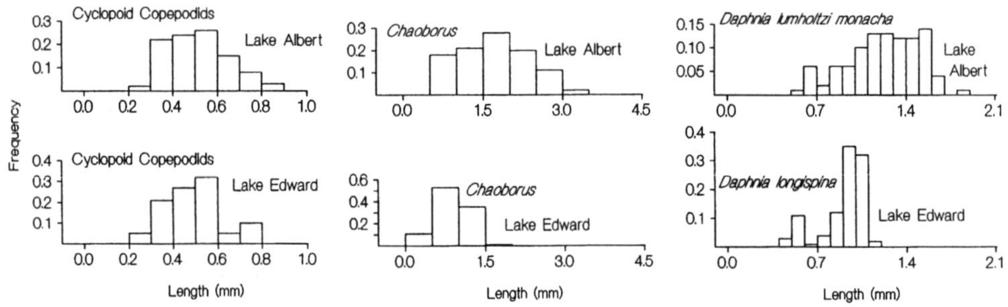

Figure 8. Size frequency distributions of cyclopoid copepodids, *Chaoborus* larvae, and *Daphnia* from offshore stations of Lake Albert and Lake Edward.

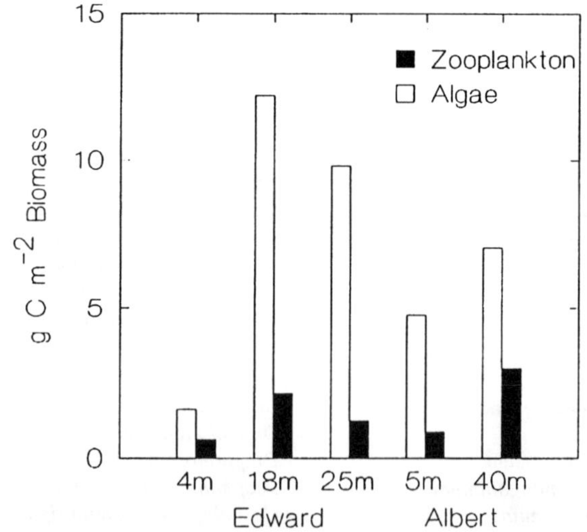

Figure 9. Total biomass of phytoplankton and zooplankton at nearshore and offshore stations in Lakes Albert and Edward.

4.3. PHOSPHATE UPTAKE

Uptake kinetics for phosphate are reported in Table 5, together with comparison data for Lake Victoria. Maximum uptake rates scaled to chlorophyll are inversely related to ambient phosphate and appear to be a useful comparative index. Initial slopes of the uptake functions (V_m/k_s) vary inversely with lake phosphate as well.

TABLE 5. Phosphate uptake parameters for Lake Edward and nearshore Lake Victoria, March 1995. ED1 and NPG (Napoleon Gulf) are nearshore sites. ED3 and BG (Bugaia) are offshore sites. Data for BG are from Lehman and Branstrator (1994). Data for NPG are from Lehman et al. (1998).

Lake	Site	Chl *a* mg m^{-3}	SRP μM	V_m nmol P L^{-1} h^{-1}	V_m nmol P (μg Chl)$^{-1}$ h^{-1}	k_s nM P	V_m/k_s h^{-1}
Edward	ED1	7.2	0.20	28 (17)	3.9 (2.4)	841 (1185)	0.033
Edward	ED3	6.9	0.57	18 (7)	2.6 (1.0)	664 (855)	0.027
Victoria	NPG	18.8	0.04	180 (23)	9.6 (1.2)	447 (168)	0.403
Victoria	BG	11.3	1.65	23 (4)	2.0 (0.4)	1200 (600)	0.019

5. Discussion

Zooplankton in Lake Albert and Lake Edward are superficially very similar. Both lakes are dominated overwhelmingly by a suite of cyclopoid species, including the predator *Mesocyclops*. Both lakes have a species of *Daphnia*, two *Ceriodaphnia*, and the same two *Diaphanosoma*. Coexistence of *Diaphanosoma* species is not uncommon (N. M. Korovchinsky, personal communication), and is different from the situation in Lake Victoria, where only one species is present (Lehman 1996). The communities of these lakes are otherwise simpler than Lake Victoria, where five cyclopoid species and three diaptomids are present offshore (Branstrator et al. 1996). Rotifers are rare, which is the typical situation in large African lakes dominated by cyclopoid species (Lehman 1996).

Despite superficial similarities, the smaller size of non-cyclopoid zooplankton in Lake Edward compared with Lake Albert suggests that visual planktivory is more intense in Edward. That notion is reinforced by the lower ratio of zooplankton biomass to phytoplankton biomass in Lake Edward. Presence of an effective piscivore, Nile perch, in Lake Albert may cause reductions in the numbers of planktivorous fish that occur in that lake. As early as the 1960s, Green (1967) argued that there was reduced intensity of predation by fish on zooplankton in mid-lake regions of Lake Albert.

Zooplankton biomass at offshore stations in both Lake Albert and Lake Edward exceed the seasonal mean values reported for Lake Victoria by Branstrator et al. (1996), but the values are within the range of annual variation observed for Lake Victoria. In all three of the lakes, zooplankton biomass falls in a range from approximately 2 to 6 g DW m^{-2}. Water column depths that were sampled differed among the lakes, however, and ranged from 25 m in Edward to 40 m in Albert, to 50 m in Victoria. As a consequence of averaging the areal abundances over different depth ranges, computed volumetric abundances of zooplankton are somewhat higher in Edward than in the other two lakes. We did not investigate the vertical distribution of the plankton in the lakes, and thus actual in situ concentrations are only approximated by water column averages.

Previous literature reports that *Diaptomus* is not present in Lake Albert are erroneous. It is unlikely that our discovery represents a new colonization event. The animals are extremely rare, on the order of one individual per cubic meter. We confirmed their presence only by diligent and sustained search of large sample volumes. An immature diaptomid copepod had already been discovered in a Lake Albert collection during 1993 with an identical net (V. Kiggundu and J. Lehman, personal observation), but its species identity could not be established at that time. The fact that calanoids are so scarce in these lakes, compared with Lake Victoria, points again to higher levels of visual planktivory in the rift lakes.

Lake Albert seems to be more transparent now than 30 years ago, evidently because of reduced algal chlorophyll. Phosphate concentrations are also lower, but that is probably not the cause of increased transparency. Talling had reported extreme phosphate concentrations (Table 1), and stoichiometric considerations dictate that P could not be a limiting factor in biomass production, then or now. The elevated temperature of the lake water is intriguing, because it is reminiscent of similar warming of Lake Victoria (Hecky 1993; Lehman 1996). Whether or how altered mixing in response to the thermal regime may have affected lake production, nutrient supplies to the epilimnion, and algal biomass cannot be answered by our limited temporal data.

We found lower values of Total P in Lake Edward than previous reports had indicated. We cannot yet determine if the reduction is paralleled by chlorophyll because historical data are not readily available for comparison. All three lakes exhibit modern values of sulfate lower than historical reports. This apparent reduction is a regional feature in common with Lake Victoria. It is not yet clear how much of this difference is real, a consequence of lessened inputs or possibly increased chemical reduction in the sediments, or how much is the result of improvements in analytical methodology.

Kinetic characteristics of the algae in Lake Edward suggest that phosphate is not growth limiting. Inverse variation of both V_m and V_m/k_s with SRP in these tropical lakes is consistent with expectations from temperate lakes. Values for Lake Edward are intermediate between inshore and offshore Lake Victoria, where bioassay experiments (Lehman and Branstrator 1993, 1994) demonstrated that P is not the proximate limitation for biomass production.

5.1. CONDITION OF LAKE GEORGE

Lake George, as well as much of eastern Lake Edward, lies within Queen Elizabeth (Ruwenzori) National Park. Restoration of political stability to Uganda in the late 1980s after a decade of strife has stimulated conservation efforts for wildlife within the National Park. In addition, copper mining has recently resumed along the Nyamwamba River, which flows into Lake George. The Kilembe mines first began operation in 1956, and shortly before they closed in 1979 elevated copper concentrations were measured along a gradient extending from Lake George, along the Kazinga Channel which links George to Edward, and then across northern Lake Edward to its outflow as the Semliki River (Bugenyi 1979).

The IBP investigations of the 1960s preceded evidence of metal pollution. They established that nitrogen was a limiting resource to the algae of the lake, and that total community respiration approximately balanced photosynthesis on a 24-h basis (Ganf and Viner 1973). Nitrogen income was dominated by fluvial inputs and nitrogen fixation by diazotrophic cyanobacteria. Hydrologic input is dominated by flow from the Ruwenzoris (Viner and Smith 1973), and the chemistry of the streams is representative of headwaters draining predominant silicate rocks (Kilham 1984).

The high pH of Lake George is the result of photosynthesis by suspended phytoplankton (Ganf and Horne 1975). Withdrawal of CO_2 from the water by photosynthesis elevates the pH. During the 1960s, the lake exhibited a diurnal pH variation from approximately 8.75 to 9.98 (Ganf 1972). More recently, pH as high as 11.2 has been reported from nearshore waters, with open lake values consistently near pH 10 (Denny et al. 1995). By elevating the pH, photosynthesis can accelerate net chemical weathering of silica from sediments, and there appears to be a trend toward higher silica concentrations in recent years (Table 3). Biogenic precipitation of amorphous silica by diatoms, a process that typically reduces silica levels in lakes, may become partly inhibited by toxic effects of the polluting metals, especially copper. Copper is known to inhibit silica uptake and silicification by diatoms (Reuter 1983).

TABLE 3. Chemical composition of Lake George surface water. Samples for 24 March 1995 were collected mid-lake at 0° 0.30' N, 30° 13.27' E (2 m depth). Values are μM unless indicated.

	1961[1]	1967[2]	1969[3]	1995
K_{25} (μS)	224	234-262	234-338	240
pH	9.6		8.2	10.0
Na	587	848-892	674-1240	620
K	108	97-118	243-271	90
Mg	333	300-321	300-412	300
Ca	505	362-500	520-565	450
F			25	18
Cl	248	214-259	364-567	240
SO_4	116	150-188	131-200	45
Alk (μEq)	1910	1250-2000	1780-2290	1730
SiO_2	300	330-340	375-410	370
SRP	<0.6	1.8-4.2		1.3
DP				1.3
TP	13.3			
NO_3			n.d.[4]	<0.1
Fe		4.5		0.10
Mn		4.7		

[1]Talling and Talling (1965)
[2]Viner (1969)
[3]Kilham (1972)
[4]Viner (1975) reports NO_3 as not detectable

Copper also inhibits the nitrogen-fixation processes of diazotrophic, heterocyst-forming cyanobacteria (Elder and Horne 1975), including the taxa *Anabaena* and *Aphanizomenon*, which formerly were important elements of the flora (Ganf 1974b). Our recent samples were dominated by colonial coccoid cyanobacteria (Chroococcales), which do not produce heterocysts, but nonheterocystous cyanobacteria can fix nitrogen in the dark when there is no production of nitrogenase-disabling oxygen by photosynthesis, or in the light either by spatial separation between aerobic and anerobic portions of colonies or by temporal separation of photosynthesis and nitrogen fixation in different phases of the cell cycle (Carpenter and Price 1976; Stal and Krumbein 1985; Mitsui et al. 1986). Small pennate diatoms are present, but not as biomass dominants. Centric diatoms, especially *Aulacoseira* (formerly *Melosira*) species, were rare. A similar situation was reported for Lake Edward during 1972 (Hecky and Kling 1987); Chroococcales dominated the phytoplankton and diatoms were significantly abundant only at a southern site, remote from the metal contamination later identified in the north.

Nitrogen fixation formerly provided nearly half of the nitrogen income for Lake George (Horne and Viner 1971). Unlike regenerated nutrients, such "new" nutrient permits net biological production and export of biogenic material from the water column to the sediments. The prospect that metal pollution may increase with mining development suggests the need for additional analyses of the productive ecosystem.

6. Conclusions

The most conspicuous changes to lakes Albert and Edward during the previous 30 years appear to have been (1) reductions in algal biomass, measured as particulate chlorophyll, and (2) reduced water column inventories of P and inorganic N. These changes are consistent with increased temperature of Lake Albert, and suggest reduced mixing of the surface layer, possibly under influence of reduced wind stress. Given sufficient silica to support fast-sinking diatoms, increased physical stability can lead to accelerated sedimentation of particulate matter, including Chl, P, and N, from surface layers. This tendency is opposed only when there is a floristic change to more buoyant taxa, as in modern Lake Victoria. We regard climatic variation to be a master controlling variable for lakes of East Africa, and its signal may be registered through changes in primary production and algal biomass as well as nutrient conditions.

7. Acknowledgments

We thank the government of Uganda for its courtesy and permission to investigate these lakes. We are grateful for assistance of M. Mugumba, V. Kiggundu, and staff of the Ecology Institute at Mweya, Uganda and the Fisheries Research Institute (FIRI) at Jinja, Uganda. Our preliminary identifications of *Diaphanosoma* species were confirmed by Prof. N. M. Korovchinsky, and our identifications of *Ceriodaphnia* were confirmed by Dr. D. Berner. A. J. Horne and C. D. Sandgren provided comments about nutrient dynamics that improved the paper. This research was supported in part by the U. S. National Geographic Society and the U. S. National Science Foundation.

8. References

Beadle, L. C. 1966. Prolonged stratification and deoxygenation in tropical lakes. I. Crater lake Nkugute, Uganda, compared with Lakes Bunyoni and Edward. Limnology and Oceanography **11**: 152-163.

Branstrator, D. K., Ndawula, L. M. and Lehman, J. T. 1996. Zooplankton dynamics in Lake Victoria, p. 337-355. *In* T. C. Johnson and E. O. Odada [eds.], The limnology, climatology and paleoclimatology of the East African lakes. Gordon and Breach Publishers.

Bugenyi, F. W. B. 1979. Copper ion distribution in the surface waters of Lakes George and Idi Amin. Hydrobiologia **64**: 9-15.

Burgis, M. J., L. G. Dunn, G. G. Ganf, L. M. McGowan, and A. B. Viner. 1972. Lake George, Uganda. Studies on a tropical freshwater ecosystem, p. 301-309. *In* K. Kajak and A. Hillbricht-Ilkowska [eds.], Productivity Problems in Freshwaters. Polish Scientific Publications.

Carpenter, E. J. and C. C. Price. 1976. Marine *Oscillatoria* (*Trichodesmium*): explanation for aerobic nitrogen fixation without heterocysts. Science **191**: 1278-1280.

Denny, P., R. Bailey, E. Tukahirwa, and P. Mafabi. 1995. Heavy metal contamination of Lake George (Uganda) and its wetlands. Hydrobiologia **297**: 229-239.

Dussart, B. 1989. Crustacés copépodes, calanoides des eaux intérieures Africaines. Crustaceana, supplement 15.

Elder, J. F., A. J. and Horne. 1978. Copper cycles and $CuSO_4$ algicidal capacity in two California lakes. Environmental Management **2**: 17-30.

Ganf, G. G. 1972. The regulation of net primary production in Lake George, Uganda, East Africa, p. 693-708. *In* K. Kajak and A. Hillbrecht-ilkowsak [eds.], Productivity Problems in Freshwaters. Polish Scientific Publications.

Ganf, G. G. 1974. Incident solar irradiance and underwater light penetration as factors controlling the chlorophyll *a* content of a shallow equatorial lake (Lake George, Uganda). Journal of Ecology **62**: 593-609.

Ganf, G. G. 1974b. Phytoplankton biomass and distribution in a shallow eutrophic lake (Lake George, Uganda). Oecologia **16**: 9-29.

Ganf, G. G., and A. J. Horne. 1975. Diurnal stratification, photosynthesis and nitrogen-fixation in a shallow, equatorial Lake (Lake George, Uganda). Freshwater Biology **5**: 13-39.

Ganf, G. G., and A. B. Viner. 1973. Ecological stability in a shallow equatorial lake (Lake George, Uganda). Proceedings of the Royal Society of London **B 184**: 321-346.

Green, J. 1967. The distribution and variation of *Daphnia lumholtzi* (Crustacea: Cladocera) in relation to fish predation in Lake Albert, East Africa. Journal of Zoology **151**: 181-197.

Green, J. 1971. Associations of Cladocera in the zooplankton of the lake sources of the White Nile. Journal of Zoology **165**: 373-414.

Hecky, R. E. 1993. The eutrophication of Lake Victoria. Verhandlungen der Internationale Vereinigung für Limnologie **25**: 39-48.

Hecky, R. E., and H. J. Kling. 1987. Phytoplankton ecology of the Great Lakes in the rift valleys of central Africa. Archiv für Hydrobiologie Ergebnisse Limnologie **25**:197-228.

Hecky, R. E. and F. W. B. Bugenyi. 1992. Hydrology and chemistry of the African Great Lakes and water quality issues: problems and solutions. Mitteilungen Vereinigung Internationale Limnologie **23**: 45-54.

Horne, A. J., and A. B. Viner. 1971. Nitrogen fixation and its significance in tropical Lake George, Uganda. Nature **232**: 417-418.

Kilham, P. 1972. Biogeochemistry of African lakes and rivers. Ph.D. thesis, Duke University.

Kilham, P. 1984. Sulfate in African inland waters: Sulfate to chloride ratios. Verhandlungen Internationale Vereinigung Limnologie **22**: 296-302.

Lehman, J. T. 1996. Pelagic food webs of the African Great Lakes, p. 281-301. *In* T. C. Johnson and E. O. Odada [eds.], The limnology, climatology and paleoclimatology of the East African lakes. Gordon and Breach.

Lehman, J. T., and D. K. Branstrator. 1993. Effects of nutrients and grazing on the phytoplankton of Lake Victoria. Verhandlungen Internationale Vereinigung Limnologie **25**: 850-855.

Lehman, J. T., and D. K. Branstrator. 1994. Nutrient dynamics and turnover rates of phosphate and sulfate in Lake Victoria, East Africa. Limnology and Oceanography **39**: 227-233.

Mitsui, A., S. Kumazawa, A. Takahashi, H. Ikemoto, S. Cao, and T. Arai. 1986. Strategy by which nitrogen-fixing unicellular cyanobacteria grow photoautotrophically. Nature **323**: 720-722.

Reuter, J. G. Jr. 1983. Effect of copper on growth, silicic acid uptake and soluble pools of silicic acid in the marine diatom, *Thalassiosira weisflogii* (Bacillariophyceae). Journal of Phycology **19**: 101-104.

Stal, L., and W. E. Krumbein. 1985. Nitrogenase activity in the nonheterocystous cyanobacteriium *Oscillatoria* sp. grown under alternating light-dark cycles. Archiv für Microbiologie **143**: 67-71.

Talling, J. F. 1963. The origin of stratification in an African rift lake. Limnology and Oceanography **8**: 68-78.

Talling, J. F. 1965. The photosynthetic activity of phytoplankton in the East African lakes. Internationale Revue gesamten Hydrobiologie **50**: 1-32.

Talling, J. F., and Talling, I. B. 1965. The chemical composition of african lake waters. Internationale Revue der gesamten Hydrobiologie **50**: 421-463.

Verheye, H. M., and H. J. Dumont. 1984. Calanoid copepods of the Nile system. Hydrobiologia **110**: 191-212.

Viner, A. B. 1969. The chemistry of the water of Lake George, Uganda. Verhandlungen der Internationale Vereinigung für Limnologie **17**: 289-296.

Viner, A. B. 1975. Relationships of nitrogen and phosphorus to a tropical phytoplankton population. Hydrobiologia **52**: 185-196.

Viner, A. B., and I. R. Smith. 1973. Geographical, historical and physical aspects of Lake George. Proceedings of the Royal Society of London **B 184**: 235-270.

Worthington, E. B. 1930. Observations on the temperature, hydrogen-ion concentration, and other physical conditions of the Victoria and Albert Nyanzas. Internationale Revue der gesamten Hydrobiologie **24**: 328-357.

COMPARISON OF ISOTOPIC RECORDS IN ABIOGENIC AND BIOGENIC CALCITE FROM LAKE TURKANA, KENYA

P. NG'ANG'A
Department of Geology, Duke University
Durham, NC 27706, USA

M. W. MUCHANE
Department of Geological Sciences, University of Tennessee
Knoxville, TN USA

THOMAS C. JOHNSON
Large Lakes Observatory, University of Minnesota
Duluth, MN 55812, USA

KEITH STURGEON
Stable Isoptope Laboratory, North Carolina State University
Raleigh, NC 27695, USA

1. Abstract

The oxygen isotopic composition of abiogenic (micrite) and biogenic calcite (ostracods) in modern Lake Turkana sediments varies greatly around the lake. The carbonates appear to form out of equilibrium with the modern lake waters. The micrite is precipitating at 0.6‰ to 2‰ below equilibrium values whereas the ostracods are forming their shells at 0.3‰ to 2.0‰ above equilibrium. This seems to be caused by the extent of interaction between inflow and basin waters. Covariance of $\delta^{18}O$ values in the micrite and in ostracods both in the modern sediments and in a 12 meter core from the lake suggests that both are responding to environmental forcing and are useful for paleoenvironmental interpretation at least on a time scale of centuries to millennia.

Covariance of $\delta^{18}O$ values in the two types of carbonates was strongest prior to 4,000 ybp when Lake Turkana was hydrologically open. Although covariance persisted after 4,000 ybp, a decrease in the degree of covariance suggests more stratification in the water column than when the lake was hydrologically open. Modern $\delta^{13}C$ values of the carbonates are depleted relative to equilibrium values and vary appreciably around the lake. The preferred explanation for this observation is chemical enhancement of CO_2 invasion due to high rates of photosynthesis associated with algal blooms. Since Lake Turkana has a large volume, covariance of $\delta^{18}O$ and $\delta^{13}C$ values of the micrite will be achieved only on a long time scale, whether lake volume increases or decreases.

J.T. Lehman (ed.), Environmental Change and Response in East African Lakes, 173–190.

2. Introduction

Stable isotope geochemistry of light elements (carbon, hydrogen, nitrogen, oxygen and sulfur) provides a powerful tool for tracing environmental changes through time. $\delta^{18}O$ records from lake sediments typically come from ostracod valves or inorganic calcite precipitated from the water column (e.g., Stuiver 1970; Owen et al. 1982; Abell et al. 1982; Hodell et al. 1991; Johnson et al. 1991; Lister et al. 1991; Phillips et al. 1992; Mckenzie and Hollander 1993). In closed-basin lakes of low latitudes, effects of interannual variability in temperature on $\delta^{18}O$ of these carbonates is relatively low compared to effects caused by interannual differences in $\delta^{18}O$ of the lake waters resulting from hydrological changes driven by climate forcing. However, calcite precipitated either abiogenically or biogenically may or may not be in isotopic equilibrium with lake water due to kinetic or biological effects (Epstein et al. 1953; Mook and Vogel 1968; Tarutani et al. 1969; Grossman and Ku 1981). Thus the value of isotopic signals of lacustrine carbonates as proxies of past climate change must be questioned.

We are faced with several questions in the interpretation of paleorecords regarding the initial controls on the isotopic values of biogenic and abiogenic calcite. In closed and open basins for example, what is the spatial and temporal variability in the isotopic values of biogenic and abiogenic calcite? Are the carbonates formed in isotopic equilibrium with the lake waters? If so, they clearly are useful as paleoclimatic indicators. If not, are there consistent relationships among the isotopic and environmental parameters that still allow a valid paleoenvironmental interpretation?

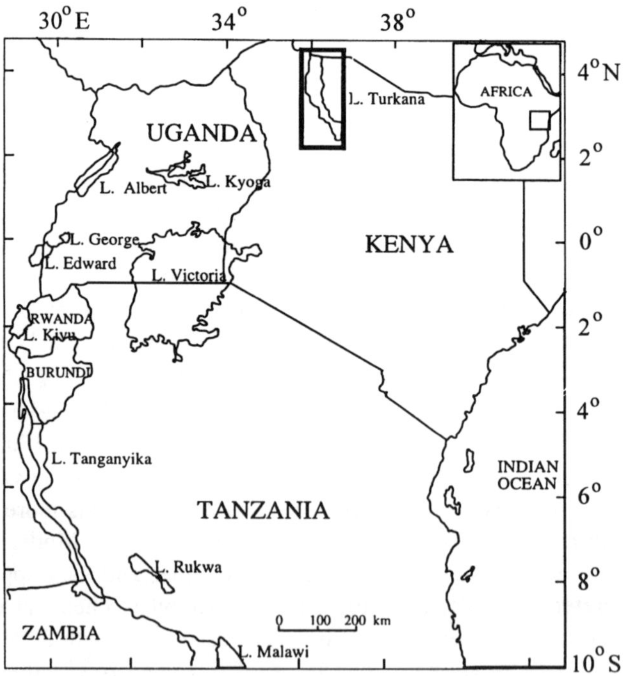

Figure 1. Location map of Lake Turkana.

In this study, we examine the spatial and temporal variability of the isotopic values of biogenic and abiogenic carbonates from Lake Turkana, Kenya. We infer the isotopic and environmental parameters that control these values. We then compare and interpret isotopic values in biogenic and abiogenic calcite in a 4400 yr record in a 12 m core from Lake Turkana, Kenya. This is the first study in a low latitude tropical lake to compare isotopic values in biogenic and inorganic carbonates from the same sediment for paleoenvironmental interpretation.

3. Site description

Lake Turkana is the largest lake in the eastern arm of the East African Rift Valley (Fig. 1). It lies between 2° 27'N and 4° 30'N in arid northern Kenya. It is elongate in shape with a mean length and width of 260 km and 30 km respectively and a mean depth of 31 m. The waters are constantly mixed by strong diurnal southeast winds (Ferguson and Harbott 1982). The modern sediments are mostly faintly-laminated sandy and silty clays (Yuretich 1979, 1986; Halfman 1987). The relative abundance of biogenic components increases with increasing distance from the Omo delta at the northern margin of the lake, resulting in relatively well preserved ostracod-diatom clays in the south basin.

Strong southeasterly winds dominate the climate at Lake Turkana. Daily mean wind speeds up to 25 km h^{-1} were recorded and averaged 14.5 km h^{-1} during a period of 2.5 yr (Ferguson and Harbott 1982). Daily averages are misleading because the winds are diurnal, typically exceeding 80 km h^{-1} during early morning hours and calm in the afternoon. The winds are slightly stronger and more persistent in direction at the south end of the lake. The seasonal pattern of winds shows a slight tendency to be more intense and more consistent in direction during October to March (Griffiths 1972).

The Lake Turkana basin has a mean annual rainfall of about 155 mm yr^{-1} supporting desert-like vegetation (East African Meteorological Dept. 1975). Most of the lake environs, and in particular the southern basin, is desert. Mean annual rainfall increases from just north of the lake to regions farther north, especially at elevations above 1,000 m in the Ethiopian Highlands, with values from 300 mm yr^{-1} to 2000 mm yr^{-1}, respectively. The lack of rainfall at Lake Turkana and other parts of northern Kenya and eastern Ethiopia is caused by widespread divergence of the southeast monsoon to the north of the Equator due, in part, to the orographic influence of the Ethiopian and Kenya Plateaus, and the north/south alignment of the eastern coastline of Africa (Nieuwolt 1977; Nicholson 1996). When the monsoon flows along the shore, the frictional contrast between the shore and the water induces subsidence of the air and dryness. Kinuthia (1992) suggested that the low-level Turkana jet that flows to the northwest can produce patterns of descending air that enhance aridity.

4. Modern hydrology

The isotopic values of precipitation in East Africa reflect the regional circulation patterns (Rozanski et al. 1996). Seasonal fluctuations are well pronounced and coincide with the seasonal displacement of the ITCZ. Isotopically depleted rains are in general associated with the rainy period(s). The annual weighted mean values of $\delta^{18}O$ are -1.3‰ at Addis Ababa (within the Ethiopian Highlands), and -2.9‰ at Entebbe (close to the source area of the Turkwel).

The Turkana drainage basin covers an area of 123,300 km^2. The hydrology of the basin is dominated by the perennial Omo river, which drains the Ethiopian highlands to the north, and provides about 90% of the fresh water flowing into the lake (Ferguson and Harbott 1982). The seasonal Turkwel and Kerio rivers contribute most of the remaining fluvial input. Other streams, direct rainfall and subterranean flow are considered insignificant in the water budget (Yuretich and Cerling 1983).

The lake has a surface area of 8860 km^2 and a volume of 251 km^3 (Ferguson and Harbott 1982). The lake is divided into two major basins, north and south, with a bathymetric high between them, near the Turkwel and Kerio rivers. The north and south basins have maximum depths of just over 80 m and 115 m, respectively (Halfman 1987).

Average maximum and minimum daily air temperatures at Lake Turkana were 32.5 ± 0.2 °C and 26 ± 0.2 °C over a 2.5 year survey in the early 1970s (Ferguson and Harbott 1982). The water column is mixed by strong diurnal winds, and has a relatively uniform water temperature of 25 to 26 °C. Surface water temperatures (<1 m depth) however reveal a daily temperature cycle as great as 2.5 °C. The largest change in temperature with depth (0.2 °C m^{-1}) is within the upper 1 to 3 m (Halfman 1996). Dissolved oxygen concentrations range from about 70 to 100% saturation (Ferguson and Harbott 1982).

The lake waters have become more saline during the past 15 years of lake level decline. Salinity ranges between 2.4‰ and 2.6‰ and pH between 9.2 and 9.7 (Yuretich and Cerling 1983; Ng'ang'a and Johnson, in prep.). The principal ions are Na^+ (37 mM), HCO_3^- (23 mEq) and Cl^- (13 mM), with low concentrations of Ca^{2+}, Mg^{2+} and SO_4^{2-}. Mass balance calculations indicate that the lake water is supersaturated with respect to calcite (Yuretich and Cerling 1983).

5. Methods

Modern sediment samples were obtained using a grab sampler from several sites throughout Lake Turkana in 1994 and 1995 (Fig. 2). Surface and deep water samples were also obtained at these localities in 1994, and only surface water samples were collected in 1995. The water samples were immediately treated with mercuric chloride to prevent microbial activity. The bottles were then filled to the top making sure no air bubbles were left at the top and then tightly sealed to avoid exchange with the atmosphere. A 12 m core (LT84-2P) previously recovered from a water depth of 57 m at the south end of Lake Turkana in 1984 was subsampled at 10 cm intervals (Muchane 1994).

The core and grab samples were frozen, then freeze-dried and dispersed in 0.25% Calgon solution for deflocculation. The sediment was sieved through nested 150 μm and 63 μm sieves. The carbonate in the <63 μm fraction is dominated by euhedral crystals of low-Mg calcite (3 mole % Mg) about 10 μm in length (Halfman et al. 1989). The >150 μm fraction contains mainly ostracod carapaces. The <63 μm fraction was washed 3 times with deionized water and once with ultra-pure deionized water through a process of centrifuging and decanting. This fraction was then ground and roasted for an hour at 320 °C under vacuum to remove any organic matter. Whole adult shells of the ostracod *Sclerocypris clavularis* were picked from the >150 μm fraction and recleaned with deionized water under a microscope using a sable brush. Not all intervals had sufficient numbers of ostracods for isotope analysis.

Both biogenic and abiogenic carbonates were processed separately for oxygen and carbon isotopic analysis on a carbonate extraction line. The samples were reacted with 100% orthophosphoric acid at 90 °C to release carbon dioxide which was purified and collected using standard cryogenic techniques. Each batch of samples was processed in conjunction with Carrara Marble standards. The extracted carbon dioxide was analyzed for $\delta^{18}O$ and $\delta^{13}C$ on a VG Sira Series II isotope ratio mass spectrometer at the Duke University Botany Department. Precision of the standards was 0.04 per mil for the $\delta^{18}O$ and 0.02 per mil for $\delta^{13}C$. Both the ostracod and the abiogenic calcite (hereafter referred to as micrite) results were reported in per mil against the PDB standard. Reproducibility of selected duplicate measurements was ± 0.13‰ and ± 0.12‰ in micrite $\delta^{18}O$ and $\delta^{13}C$ respectively, ±0.12‰ and ±0.07‰ in the $\delta^{18}O$ and $\delta^{13}C$ of *Sclerocypris clavularis*, and ±0.09‰ in both $\delta^{18}O$ and $\delta^{13}C$ of *Hemicypris intermedius*. The water samples were analyzed for $\delta^{18}O$ and δD at the International Atomic Energy Agency laboratories in Vienna.

6. Results

6.1. SEDIMENTOLOGY OF LT84-2P

The sedimentology of core LT84-2P has been described in detail by Halfman (1987). Visual inspection and smear-slide analysis of the sediment reveals that it is predominantly faintly laminated clays. The clay and fine-silt fractions constitute approximately 85 to 95% of the sediment and consist primarily of clay minerals and microcrystalline carbonate grains with minor amounts of quartz and feldspar. Bulk carbonate concentrations average 10% ($CaCO_3$) of the dry weight, and range from 1 to 24% (Halfman 1987). The accessory components include detrital grains (coarse silt to sand), volcanic debris and biogenic material. Diatoms dominate the biogenic fraction, and ostracods are also abundant. The diatom population is dominated by varying proportions of *Aulacoseira nyassensis*, *Stephanodiscus astria*, *Thalassiosira rudolfi* and *Surirella* spp. (Halfman et al. 1992). Species of *Anomoeneis*, *Rhopalodia*, *Gomphocymbella*, *Cocconeis* and *Cymbella* occur in trace amounts. The ostracod population is dominated by *Limnocythere africana*, *Potamocypris producta*, *Hemicypris intermedius*, *Sclerocypris clavularis* and *Candonopsis cf. africana*. Species of *Gomphocythere angulata*, *Darwinula stevensoni* and cf. *Cetacella* species occur in trace amounts at the bottom of the core.

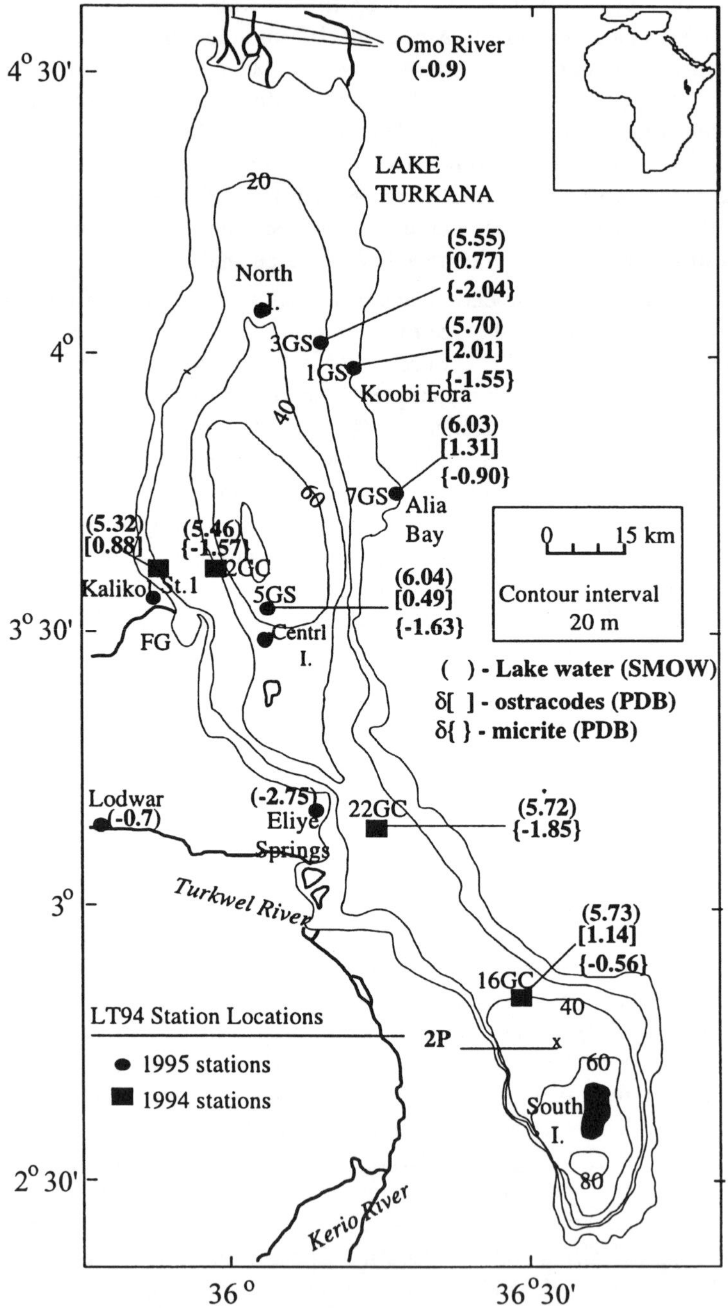

Figure 2. Lake Turkana bathymetry showing location of core LT84-2P and sample sites for modern carbonates. GC: gravity cores; GS: grab samples; (isotope values of lake waters, depths in Table 1); [deviations of ^{18}O values in modern ostracods from calculated equilibrium values]; {deviations from equilibrium values in modern micrite}.

6.2. CHRONOLOGY

A best fit line through the AMS radiocarbon dates on the ostracod shells from LT84-2P gives a core-top age of 1070 ybp and an age of 5450 ybp (radiocarbon years before present) at the bottom of the core (Halfman et al. 1994). This yields a steady sedimentation rate of 0.27 cm yr^{-1} ($r^2 = 0.997$). The relatively old date at the top of the core reflects over-penetration of the piston core and is not an artifact of old dissolved inorganic carbon in the water column. The ostracods probably reflect the true age of the sediment whereas the micrite reflects an admixture of modern calcite formed in the lake and calcite from exposed lake beds onshore, introduced into the lake by strong diurnal winds that often create intense dust storms in the Turkana basin (Halfman et al. 1994). Lacustrine sediments and soils that may contain old carbonates are exposed around the basin, although they are much more widespread in the northern part of the basin than in the south where this study is focused. In this study, we use the ostracod dates to distinguish the major periods of climate change in Lake Turkana.

6.3. ISOTOPIC COMPOSITION OF LAKE WATERS.

Ricketts and Johnson (1996) reported surface lake waters to have $\delta^{18}O$ values that range from 5.1‰ near the mouth of the Omo to 5.7‰ in the south basin. The bottom waters are slightly heavier than surface waters and vary from 5.3‰ near the Omo to 6.1‰ in the south basin. A maximum difference between surface and bottom of 0.3‰ was observed in the south basin. The isotopic composition of the Omo River was -1‰ in 1979 (measurement by T. Cerling reported by Halfman et al. 1989). In 1995 we measured a value of -0.9‰. The Turkwel river had a $\delta^{18}O$ value of -0.7‰, measured at Lodwar at the end of the flood season. Values for other inputs to the lake are given by Ricketts and Johnson (1996). In order to calculate expected $\delta^{18}O$ equilibrium values for micrite, we used $\delta^{18}O$ values of lake surface waters. Except in stations shallower than 20 m, expected $\delta^{18}O$ equilibrium values for ostracods were calculated using $\delta^{18}O$ values of the lake bottom waters (Table 1).

TABLE 1. Isotopic values for lake waters and micrite. $\delta^{18}O$ equil are calculated equilibrium values and $\Delta\delta^{18}O$ are deviations from equilibrium values. T °C meas is measured temperature. For stations equal to or deeper than 20 m, measured temperatures are maximum mean surface temperatures from Ferguson and Harbott (1982).

Station	Depth (m)	$\delta^{18}O$ water	$\delta^{18}O$ micrite	$\delta^{13}C$ micrite	$\delta^{18}O$ equil	$\Delta\delta^{18}O$	T °C meas
LT95-1GS	1	5.7	0.68	-2.57	2.23	-1.55	32.8
LT95-1GS	1	5.7	0.69	-2.45	2.23	-1.54	32.8
LT97-7GS	1	6.03	2.33	-0.67	3.28	-0.95	28.2
LT97-7GS	1	6.03	2.42	-0.75	3.28	-0.86	28.2
LT95-3GS	20	5.55	0.54	-1.91	2.58	-2.04	29.4
LT95-3GS	20	5.55	0.61	-1.99	2.58	-1.97	29.4
LT94-22GC	30	5.72	0.9	-2.39	2.75	-1.85	29.4
LT94-2GC	40	5.32	0.75	-0.99	2.36	-1.61	29.4
LT94-2GC	40	5.32	0.83	-1	2.36	-1.53	29.4
LT94-16GC	40	5.73	2.20	-0.09	2.76	-0.56	29.4
LT95-5GS	70	6.04	1.36	-1.2	3.06	-1.70	29.4
LT95-5GS	70	6.04	1.49	-1.3	3.06	-1.57	29.4

6.4. ISOTOPIC COMPOSITION OF MODERN CARBONATES

$\delta^{18}O$ values of the modern micrite vary appreciably around the lake (Table 1). Nearshore regions have $\delta^{18}O$ values around 0.7‰ except at Alia Bay which had a value of 2.3‰. Mid-lake values vary between 0.5‰ and 2.2‰. The micrite $\delta^{18}O$ values around the lake are below expected equilibrium values by 0.6‰ in the south basin and by 2‰ near the Omo delta (Fig. 2). The $\delta^{13}C$ values of the micrite are depleted close to river inputs. Stations 3GS and 1GS, closest stations to the Omo, have $\delta^{13}C$ values of -1.9‰ and -2.5‰ respectively, and station 22GC, close to the Turkwel has a value of -2.4‰. $\delta^{13}C$ values are higher in nearshore regions and away from river inputs and approach -1‰ (stations 5GS, 7GS and 16GC). Comparison of the $\delta^{18}O$ and $\delta^{13}C$ values of the modern micrite shows no significant correlation.

TABLE 2. As Table 1, but including isotopic values for *Sclerocypris clavularis* (Sc).

Station	Depth (m)	$\delta^{18}O$ water	$\delta^{18}O$ Sc	$\delta^{13}C$ Sc	$\delta^{18}O$ equil	$\Delta\delta^{18}O$	T °C meas
LT95-1GS	1	5.7	3.77	-1.79	1.91	1.86	32.8
LT95-7GS	1	6.03	3.04	-0.81	3.34	-0.30	28.2
LT95-3GS	20	5.55	3.5	-4.11	3.29	0.21	26.4
LT95-3GS	20	5.55	3.38	-4.18	3.29	0.09	26.4
LT95-St.1	20	5.32	3.73	-1.54	3.06	0.67	26.4

The ostracod $\delta^{18}O$ values around the lake do not vary as much as the micrite values. *Sclerocypris clavularis* $\delta^{18}O$ values range from 3‰ to 3.7‰ (Table 2) and those of *Hemicypris intermedius* vary from 3.9‰ to 4.6‰ (Table 3). Thus both species vary by about 0.7‰. Despite this small intraspecies variability in ostracod $\delta^{18}O$ values, there is appreciable variability around the lake and values are in most cases above expected equilibrium values (Fig. 2). Deviations from equilibrium values vary between -0.3‰ and 1.86‰ in *Sclerocypris clavularis,* and between 0.45‰ and 2.01‰ in *Hemicypris intermedius*. The distribution of $\delta^{13}C$ ratios of the ostracods around the lake does not show any pattern. A comparison of the $\delta^{18}O$ and $\delta^{13}C$ ratios of *Hemicypris intermedius* shows no correlation. $\delta^{18}O$ values for the micrite and the ostracod *Hemicypris intermedius* from five stations around the lake covary (Fig. 3). No such correlation was found between micrite and *Sclerocypris clavularis*, perhaps because there were only three sites where a comparison could be made.

TABLE 3. As Table 1, but including isotopic values for *Hemicypris intermedius* (Hi).

Station	Depth (m)	$\delta^{18}O$ water	$\delta^{18}O$ Hi	$\delta^{13}C$ Hi	$\delta^{18}O$ equil	$\Delta\delta^{18}O$	T °C meas
LT95-1GS	1	5.7	3.92	0.48	1.91	2.01	32.8
LT95-7GS	1	6.03	4.65	1.31	3.34	1.31	28.2
LT95-3GS	20	5.55	4.1	-0.25	3.29	0.81	26.4
LT95-3GS	20	5.55	4.02	-0.13	3.29	0.73	26.4
LT95-St.1	20	5.32	3.94	-0.06	3.06	0.88	26.4
LT94-16GC	40	5.73	4.61	0.39	3.47	1.14	26.4
LT95-5GS	70	6.04	4.32	0.50	3.78	0.54	26.4
LT95-5GS	70	6,04	4.23	0.41	3.78	0.45	26.4

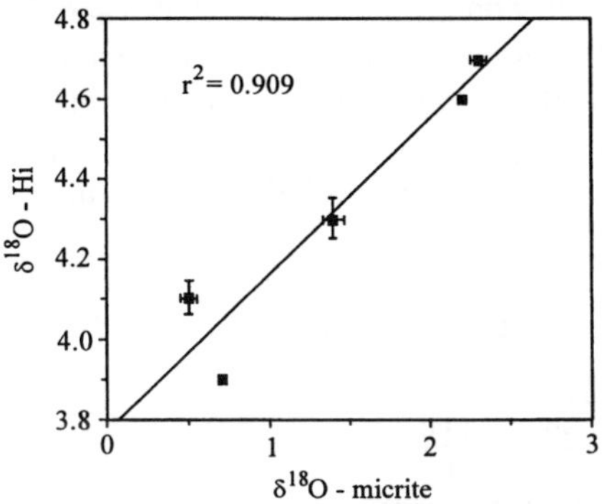

Figure 3. Covariance in ostracod and micrite $\delta^{18}O$ in the modern lake sediments in five different locations around the lake. Hi = *Hemicypris intermedius* and m = micrite.

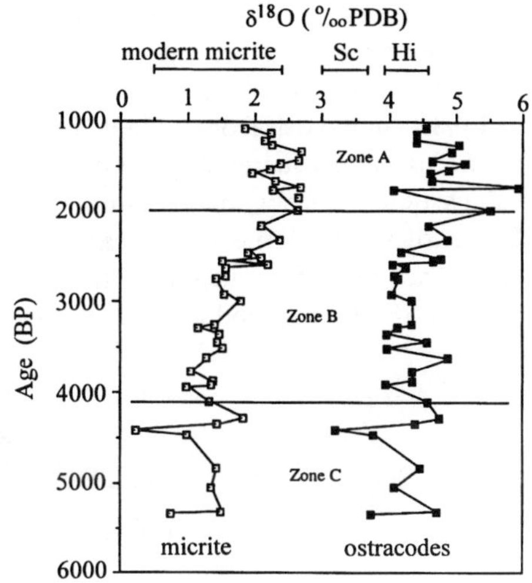

Figure 4. $\delta^{18}O$ of ostracods and micrite in the same subsamples in core LT84-2P plotted against age downcore.

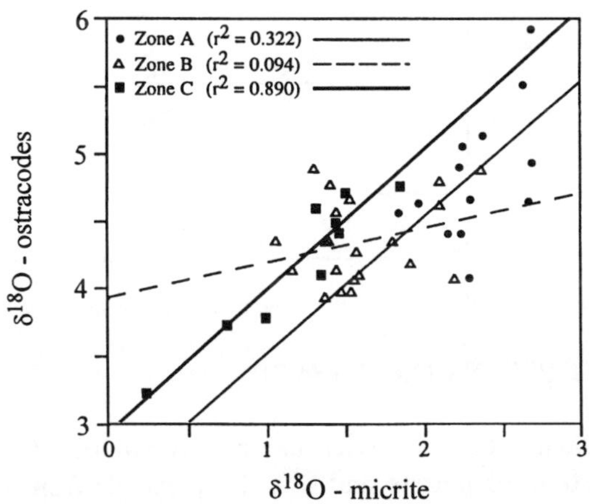

Figure 5. Covariance in ostracod and micrite $\delta^{18}O$ values at corresponding depths in core 2P. The boundaries between zones A, B and C were selected following the results of Ricketts and Johnson (1996).

6.5. OXYGEN ISOTOPES IN CORE LT84-2P

Muchane (1994) analyzed in detail oxygen and carbon isotopes in micrite in the core. Ricketts and Johnson (1996) delineated three zones in the core on the basis of the micrite $\delta^{18}O$ values. These are from 5400 ybp to 4000 ybp (Zone C), from 4000 to 2000 ybp (Zone B), and from 2000 ybp to 1100 ybp (Zone A) (Fig. 4). The $\delta^{18}O$ of the carbonates have the lightest values in Zone C. The micrite values average 1.2‰ and those of the ostracods average 4.2‰, a difference of 3‰. In Zone B, the $\delta^{18}O$ of the carbonates are slightly heavier compared to zone C with micrite values averaging 1.7‰ and those for ostracods averaging 4.4‰, a difference of 2.7‰. The average $\delta^{18}O$ shifts in both micrite and ostracods within Zone B are comparable being about 1‰ in micrite and 0.8‰ in the ostracods. Mean $\delta^{18}O$ of micrite in Zone A increases to 2.3‰ and that of the ostracods increases to 4.8‰. The $\delta^{18}O$ values of ostracods track the micrite values quite closely, maintaining values of about 3‰ heavier than micrite values. Comparison of micrite and ostracod $\delta^{18}O$ values for the whole core shows a positive covariance (Fig. 5). The correlation is strongest in Zone C ($r^2 = 0.890$), intermediate in Zone A ($r^2 = 0.322$), and lacking in Zone B ($r^2 = 0.094$).

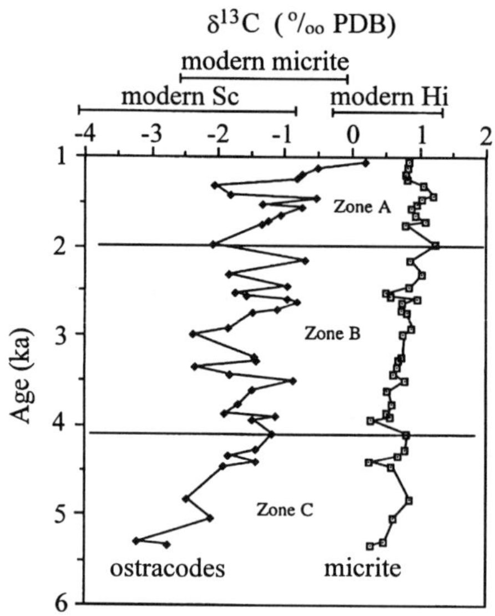

Figure 6. $\delta^{13}C$ values for micrite and ostracods in core 2P.

6.6 CARBON ISOTOPES IN CORE LT84-2P

The $\delta^{13}C$ values of the micrite are heavier than ostracod values (Fig. 6). No correlation was found between $\delta^{13}C$ of micrite and $\delta^{13}C$ in ostracods from corresponding depth intervals. Micrite values range from 0.3‰ to 0.9‰ and ostracod values range from -2.7‰ to 0.2‰ (Fig. 6). The ostracod profile has amplitude fluctuations of up to 1.5‰ compared to the relatively stable micrite values. Micrite values increase steadily throughout zone B, whereas ostracod values increase upwards through this zone in a less linear manner. The difference between the averages in micrite and ostracod $\delta^{13}C$ is least in Zone A than deeper in the core. The oxygen and carbon isotopes of micrite correlate positively (r= 0.844; Fig. 7). The correlation is strong in zones A (r^2= 0.618), and C (r^2= 0.634), and weak in zone B (r^2= 0.341). A comparison of the $\delta^{18}O$ and $\delta^{13}C$ of the ostracods shows no correlation.

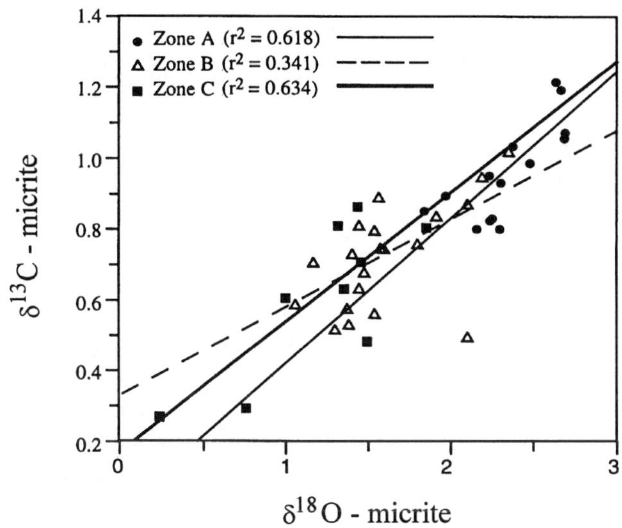

Figure 7. Covariance in micrite $\delta^{18}O$ and $\delta^{13}C$ values at corresponding depths in core 2P.

7. Discussion

7.1. OXYGEN ISOTOPES

As expected, the $\delta^{18}O$ values of the waters in Lake Turkana are significantly enriched over river values because the lake is closed and there has been extensive evaporative fractionation. Lake Turkana values of δD versus $\delta^{18}O$ plot on a line that intercepts the global meteoric water line at a $\delta^{18}O$ value of -1.7‰ (Ricketts and Johnson 1996). This is taken to be the mean composition of the source waters. The $\delta^{18}O$ values of the lake waters show no significant north-south gradient, reflecting the well-mixed waters in this relatively shallow lake that is subjected to intense winds much of the year.

O'Neil et al. (1969) determined oxygen isotope fractionation between inorganically-precipitated calcite and water. Their results over the temperature range 0 to 500 °C are described by :

$$10^3 \ln(\alpha)_{\text{calcite-water}} = (2.78 \times 10^6)T^{-2} - 2.89 \quad (1)$$

$$\text{and } 10^3 \ln(\alpha) \sim \delta^{18}O_c - \delta^{18}O_w(\text{SMOW})$$

where $\alpha = R_A/R_B$, R_A is the ratio of the heavy to the light isotope in molecule or phase A ($^{18}O/^{16}O$), R_B is the same in phase B, T is the temperature of the reaction (°K) and the isotope ratios are relative to the SMOW standard. This equation was later modified by Friedman and O'Neil (1977) using $\alpha_{CO2 - H20} = 1.0412$.

The relationship between $\delta^{18}O$ values for water and shell calcite is given by Buchardt and Fritz (1980):

$$T = 16.9 - 4.2(\delta^{18}O_c - \delta^{18}O_w) + 0.13(\delta^{18}O_c - \delta^{18}O_w)^2 \quad (2)$$

where $\delta^{18}O_w$ is the $\delta^{18}O$ value of the water in SMOW units, $\delta^{18}O_c$ is the $\delta^{18}O$ value of the shell calcite in SMOW units, and T is the water temperature at calcification in °C.

Surface water temperatures in Lake Turkana range between 27.2 and 29.4 °C and bottom water temperatures range between 25.4 and 26.4 °C (Ferguson and Harbott 1982). In the open lake, water temperatures can be as high as 32.8 °C in protected bays and lagoons, for example at station 1GS. Applying Eq. 1 indicates that the calcite values are below equilibrium values (Table 1). The difference between calculated and measured $\delta^{18}O$ values ranges from 0.6‰ to 2.0‰. This would imply carbonate precipitation temperatures that are 0.4 to 8.4 °C above average water temperatures. Johnson et al. (1991) hypothesized that micrite may form in a thin microlayer at the lake surface heated on calm, hot sunny days. It is unlikely that such a microlayer could achieve temperatures that are more than about 1 °C more than the underlying waters. This is probably a kinetic effect, perhaps influenced by biological processes such as algal blooms.

Application of Eq. 2 to the ostracod shells indicates that *Sclerocypris clavularis* shells deviate by -0.3‰ to +1.86‰ relative to equilibrium values for $\delta^{18}O$ and those of *Hemicypris intermedius* are +0.45‰ to +2.01‰ above equilibrium values. The ostracods form shells that are out of equilibrium with the lake waters but in contrast to the micrite, most ostracod values are higher than equilibrium values.

The $\delta^{18}O$ values of modern ostracods in Lake Turkana are higher than micrite by about 0.7‰ to 3.1‰ (*Sclerocypris clavularis*) and 2.4‰ to 3.2‰ (*Hemicypris intermedius*). This cannot be explained by slight differences in the $\delta^{18}O$ or temperature between the surface waters and bottom waters. Culture experiments show that when ostracods and micrite are precipitated at the same temperature, the ostracods are 2‰ heavier than equilibrium values due to vital effects during shell formation (E. Ito and R. Forester, personal communication, 1995). This would be consistent with the ~2‰ heavier values found in this study. Different species of ostracods collected from the same lake frequently have different $\delta^{18}O$ values (Lister et al. 1991).

The $\delta^{18}O$ values for micrite and ostracods correlate positively both in modern sediments and in core LT84-2P (Figs. 3 and 5). Although neither form of carbonate forms in isotopic equilibrium with the lake water, the profiles track one another through time. This suggests that both are responding to environmental forcing and are useful for paleoenvironmental interpretation, at least on a time scale of centuries to millennia. We noted that the largest deviations for modern calcite occur close to sources of river input. Thus on a short time scale (months to years), the interaction between inflow and basinal waters adversely affects the reliability of the isotopic signal of climate change.

In this low latitude setting, past variations in the isotopic composition of the lake waters is the most likely cause of temporal change in the $\delta^{18}O$ of carbonates. Oxygen isotopic values are lightest in zone C, corresponding to fresh water and open conditions prior to 4000 ybp as deduced from diatom assemblage analyses (Fig. 4) (Halfman et al. 1992), and ostracod assemblages (Ng'ang'a and Johnson, in prep). The steady increase in $\delta^{18}O$ values observed in zone B, representing the period from 4000 to 2000 ybp, reflects a long-term steady change in climatic conditions during that interval (Ricketts and Johnson 1996). The relatively heavy $\delta^{18}O$ values of Zone A are consistent with results of previous studies that show lake level to have remained relatively low during the past 2000 yr (Owen et al. 1982; Johnson et al. 1991; Halfman et al. 1992; Ricketts and Johnson 1996).

7.2. CARBON ISOTOPES

The isotopic equilibrium reaction involving transfer of CO_2 across the air-water interface and dissociation of dissolved CO_2 to HCO_3^- may be written as a composite of two processes, gas exchange and dissociation of carbonic acid:

$$^{13}CO_2(g) + H^{12}CO_3^-(aq) = {}^{12}CO_2(g) + H^{13}CO_3^-(aq) \qquad (3)$$

where the equilibrium constant, K

$$K = [^{12}CO_2(g)][H^{13}CO_3^-(aq)]/[^{13}CO_2(g)][H^{12}CO_3^-(aq)]$$

$$= (^{13}C/^{12}C)_{HCO3-}/(^{13}C/^{12}C)_{CO2} = \alpha \qquad (4)$$

is equal to the isotope fractionation factor α (Herczeg and Fairbanks 1987). Enrichments and depletions are generally reported as: $\varepsilon = (\alpha\text{-}1) \times 1000‰$.

Table 4 compares theoretical and observed $\delta^{13}C$ values of carbonates relative to atmospheric CO_2. The $CaCO_3$ that forms in equilibrium with atmospheric CO_2 should have a $\delta^{13}C$ value of 1.75 ‰. The $\delta^{13}C$ values of modern micrite around the lake vary from -0.09‰ to -2.6‰ implying $\delta^{13}C$ values that are lighter than $CO_{2(aq)}$ in equilibrium with the atmosphere by -1.8‰ to -4.3‰. The $\delta^{13}C$ values for ostracods are lighter than those for $CO_{2(aq)}$ in equilibrium with the atmosphere by -1.88‰ to -0.44‰.

Table 4. (a). Carbon isotope fractionation (‰) in natural waters (Herczeg and Fairbanks 1987). The value ε (enrichment or depletion) is defined as $\varepsilon = (\alpha\text{-}1) \times 1000$ where α is the fractionation factor for the given reaction or process. (b). Expected (*), observed (†) and calculated (‡) $\delta^{13}C$ values for dissolved carbon species in water and carbonate in equilibrium with atmospheric $CO_2(g)$. m = micrite; Hi = *Hemicypris intermedius*.

a	Reaction		ε	
	CO_2 (aq) = CO_2 (g)		+1.1	
	CO_2 (aq) = HCO_3^-		+9 to +12 (25 to 0 °C)	
	CO_2 (aq) + OH^- = HCO_3^-		~ -15	
b	CO_2 (atm)	CO_2 (aq)	HCO_3^- (aq)	$CaCO_3$
	-8	-9.1*	-0.1*	1.75*
	-12.04 to -14.5†	-10.94 to -13.4†	-1.9 to -4.4†	-0.09 to -2.6 (m)‡
	-12.8 to -10.64†	-10.98 to -9.54†	-1.98 to -0.54†	-0.25 to 1.31 (Hi) ‡

If the partial pressure of carbon dioxide is very low in surface waters as a result of algal productivity, there is a large net negative carbon isotope fractionation between atmospheric CO_2 and the CO_2 absorbed by lake water at pH greater than 8 (Herczeg and Fairbanks 1987). In many freshwater systems where pH exceeds 9 during algal blooms, direct hydroxylation of CO_2 by OH^- at the air-water interface, and the isotope fractionation associated with that reaction, may also be important. High rates of photosynthesis not only cause chemically enhanced invasion but also promote the predominant role of invading CO_2 highly depleted in ^{13}C.

Under ideal conditions, micrite that forms in equilibrium with atmospheric CO_2 ($\delta^{13}C$= -8‰) will have a $\delta^{13}C$ value of 1.75‰ (Table 4b). A calculation using $\delta^{13}C$ values of the modern micrite in Lake Turkana gives an atmosphere with a $\delta^{13}C$ value of -12.04‰ to -14.5‰. Since the present day atmosphere has a $\delta^{13}C$ value of -8‰ and the pH in Lake Turkana is greater than 9, we conclude that the reaction between CO_2 and OH^- at the air-water interface and algal blooms in the lake are important factors that cause the modern micrite and the ostracods to be depleted in $\delta^{13}C$ relative to what would be expected if the lake waters were in equilibrium with the atmosphere. The relative importance of each of these effects on the carbon isotopic composition of the carbonates forming in the modern lake is however unknown.

7.3. COMPARISON BETWEEN CARBON AND OXYGEN ISOTOPES

The $\delta^{18}O$ and $\delta^{13}C$ for micrite correlate positively as is commonly observed in lacustrine carbonates (Fig. 7) (Stuiver 1970; Botz et al. 1988; Gasse et al. 1987; Talbot and Kelts 1990; Johnson et al. 1991; Li 1995). The covariance in each lake has a unique trend (Talbot and Kelts 1990). The observed $\delta^{13}C$-$\delta^{18}O$ covariance depends on the relative effects of total CO_2 buffering, hydrologic dilution, vapor exchange, enhanced evaporation and productivity (Li 1995). Covariance of $\delta^{13}C$-$\delta^{18}O$ can be achieved in several ways. As lake volume rapidly decreases due to evaporation, its $\delta^{18}O$ will increase. It is possible that nutrient flux to surface waters may increase due to a shallower water column leading to increased productivity which elevates $\delta^{13}C$ in the surface waters of the lake. This is because ^{12}C is preferentially assimilated into organic matter resulting in an enrichment of the heavier isotope ^{13}C in the remaining dissolved inorganic carbon (McKenzie 1985). For a highly productive lake, the $\delta^{13}C$ value can reach as high as 6‰ (Li 1995). Therefore, a rapidly declining lake will have its $\delta^{13}C$ value enriched due to increased productivity and its $\delta^{18}O$ enriched due to evaporation.

Persistence of $\delta^{13}C$-$\delta^{18}O$ covariance in micrite indicates a long-term response that probably reflects a change in the isotopic composition of inflowing waters. Generally, the isotopic composition of a closed-basin lake and its input waters differ largely. Therefore, when large amounts of fresh water discharge to the lake, the ^{13}C and ^{18}O of the lake will both decrease, but on different time scales, along with carbonate alkalinity and salinity. When lake volume shrinks under dry and warm climates, alkalinity and salinity will increase (Forester 1987; Gasse et al. 1987; Benson et al. 1990). Li (1995) has shown that in lakes where carbonate alkalinities are much higher than in the inputs, the dilution effect on $\delta^{13}C$ is much smaller than that on the $\delta^{18}O$. As the carbonate alkalinity decreases, the dilution effect on $\delta^{13}C$ becomes more important.

For a large closed-basin lake with moderately high alkalinity the $\delta^{13}C$ of dissolved inorganic carbon (DIC) will not vary rapidly with fluctuations in input water. Since Lake Turkana has a large volume, covariance of $\delta^{18}O$ and $\delta^{13}C$ will be achieved only on a long term scale, whether lake volume increases or decreases. Depending on the size of the basin, there exists a time lag in the response of isotopic signals in carbonates to a lake volume (or climate) change (Ricketts and Johnson 1996). This makes it hard to draw conclusions of changing hydrological conditions in the drainage basin without using other independent proxies, for example diatoms and ostracod assemblages (Ng'ang'a and Johnson, in preparation). Results of this study strongly suggest that the degree of covariance in the $\delta^{18}O$ of abiogenic and biogenic calcite can be used to infer past hydrologic conditions of a lake. Strong covariance in oxygen isotope profiles of biogenic and abiogenic calcite in a core may be an indication of open, fresh water conditions whereas low covariance may be an indication of stratification or a closed basin.

8. Conclusions

After investigating the modern relationships between $\delta^{18}O$ of the lake waters and $\delta^{18}O$ of micrite and of two species of ostracods, we conclude that:

(1) Modern micrite and ostracods in Lake Turkana are forming out of equilibrium with the lake waters. On the short term, the isotopic composition of inflowing waters affects disequilibrium only locally in the precipitating carbonates. The large reservoir of basin waters controls the isotopic values on the long term via isotope exchange.

(2) Although there is lake-wide variability in the $\delta^{18}O$ values of micrite and ostracods, covariance between the two both in the modern sediments and in the core suggests that both are responding to environmental parameters (temperature, $\delta^{18}O$ of the lake waters, primary production, etc.), and are useful for paleoenvironmental interpretation at least on a time scale of centuries to millennia.

(3) Covariance of $\delta^{18}O$ values in micrite and in ostracods was strongest prior to 4,000 ybp when Lake Turkana was hydrologically open. Although covariance persisted after 4,000 ybp, a decrease in the degree of covariance suggests more stratification in the water column than when the lake was hydrologically open.

(4) Modern $\delta^{13}C$ values of the carbonates are depleted relative to equilibrium values and suggest inputs of isotopically light carbon into the lake. This may result from light fractionation of atmospheric CO_2 invading the lake during periods of high productivity.

9. Acknowledgments

We thank the Kenya Government for permission to undertake this research. We are grateful to the Kenya Wildlife Services and the National Museums of Kenya for their assistance in field logistics. Larry Giles provided invaluable assistance with isotopic analyses of carbonates and K. Rozanski supervised analyses of waters at IAEA laboratories in Vienna. This project was funded by the Leakey Foundation and NOAA grant NA56GP0075 to TCJ.

10. References

Abell, P.I., S. M. Awramik, R. H. Osborne, and S. Tomellini. 1982. Plio-Pleistocene lacustrine stromatolites from Lake Turkana, Kenya: Morphology, Stratigraphy and Stable Isotopes. Sedimentary Geology **32**: 1-26.

Benson, L.V., D. R. Currey, R. I. Dorn, K. R. Lajoie, C. G. Oviatt, S. W. Robinson, G. I. Smith, and S. Stine. 1990. Chronology of expansion and contraction of four Great Basin lake systems during the past 35,000 years. Palaeogeography, Palaeoclimatology, Palaeoecology **78**: 241-286.

Botz, R., P. Stoffers, E. Faber, and K. Tietze. 1988. Isotope geochemistry of carbonate sediments from lake Kivu (East-Central Africa). Chemical Geology **69**: 299-308.

Buchardt, B., and P. Fritz. 1980. Environmental Isotopes as Environmental and Climatological Indicators, p. 473-504. *In* P. Fritz and J. C. Fonte [eds.], Handbook of Environmental Geochemistry. Elsevier.

East African Meteorological Department. 1975. Climatological records for East Africa. East Africa Community, Nairobi.

Epstein, S., R. Buchsbaum, H. A. Lowenstam, and H. C. Urey. 1953. Revised carbonate-water isotopic temperature scale. Geological Society of America Bulletin **64**: 1315-1326.

Ferguson, A. J. D., and B. J. Harbott. 1982. Geographical, physical and chemical aspects of Lake Turkana. p. 1-107. *In* A. J. Hopson [ed.], Lake Turkana : A report on the findings of the Lake Turkana Project, 1972-1975. Overseas Development Administration, London.

Forester, R.M. 1987. Late Quaternary paleoclimate records from lacustrine ostracodes, P. 261-276. *In* W. F. Ruddiman and H. E. Wright, Jr. [eds.], North America and adjacent oceans during the last deglaciation. Geological Society of America.

Friedman, I., and J. R. O'Neil. 1977. Compilation of stable isotope fractionation factors of geochemical interest, p. KK1-KK12. *In* M. Fleischer [ed.], Data of Geochemistry, 6th ed. U.S. Geological Survey Professional Paper 440-KK.

Gasse, F., J. C. Fontes, J. C. Plaziat, P. Carbonel, I. Kaczmarska, P. DeDeckker, I. Soulié-Marsche, Y. Callot, and P. A. Dupeuble. 1987. Biological remains, geochemistry and stable isotopes for the reconstruction of environmental and hydrological changes in the Holocene lakes from North Sahara. Palaeogeography, Palaeoclimatology, Palaeoecology **60**: 1-46.

Griffiths, J.F. 1972. Climates in Africa. Elsevier.

Grossman, E.T., and T. L. Ku. 1981. Aragonite-water isotopic paleotemperature scale based on the benthic foraminifera Hoeglundia elegans. Geological Society of America Abstracts.

Halfman, J.D. 1987. High-resolution sedimentology and paleoclimatology of Lake Turkana, Kenya. Ph.D. dissertation. Duke University.

Halfman, J. D. 1996. CTD-Transmissometer profiles from Lake Malawi and Turkana, p. 169-182. *In* T. C. Johnson and E. Odata [eds.], The limnology, climatology and paleoclimatology of the East African lakes. Gordon and Breach.

Halfman, J. D., and P. J. Hearty. 1990. Cyclic sedimentation in Lake Turkana, Kenya, p. 187-196. *In* B. Katz [ed.], Lacustrine Basin Exploration - Case Studies and Modern Analogs. American Association of Petroleum Geologists, Memoir 50.

Halfman, J. D., D. F. Jacobson, C. M. Canella, K. A. Haberyan, and B. P. Finney. 1992. Fossil diatoms and the mid to late Holocene paleolimnology of Lake Turkana, Kenya: a reconaissance study. Journal of Paleolimnology **7**: 23-35.

Halfman, J. D., T. C. Johnson, and B. P. Finney. 1994. New AMS dates, stratigraphic correlations and decadal climatic cycles for the past 4 ka at Lake Turkana, Kenya. Palaeogeography, Palaeoclimatology, Palaeoecology **111**: 83-98.

Halfman, J. D., T. C. Johnson, W. J. Showers, and G. S. Lister. 1989. Authigenic Low-Mg Calcite in Lake Turkana, Kenya, p. 533-540. *In* B. R. Rosendahl, J. J. W. Rogers, and N. M. Rach [eds.], African Rifting. Journal of African Earth Sciences Special Publication 8.

Herczeg, A.L., and R. G. Fairbanks. 1987. Anomalous carbon isotope fractionation between atmospheric CO_2 and dissolved inorganic carbon induced by intense photosynthesis. Geochimica et Cosmochimica Acta **51**: 895-899.

Hodell, D.A., J. H. Curtis, G. A. Jones, A. Higuera-Gundy, M. Brenner, M. W. Binford, and K. T. Dorsey. 1991. Reconstruction of the Caribbean climate change over the past 10,500 years. Nature **352**: 790-793.

Hopson, A.J.[ed.]. 1982. Lake Turkana : A report on the findings of the Lake Turkana Project, 1972-1975, 6 vols. Overseas Development Administration.

Johnson, T.C., J. D. Halfman, and W. J. Showers. 1991. Paleoclimate of the past 4000 years at Lake Turkana, Kenya based on isotopic composition of authigenic calcite. Palaeogeography, Palaeoclimatology, Palaeoecology **85**: 189-198.

Kinuthia, J. K. 1992. Horizontal and vertical structure of the Lake Turkana jet. Journal of Applied Meteorology **31**: 1248-1274.

Li, H.-C. 1995. Isotope geochemistry of Mono Basin, California: Applications to Paleoclimate and Paleohydrology. Ph.D. dissertation. University of Southern California.

Lister, G.S., K. Kelts, C. K. Zao, J-Q. Yu, and F. Niessen, 1991. Lake Qinghai, China: closed-basin lake levels and the oxygen isotope record for Ostracodea since the latest Pleistocene. Palaeogeography, Palaeoclimatology, Palaeoecology **84**: 141-162.

McKenzie, J. A. 1985. Carbon isotopes and productivity in the lacustrine and marine environment, p. 99-118. *In* W. Stumm [ed.], Chemical processes in lakes. Wiley.

McKenzie, J. A., and D. J. Hollander. 1993. Oxygen-Isotope record in recent carbonate sediments from Lake Greifen, Switzerland (1750-1986): Application of continental isotopic indicator for evaluation of changes in climate and atmospheric circulation patterns, p. 101-111. *In* P. K. Swart, K. C. Lohmann, J. McKenzie, and S. Savin, Climate change in Continental Isotopic records. Geophysical Monograph 78.

Mook, W.G., and J. C. Vogel. 1968. Isotope equilibrium between shells and their environment. Science **159**: 874-875.

Muchane, M.W. 1994. Stable isotope analyses of authigenic calcite from Lake Turkana, Kenya: High-Resolution paleoclimatic iplications for the past 5000 years. Ph.D. dissertation. Duke University.

Ng'ang'a, P., and T. C. Johnson. (in prep). The paleohydrochemistry of Lake Turkana in the last 5400 yrs based on ostracode species variability.

Nicholson, S. E. 1996. A review of climate dynamics and climate variability in Eastern Africa, p. 25-56. *In* T. C. Johnson and E. Odada [eds.], The limnology, Climatology and Paleoclimatology of the East African Lakes. Gordon and Breach.

Nieuwolt, S. 1977. Tropical Climatology. Wiley.

O'Neil, J. R., R. N. Clayton, and T. K. Mayeda. 1969. Oxygen isotope fractionation in divalent metal carbonates. Journal of Chemistry and Physics **51**: 5547-5548.

Owen, R. B., J. W. Barthelme, R. W. Renaut, and A. Vincens. 1982. Palaeolimnology and archaeology of Holocene deposits north-east of Lake Turkana, Kenya. Nature **298**: 523-528.

Phillips, F.M., A. R. Campbell, C. Kruger, P. Johnson, R. Randall, and E. Keyes. 1992. A reconstruction of the response of the water balance in Western United States lake basins to climate change. New Mexico Water Resources Research Institute. Technical Completion Report.

Ricketts, R.D., and T. C. Johnson. 1996. Climatic change in the Turkana Basin as deduced from a 4000 year long 18O record. Earth and Planetary Science Letters **142**: 7-17.

Rozanski, K., L. Araguas-Araguas, and R. Gonfiatini. 1996. Isotope patterns of Precipitation in the East African Region, p. 79-93. *In* T. C. Johnson and E. Odada [eds.], The limnology, climatology and paleoclimatology of the East African Lakes. Gordon and Breach.

Stuiver, M. 1970. Oxygen and carbon isotope ratios of fresh water carbonates as climate indicators. Journal of Geophysical Research **75**: 5247-5257.

Talbot, M. R., and K. Kelts. 1990. Paleolimnological signatures from carbon and oxygen isotopic ratios carbonates from organic carbon-rich lacustrine sediments, p. 99-112. *In* B. Katz [ed.], Lacustrine basin exploration: case studies and modern analogs. American Association for Petroleum Geologists Memoir 50.

Tarutani, T., R. N. Clayton, and T. K. Mayeda. 1969. The effect of polymorphism and magnesium substitution of oxygen isotope fractionation between calcium carbonate and water. Geochimica et Cosmochimica Acta **33**: 987-996.

Yuretich, R. F. 1979. Modern sediments and sedimentary processes in Lake Rudolf (Lake Turkana), eastern Rift Valley, Kenya. Sedimentology **26**: 313-331.

Yuretich, R. F. 1986. Controls on the composition of modern sediments, Lake Turkana, Kenya, p. 141-152. *In* L. E. Frostick, et al. [eds.], Sedimentation in the African Rifts. Geological Society of London Special Publication #25.

Yuretich, R. F., and T. E. Cerling. 1983. Hydrogeochemistry of Lake Turkana, Kenya: Mass balance and mineral reactions in an alkaline lake. Geochemica et Cosmochimica Acta **47**: 1099-1109.

A COMPARISON BETWEEN THE STABLE ISOTOPIC COMPOSITION OF EARLY HOLOCENE AND LATE PLEISTOCENE CARBONATES FROM LAKE MALAWI, EAST AFRICA

R. D. RICKETTS
Large Lakes Observatory, University of Minnesota
Duluth, MN 55812 USA

1. Abstract

Late Pleistocene carbonates in two cores from Lake Malawi were analyzed for stable isotopic composition. A comparison is made between carbonates deposited during the late Pleistocene (ca. 31-40 ka) and the early Holocene (ca. 5-10 ka). Differences in $\delta^{13}C$ values record changes in the composition of atmospheric CO_2 and lake productivity. The $\delta^{18}O$ values of the late Pleistocene carbonates record either spatial variability in the composition of lake water or a change in climatic conditions in the basin. This change could have been caused by at most a 3.7 °C decrease in temperature, but other climatic parameters may also have contributed to the observed change in the isotopic composition of the carbonates.

2. Introduction

There is disagreement about the extent of temperature change in the tropics during the last glacial period. CLIMAP (1976) and more recent authors (e.g., Mix et al. 1986; Sikes and Keigwin 1994; Billups and Spero 1996) hypothesize that sea surface temperatures in the equatorial ocean fell by 1 to 3 °C during the last glacial maximum. Other oceanographers and terrestrial scientists find a larger drop in temperatures in the equatorial regions (e.g., Livingstone 1975; Vincens et al. 1993; Guilderson et al. 1994; Webb et al. 1997). Cores from Lake Malawi, a large tropical lake in East Africa, contain sediment which spans at least the last 40,000 years. The cores allow comparison between environmental conditions in the Lake Malawi basin during the late Pleistocene and early Holocene.

There is evidence for two closed-basin periods each longer than 4 kyr duration in Lake Malawi during the early Holocene and late Pleistocene. This evidence consists of erosional unconformities, intervals of carbonate deposition in otherwise carbonate-free sediments, variations in grain size, the abundance of shallow water diatoms, and the C:N ratio of organic matter (Finney and Johnson 1991; Finney et al. 1996). The early Holocene period of lower lake level occurred from approximately 5 to 10 ka and was a

J.T. Lehman (ed.), Environmental Change and Response in East African Lakes, 191–206.

100 to 150 m drop (Finney and Johnson 1991; Ricketts and Johnson 1996a). The late Pleistocene fall in lake level was greater, perhaps 250 m, and spanned from 28 to 40 ka (Finney et al. 1996). The timing of these lowstands is out of phase with the timing of lowstands in African lakes to the north (Finney et al. 1996; Johnson 1996).

In this study, we compare the stable isotopic composition of the authigenic carbonate deposited during the two lowstands. The isotopic composition of carbonate is dependent on the temperature at which it formed and the isotopic composition of the water in which it formed. Similarities and differences between carbonates deposited during the two periods are assessed using a quantitative interpretation of the oxygen isotope mass balance of the lake. This and a comparison of the $\delta^{13}C$ values between time periods lead to some hypotheses about the different environments present in the lake basin prior to and after the last glacial maximum.

Lake Malawi is located between 9° and 15° S in the western branch of the East African Rift System (Fig. 1). The lake is 560 km long, has a maximum width of approximately 70 km, and a maximum depth of 700 m (Eccles 1974; Johnson and Ng'ang'a 1990). The half-graben pattern of continental rifting prevalent in East Africa accounts for the shape of the lake (Rosendahl 1987; Flannery 1988; Johnson and Ng'ang'a 1990). The Shire River, which flows from the southern end of the lake, has a shallow outlet so that a 6 m drop in present lake level would close the basin (Beadle 1981). Eighty-two percent of the total water loss from the lake is by evaporation (Owen et al. 1990). The shallow outlet and high evaporation rate increase the likelihood of the lake becoming a closed basin.

Lake Malawi is a fresh water lake (0.2‰ salinity) dominated by calcium, magnesium, sodium, and carbonate ions. It is slightly alkaline with an average pH of 7.6 (Gonfiantini et al. 1979). The lake is meromictic with surface temperatures ranging seasonally from 28 °C to 23.5 °C and temperatures in the monimolimnion remaining constant at 22.7 ±0.03 °C (Eccles 1974; Halfman 1993, 1996; Patterson and Kachinjika 1995). A thermocline exists at depths of up to 100 m, and the lake is anoxic below the chemocline (200-250 m depth; Eccles 1974).

3. Methods

Thirty-three piston cores were recovered from Lake Malawi in 1986 using a Kullenberg piston corer (corer described by Kelts et al. 1986). Carbonate-rich intervals observed in five of the cores were analyzed for their isotopic composition (Table 1). The carbonate in three of the cores (12P, 22P, and 24P) deposited between 5 and 10ka have been described previously (Finney and Johnson 1991; Ricketts and Johnson 1996a).

The carbonate-rich intervals in the remaining two cores (18P and 27P) were deposited during the late Pleistocene, from about 30 to 40 ka (Finney et al. 1996). Sediments in these intervals were sampled at 5 cm spacing, freeze-dried, and analyzed for calcium carbonate content using a vacuum-gasometric method (Jones and Kaiteris 1983). The standard deviation of percent calcium carbonate for replicate samples was approximately 0.5%.

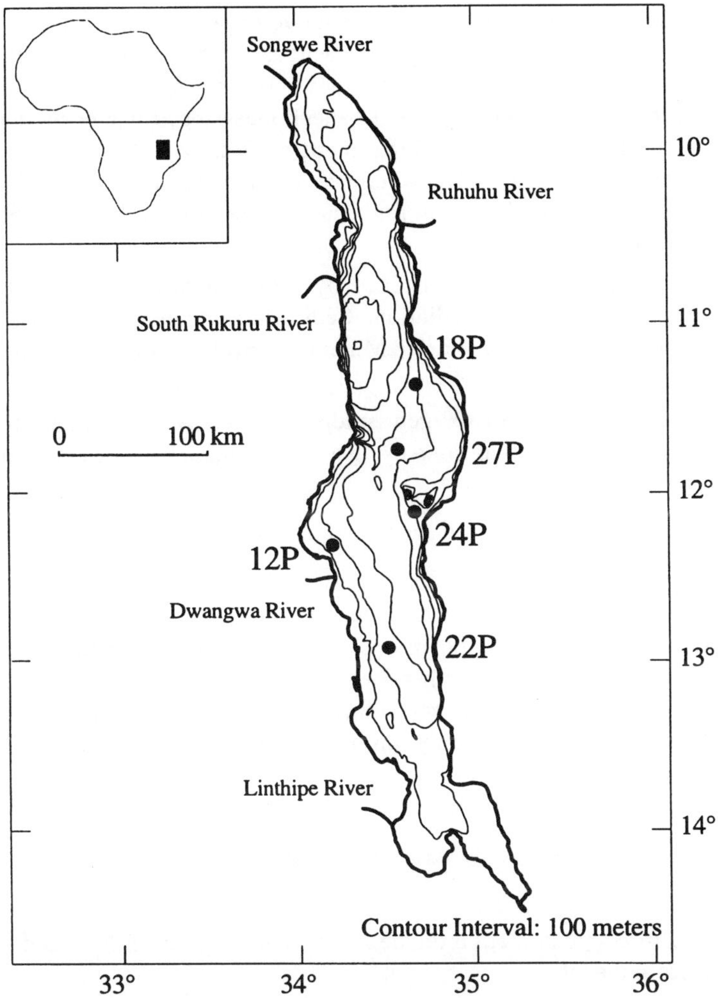

Figure 1. Lake Malawi bathymetry, major influent rivers, and core sites.

TABLE 1. Core Descriptions

	M86-12P	M86-18P	M86-22P	M86-24P	M86-27P
latitude	12°19.5′ S	11°23.6′ S	12°54.2′ S	12°09.0′ S	11°47.0′ S
longitude	34°08.8′ E	34°40.4′ E	34°31.2′ E	34°39.4′ E	34°34.0′ E
water depth	143 m	298 m	265 m	183 m	324 m
core length	555 cm	1028 cm	952 cm	733 cm	563 cm
depth interval of $CaCO_3$ (cm)	80-410	510-620 and 955-1015	810-952	200-410	130-175 and 355-505

In preparation for stable isotope analyses, 20-40 mg of powdered bulk sediment were roasted under vacuum at 350 °C to remove organic matter. The sediment was then reacted with 100% ortho-phosphoric acid at 90 °C, and the evolved CO_2 gas collected using standard cryogenic techniques. The stable isotope ratios for oxygen and carbon were determined using a VG Isotech mass spectrometer model Sira Series II located in the Duke University Department of Botany. They are reported in the standard δ notation using the PDB reference standard (Craig 1957). The analytical precision of this technique, based on replicate analyses of the same sample, is 0.10‰ for ^{18}O and 0.05‰ for ^{13}C.

Carbonate mineralogy was determined by x-ray diffraction using a Philips XRG diffractometer with Cu-K alpha radiation. Sediment samples were ground, mixed with acetone, and smeared on glass slides. Minerals were identified from the XRD peaks using standard tables (Chao 1969).

Modeling the sensitivity of oxygen isotopes in Lake Malawi waters to changes in various climatic parameters was done using a general mass balance equation:

$$\frac{dV\delta_{lake}}{dt} = \delta_{inputs}Q_{inputs} - \delta_{outputs}Q_{outputs} \quad (1)$$

in which V equals the volume of the lake, t is time, Q is the volume of the various fluxes, and δ is the isotopic composition of the various water masses. Since carbonate deposition occurs only when Lake Malawi is a closed-basin lake (Finney and Johnson 1991; Finney et al. 1996; Ricketts and Johnson 1996a) and groundwater inflow (Owen et al. 1990) and outflow are probably minimal compared to the rest of the hydrological budget, Eq. 1 can be simplified to:

$$\frac{dV\delta_{lake}}{dt} = \delta_{rain}Q_{rain} + \delta_{river}Q_{river} - \delta_{evap}Q_{evap} \quad (2)$$

"rain" denotes direct precipitation on the lake, "river" is river inflow, and "evap" is net evaporation from the lake.

Eq. 2 can be simplified by combining the input terms and generating a weighted mean input composition ($\delta_{river+rain}$) and a total input flux ($Q_{river+rain}$). Taking the simplified Eq. 2, substituting in the Benson and White (1994) formulation of δ_{evap}, and integrating the equation gives:

$$\delta_{lake} = \left(\delta_{lake(0)} - \frac{\delta_{river+rain} - a}{b}\right)e^{-b\frac{t}{R}} + \frac{\delta_{river+rain} - a}{b} \quad (3)$$

$$a = \frac{1000\left[\frac{1-k}{\alpha_{eq}} - hf(1-k)\right] - hf\delta_A[1-k]}{(1-h(k+f(1-k)))} - 1000$$

$$b = \frac{1 - k}{\alpha_{eq} (1 - h (k + f (1 - k)))}$$

in which h is humidity over the lake, and f is the fraction of atmospheric moisture over the lake that is not derived from the lake evaporate (Benson and White 1994). α_{eq} and k are the equilibrium and kinetic fractionation factors, and δ_A is the isotopic composition of atmospheric moisture. δ_{lake} is the final lake composition, $\delta_{lake(0)}$ is the initial lake composition, and R is the residence time of the lake ($V/Q_{river+rain}$). The integration assumes that variations in the climatic parameters (h, k, δ_A, $\delta_{river+rain}$, or α_{eq}) were instantaneous changes, that lake volume remained constant ($Q_{evap}/Q_{river+rain} = 1$), and that the lake was a closed-basin lake.

The term:

$$\frac{\delta_{river + rain} - a}{b} \tag{4}$$

is the steady state composition of the lake (δ_{Lss}) under the new climatic parameters. The exponential term in Eq. 3 is a rate term. Under modern conditions in the lake basin, 99% of the change to any new isotopic steady state value will occur in approximately 3 residence times of the lake.

Modern conditions in the lake basin were used to determine the sensitivity of the isotopic composition of lake water to changes in climatic parameters. Relative humidity has a mean annual value of 66-78% at 4 stations around the lake (Patterson and Kachinjika 1995). A humidity value of 70% was used in the sensitivity calculations. The equilibrium fractionation factor (α_{eq}) between liquid water and water vapor is dependent on temperature (Friedman and O'Neil 1977). Surface water temperatures range from 28 °C to 23.5 °C (Eccles 1974; Halfman 1993, 1996; Patterson and Kachinjika 1995), and a value of 27 °C was used to calculate α_{eq}. The kinetic fractionation factor (k) between liquid water and water vapor is dependent on wind speed above the body of water (Merlivat and Jouzel 1979). Mean annual wind speeds near the shores of Lake Malawi are between 1.7 m s^{-1} and 3.8 m s^{-1}, although these values may be low due to station location (Patterson and Kachinjika 1995). Sustained winds of 11 m s^{-1} have been noted during the dry season (Eccles 1974). A wind speed value between 6 and 7 m s^{-1} was used to determine the kinetic fractionation factor for Lake Malawi ($k \approx 0.002$; Merlivat and Jouzel 1979).

The isotopic composition of precipitation at the two closest IAEA/WMO (International Atomic Energy Agency/ World Meteorological Organization) network stations, Ndola and Harare, have weighted mean values of –6.6‰ and –6.1‰ (Rozanski et al. 1996). A value of –6.5‰ was chosen as the composition of precipitation (δ_{rain}) on Lake Malawi. The composition of atmospheric moisture (δ_A) was assumed to be in equilibrium with rainfall at 27 °C. The $\delta^{18}O$ value of river input to Lake Malawi (δ_{river}) is unknown. δ_{river} was set at –5‰ to reflect evaporative concentration of the local precipitation as water flows down river to the lake. The mean composition of the input ($\delta_{river+rain}$) was calculated using the steady state fluxes discussed by Ricketts and Johnson (1996a). Since there is no measure of the isotopic composition of the water vapor above the lake, the value of f, the fraction of air advected into the basin (Benson and White 1994), was calculated using the values of parameters mentioned above and a version of Eq. 2 in which δ_{evap} and therefore f were unknown:

$$f = \frac{\left(\frac{\delta_{rain + river}}{1000}+1\right)\left(\frac{1-h}{1-k}\right)+\left(\frac{\delta_{rain + river}}{1000}+1\right)h - \frac{\frac{\delta_{lake}}{1000}+1}{\alpha_{eq}}}{\left(\frac{\delta_{rain + river}}{1000}\right)h - h - h\left(\frac{\delta_{A}}{1000}\right)} \tag{5}$$

The value of f is approximately 45% using the composition of modern surface water (0.7‰).

4. Results

Sediment ages within the carbonate intervals of cores 18P and 27P were assigned using a least squares linear fit between ^{14}C dates determined from calcium carbonate and organic matter (Table 2; Finney and Johnson 1991; Finney et al. 1996). Several dates in core 18P are not stratigraphically consistent. The date of 33,250 yrBP at 557-558 cm depth is from the upper carbonate horizon in core 18P. This date is very similar to the dates found in the upper carbonate horizon in core 27P. Finney et al. (1996) have noted that a minor amount of modern carbon (1 to 5%) would make a radiocarbon age of 33,250 yrBP appear to be between 22,000 and 28,000 yrBP. Accepting the date of 33,250 yrBP and ignoring the dates of 28,100 yrBP (557-558 cm depth) and 22,050 yrBP (560-600 cm depth) in core 18P produces agreement in the timing of changes in lithology, porosity, $\%CaCO_3$, and diatom assemblages between cores 18P and 27P (Finney et al. 1996).

TABLE 2. ^{14}C dating results ('Beta' - Beta Analytical Inc.; 'CAMS' - Center for Accelerator Mass Spectrometry, Lawrence Livermore National Lab; 'GU' - Scottish Universities Research and Reactor Centre; 'AA' - University of Arizona NSF Accelerator Facility) (Finney and Johnson 1991; Finney et al. 1996).

Lab I.D.	Core	Depth (cm)	Material Dated	Age (yrBP)
GU 2626	M86-18P	5-25	TOC	9260 ±160
CAMS		557-558	TOC	33250 ±790
CAMS		557-558	$CaCO_3$	28110 ±420
Beta 33164		560-600	$CaCO_3$	22050 ±510
GU 2627		610-620	TOC	>27,100
AA 5589		980-981	$CaCO_3$	37150 ±1100
CAMS 1756	M86-27P	168-170	$CaCO_3$	34440 ±940
CAMS 1978		168-170	TOC	31860 ±680
CAMS		170-171	$CaCO_3$	33470 ±1800
CAMS 1757		361-363	$CaCO_3$	34950 ±1160
CAMS 1977		361-363	TOC	37040 ±1390
CAMS		390-391	$CaCO_3$	46600 ±2000

Sedimentation rates in core 18P are 0.2 mm yr^{-1} above 557 cm (younger than about 33 ka) and 1.1 mm yr^{-1} between 557 and 981 cm (about 33 to 37ka). In core 27P the sedimentation rate is 0.7 mm yr^{-1} between 170 and 362 cm (about 33 to 36 ka). Finney et al. (1996) noted that core 27P is missing a distinct Holocene unit. The lack of the Holocene unit and the ^{14}C age of 47ka at 390 cm depth, which is close to the age limit of the ^{14}C dating method, makes a meaningful discussion of sedimentation rates for core 27P above 170 cm and below 362 cm difficult. The sedimentation rate of 0.7 mm yr^{-1} found for core 27P between 170 and 362 cm was applied to the 40 cm of carbonate-bearing sediment above 170 cm in order to facilitate comparisons between the two cores.

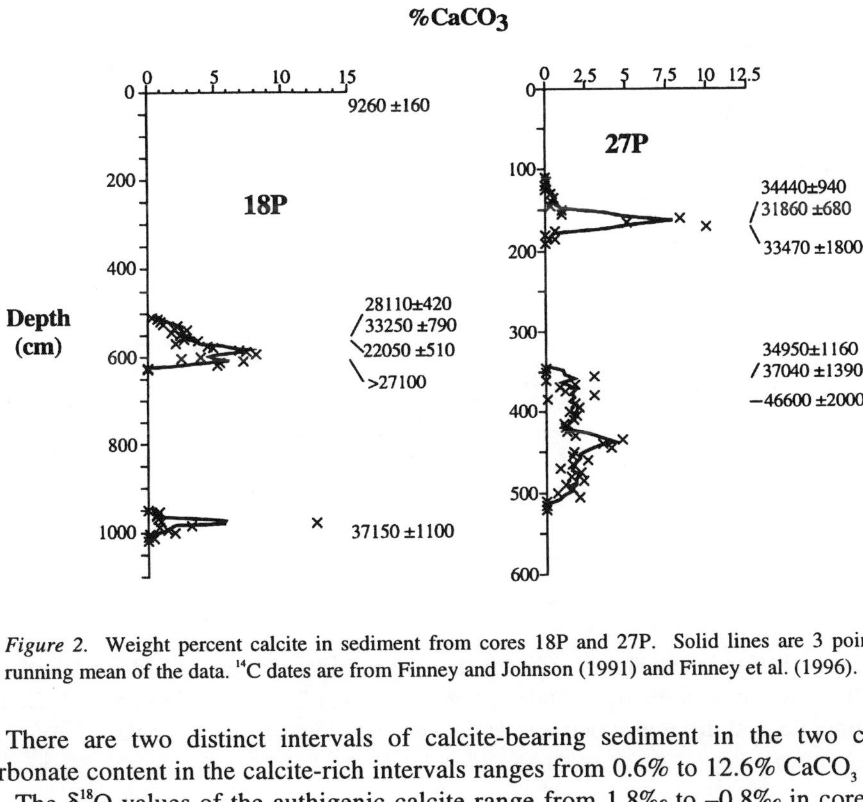

Figure 2. Weight percent calcite in sediment from cores 18P and 27P. Solid lines are 3 point running mean of the data. ^{14}C dates are from Finney and Johnson (1991) and Finney et al. (1996).

There are two distinct intervals of calcite-bearing sediment in the two cores. Carbonate content in the calcite-rich intervals ranges from 0.6% to 12.6% $CaCO_3$ (Fig. 2). The $\delta^{18}O$ values of the authigenic calcite range from 1.8‰ to –0.8‰ in core 18P and from 1.8‰ to –1.6‰ in core 27P (Fig. 3). The standard deviation of at least four replicate samples from the same depth interval for $\delta^{18}O$ is, on average, 0.13‰ for 18P, and 0.12‰ for 27P. The upper carbonate intervals in both cores have similar $\delta^{18}O$ values. The values in the lower interval in core 18P are somewhat higher on average than the values found in the lower interval in core 27P. The hiatus in calcite deposition occurred between 34 and 36 ka and calcite deposition seems to have ended completely at approximately 31ka. The younger calcite interval in core 18P has a coherent peak, whereas the few values present for the upper interval in core 27P and the lower interval in core 18P do not. The older calcite interval in core 27P has rather poorly defined peaks at 420 cm and at 480 cm depth.

The $\delta^{13}C$ values in the two cores vary between –5‰ and 2‰ (Fig. 3). The standard deviation of at least four replicate samples from the same depth interval for $\delta^{13}C$ is, on average, 0.03‰ for both cores. The older calcite interval in core 27P has a slight long term trend towards more positive values. The younger calcite intervals exhibit decreasing $\delta^{13}C$ values.

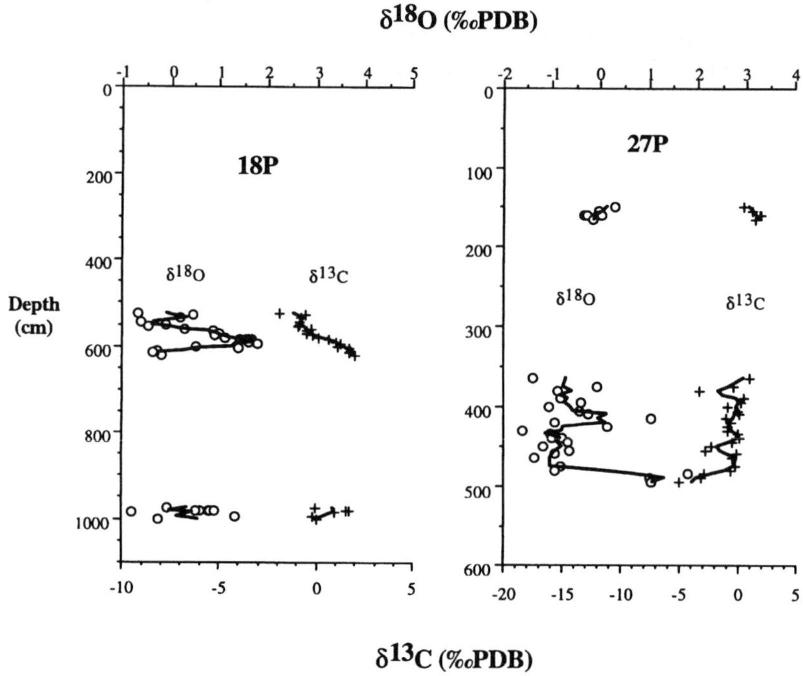

Figure 3. $\delta^{18}O$ (o) and $\delta^{13}C$ (+) values from cores 18P and 27P. Solid lines are 3 point running mean of the data.

X-ray diffraction analyses indicate that magnesium content (Goldsmith and Graf 1958) of calcite in cores from Malawi varies between 2 and 5 mole percent. There is no mineralogical difference between Holocene and Pleistocene carbonate.

5. Discussion

The carbonates from Lake Malawi generally do not exhibit a strong correlation between $\delta^{18}O$ and $\delta^{13}C$ values. Talbot (1990) hypothesized that correlation coefficients (r values) higher than 0.7 were indicative of closed-basin lakes. Only core 22P, which has $\delta^{13}C$ values indicating diagenetic formation of calcite coincident with methane production in pore waters (Ricketts and Johnson 1996a), has an r value higher than 0.7. The unaltered carbonate in the remaining four cores all have r values below 0.7.

The $\delta^{13}C$ values of carbonate in the 5 cores can be broken into 3 groups (Fig. 4). Relatively low values and high variability characterize late Pleistocene carbonates (18P and 27P: mean= 0.26‰ and –0.64‰; SD= 1.07‰ and 1.56‰). Early Holocene carbonate intervals in cores 12P and 24P have low $\delta^{13}C$ variability and higher values (12P and 24P: mean= 1.91‰ and 1.53‰; SD= 0.5‰ and 0.17‰). Core 22P has the highest $\delta^{13}C$ values of all cores (mean= 5.02‰) and the highest variability (SD= 2.69‰).

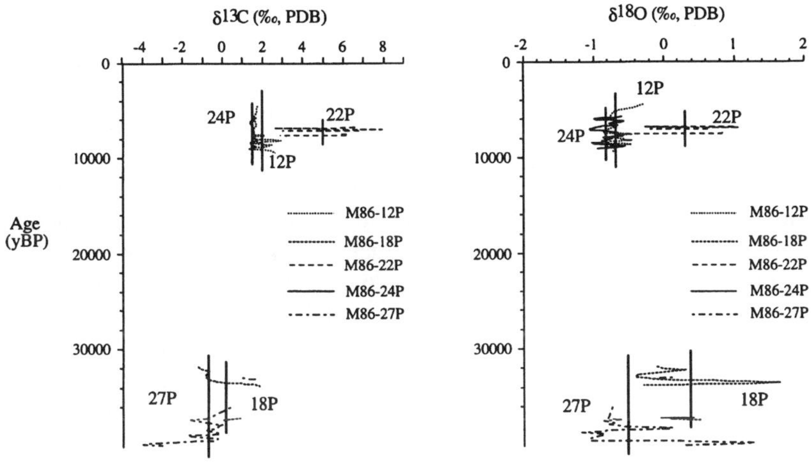

Figure 4. $\delta^{18}O$ and $\delta^{13}C$ values from the Lake Malawi cores. Thin lines are 3 point running mean of original data. Thick lines denote the mean value of all data for an individual core.

As noted above, high $\delta^{13}C$ values in 22P indicate diagenetic alteration of calcite. The $\delta^{13}C$ values of cores 12P, 18P, 24P, and 27P are lower than values seen in 22P. Values for cores 18P and 27P are also lower than 1.85‰, the value they would have if they were in equilibrium with modern surface water DIC (surface water $\delta^{13}C_{DIC}$ value for Lake Malawi surface water is 0.85‰; fractionation factor = 1.0 ± 0.2‰, Romanek et al. 1992). Carbonate in core 12P and 24P is in approximate equilibrium with modern DIC while the Pleistocene carbonate is 1.5 to 2.5‰ lower than equilibrium (Table 3).

The $\delta^{13}C$ values from Pleistocene cores are approximately 1.9‰ lower than values from unaltered Holocene cores. This difference may be explained by non-equilibrium precipitation of carbonate or a change in isotopic composition of DIC in Lake Malawi. Increased precipitation rates may introduce non-equilibrium precipitation of carbonates due to kinetic isotopic fractionation effects between DIC and calcite (Turner 1982), although there is some argument about whether or not this occurs (Romanek et al. 1992). In lacustrine systems where non-equilibrium precipitation of carbonates is hypothesized to takes place, oxygen isotopic composition of carbonate is much lower than equilibrium precipitation would predict (Fronval et al. 1995). In Lake Malawi, Pleistocene and Holocene carbonates have $\delta^{18}O$ values which are higher then expected assuming equilibrium precipitation from modern surface waters (Table 3). This implies that non-equilibrium precipitation of carbonates did not take place.

TABLE 3. Mean $\delta^{13}C$ and $\delta^{18}O$ values for Malawi carbonates and isotopic composition for hypothetical source waters at 27°C. Calculations assume only equilibrium fractionation between calcite and water or DIC. Fractionation factors are from Friedman and O'Neil (1977) and Romanek et al. (1992). Modern surface waters have $\delta^{18}O$ of 0.7‰ (SMOW) and $\delta^{13}C_{DIC}$ of 0.9‰ (PDB).

Core	Calcite Mean $\delta^{13}C$ value (PDB)	$\delta^{13}C$ value of DIC in equilibrium with calcite (PDB)	Calcite Mean $\delta^{18}O$ value (PDB)	$\delta^{18}O$ value of water in equilibrium with calcite (SMOW)
12P	1.9‰	0.9‰	-0.7‰	1.8‰
18P	0.3‰	-0.7‰	0.4‰	2.8‰
22P	5.0‰	4.0‰	0.3‰	2.8‰
24P	1.5‰	0.5‰	-0.8‰	1.6‰
27P	-0.6‰	-1.6‰	-0.4‰	2.0‰

Various factors, such as changes in sources of DIC or lake productivity could impact the composition of lake DIC. The isotopic composition of DIC in modern Lake Malawi and other large lakes in East Africa, such as Lakes Tanganyika and Turkana, is in equilibrium or close to equilibrium with atmospheric CO_2 (Craig 1974; Johnson et al. 1991; Cerling 1996). These equilibrium conditions might have existed in the Holocene and Pleistocene as well. The composition of atmospheric CO_2, as determined from gas trapped in ice cores, was approximately 0.3‰ lighter during the last glacial period than during the current interglacial period (Leuenberger et al. 1992). Thus, changes in composition of atmospheric CO_2 could explain a portion, but not all, of the difference in $\delta^{13}C$ between Holocene and Pleistocene carbonates.

Measurements of total organic C are higher in sediment containing Pleistocene carbonate than sediment containing Holocene carbonate (5-12% vs. 1-5% TOC; Johnson et al. 1988), perhaps indicating higher productivity during the Pleistocene. Very high rates of photosynthesis have been hypothesized to enhance invasion of surface waters by isotopically light CO_2, due to kinetic fractionation effects between $CO_{2(aq)}$ and $CO_{2(g)}$. Isotopically altered DIC may affect the isotopic composition of $CaCO_3$ precipitating from surface waters (Herczeg and Fairbanks 1987). This process may explain some of the difference between $\delta^{13}C$ values in Early Holocene and Late Pleistocene carbonates.

Carbon-to-nitrogen ratios in sediment deposited during the late Pleistocene are higher than in sediment deposited during the early Holocene (Johnson et al. 1988), possibly a sign of a higher supply of organic matter from nearshore macrophytes or terrestrial sources (Kemp et al. 1977; Finney and Johnson 1991). The input of terrestrial organic matter dominated by C_3 plants, with its relatively low $\delta^{13}C$ composition, could have an impact on the isotopic composition of DIC of the lake and therefore of the carbonate which precipitates from lake waters. It is not readily apparent from pollen analyses that there was more C_3 vegetation in the Malawi Basin during the time the Pleistocene carbonates were deposited (DeBusk 1994).

The relatively low $\delta^{13}C$ values of late Pleistocene carbonates may result from changes in lake productivity or, possibly, changes in the source materials of DIC for the lake. This, in combination with the lighter $\delta^{13}C$ composition of atmospheric CO_2

during the glacial period, may explain the lighter $\delta^{13}C$ values seen in the late Pleistocene carbonates compared to Holocene carbonates.

The $\delta^{18}O$ values in cores are more difficult to group than $\delta^{13}C$ values (Fig. 4). Cores 12P and 24P have similar means (–0.65‰ and –0.79‰) and low variability (SD= 0.18‰ and 0.24‰). Core 22P has a higher mean (0.35‰) and variability (SD= 0.80‰). 18P seems to have characteristics similar to 22P (mean= 0.41‰, SD= 0.80‰) but, while the carbonate in 22P has been altered by methanogenesis, as is apparent from its $\delta^{13}C$ values, 18P is composed of unaltered carbonate. The lower horizon in 27P has a mean similar to 12P and 24P (–0.48‰) but is highly variable (SD= 0.87‰). The upper horizon in 27P appears to be similar to 18P (mean= –0.02‰) but is less variable (SD= 0.21‰) possibly because it is defined by few data points (n= 4). The few values present in the lower horizon in core 18P (n= 5) and upper horizon in 27P may be depicting one of the transitory swings to higher $\delta^{18}O$ values which are seen in the $\delta^{18}O$ composition of Pleistocene carbonates (Fig. 4).

Variations in $\delta^{18}O$ values reflect past changes in lake water temperature or oxygen isotopic composition. Variations in seasonal timing of calcite formation, temporal or spatial changes in lake water composition, or long term changes in various climatic parameters may influence the $\delta^{18}O$ composition of lake water.

The carbonate in cores 12P and 24P would have precipitated from water with a temperature range of 20 to 25 °C assuming equilibrium precipitation between water and calcite (Friedman and O'Neil 1977) and a $\delta^{18}O$ water composition equal to the modern surface composition (0.7‰ SMOW; Ricketts and Johnson 1996a). While this temperature range is approximately equal to the seasonal surface water range of 4 °C found today, the temperatures are up to 9 °C lower than modern surface values (Eccles 1974; Halfman 1993, 1996; Patterson and Kachinjika 1995). The range of possible temperatures calculated from carbonate of Pleistocene age is 11 to 26 °C, which is larger than the modern seasonal range seen in Lake Malawi. These two observations imply that while part of the variation seen in the cores could be due to seasonal changes in temperature or changes in the period of calcite formation, part is due either to cooler water temperatures or heavier isotopic composition of lake water in the late Pleistocene and early Holocene compared to modern conditions.

Debusk (1994) found no indication in pollen spectra of cooler conditions during the early Holocene. Water in isotopic equilibrium with Holocene carbonates had a higher $\delta^{18}O$ composition than modern waters (Table 3) due to evaporative concentrations in the closed-basin lake. Pollen spectra indicate that the Pleistocene was slightly cooler than present conditions; therefore evaporative concentration (Table 3) and lower temperatures could explain the relatively high Pleistocene $\delta^{18}O$ values.

The $\delta^{18}O$ values from 27P are about 1‰ lower than the record from core 18P. The difference in $\delta^{18}O$ values between the two late Pleistocene cores could be indicative of different local water $\delta^{18}O$ compositions. This situation may at least partially be caused by distance from fresh water input, similar to the situation in Lake Turkana (Ricketts and Anderson, in press). The 18P and 27P core sites are closer to the eastern shore of the lake than the western shore and would be almost 100 km from any major rivers even after a major fall in lake level (Fig. 1). The eastern shore of the lake contributes less river input than the western shore (Crossley 1984), and during a lake lowstand river input would be expected to be lower. The distance from possible river input implies that spatial variability in water composition caused by fresh water input may not be an explanation for the differences between cores 18P and 27P.

As stated earlier, the few $\delta^{18}O$ values in the lower horizon in core 18P and upper horizon in core 27P could be indicative of a positive swing in values, similar to the short duration swings seen in the lower horizon in 27P and the upper horizon in 18P. If that is true, we may want to limit our discussion of changes in the Pleistocene $\delta^{18}O$ records to the two longer term intervals, the lower interval in 27P and the upper interval in 18P. Using these two intervals, late Pleistocene carbonates may be depicting a long term trend of increasing values, with high frequency fluctuations superimposed on the trend (Fig. 4). The mean $\delta^{18}O$ value of carbonate in the upper horizon in core 18P is 1.0‰ higher than the lower horizon in core 27P, therefore 1.0‰ would be a good approximation of the change in $\delta^{18}O$ values over the duration of the trend. Long term trends in isotope data generally cannot be associated solely with decreasing lake level, because an extreme change in lake level is generally required to create the trend. The trends are more likely associated with changes in other climatic parameters (Ricketts and Johnson 1996b).

To explain the long term 1‰ shift in $\delta^{18}O$ values we used Eq. 4 to determine the effect that variations in climatic parameters would have on a steady state closed-basin Lake Malawi. All other parameters were held constant as the climatic variable in question was varied. The steady state composition of water (δ_{Lss}) in Lake Malawi changes by 0.07‰ °C^{-1}. This is in addition to changes in the fractionation factor between calcite and water with changes in temperature, which in Lake Malawi is approximately 0.2‰ °C^{-1}. The total impact that temperature has on the isotopic composition of calcite is 0.27‰ °C^{-1}. Therefore, the 1.0‰ shift between carbonate older that 36 ka and carbonate deposited between 34 and 32 ka could be caused by a 3.7 °C decrease in temperature (Table 4).

TABLE 4. Change in climatic parameters necessary to cause a 1‰ increase in $\delta^{18}O$ composition of lake water (δL_{ss}) calculated using Eq. 4. All other parameters were held constant as the climatic variable in question was varied.

Climatic parameter	Change necessary
Temperature	3.7 °C decrease
Wind speed	6.7 m s^{-1} increase
Humidity	17% decrease
Fraction of air advected into the basin	15% decrease
Isotopic composition of the atmosphere	3‰ increase
Isotopic composition of inputs	1.4‰ increase

Each 1 m s^{-1} increase in wind speed increases δ_{Lss} by about 0.15‰. A wind speed increase of 6.7 m s^{-1} could explain the 1.0‰ change in the Pleistocene carbonates (Table 4).

Sensitivity of δ_{Lss} to humidity at low f values is small but increases as f increases (Benson and White 1994). At f = 45%, each 1% increase in humidity decreases δ_{Lss} by 0.06‰, while at f = 80%, δ_{Lss} decreases by 0.1‰ per 1% increase in humidity. A 17% decrease in humidity would explain the 1.0‰ increase in $\delta^{18}O$ under modern conditions (Table 4).

As with humidity, effects of composition changes of the atmosphere or inputs is dependent on the amount of air advected into the basin. At f = 45% and humidity of 70%, δ_{Lss} increases by 0.32‰ per 1‰ increase in δ_A and increases 0.69‰ per 1‰ increase in $\delta_{river+rain}$. Under modern conditions a 3‰ increase in δ_A or a 1.4‰ increase in $\delta_{river+rain}$ will increase δ_{Lss} by 1.0‰ (Table 4).

Dust concentrations in ice and deep sea cores indicate the last glacial maximum (LGM) was a period of stronger winds than the current interglacial (De Angelis et al. 1987; Petit et al. 1990). The period of inception of stronger winds was approximately 30 ka. Strongest impact of glacial ice sheets on the isotopic composition of the global ocean was also during the LGM, when ocean waters were approximately 1.25‰ heavier than at present, and 0.4‰ heavier than the glacial ocean prior to the LGM (Broecker 1993; Guilderson et al. 1994). Initiation of the 0.4‰ change also occurred at approximately 30 ka. Since ocean waters serve as the ultimate source of atmospheric moisture over a basin and precipitation in a basin, an increase in the $\delta^{18}O$ composition of ocean waters could lead to increases in δ_A and $\delta_{river+rain}$.

These are two examples of climatic changes which could explain the 1‰ increase in isotopic composition of calcite from the lower to the upper carbonate horizons in the Pleistocene cores. Sensitivity analysis demonstrates that both an increase in wind speed and an increase in the composition of δ_a and $\delta_{river+rain}$, along with a decrease in temperature, could explain the increase in calcite values. Likely changes in atmospheric circulation, changes in humidity, and changes in the fraction of advected air also probably added to the shift in values.

Large amplitude swings of up to 2‰ are present in the $\delta^{18}O$ records from both Pleistocene cores. These high frequency fluctuations are much larger than the approximately 0.4‰ high frequency swings in $\delta^{18}O$ values found in the unaltered Holocene cores. Short term fluctuations in the $\delta^{18}O$ values of carbonate records are hypothesized to be mostly an indication of fluctuations in the ratio of evaporation to freshwater input of a basin (Ricketts and Johnson 1996b). Thus, the evaporation-to-input ratio in the Malawi basin during the deposition of the two Pleistocene horizons could have fluctuated by similar amounts and these fluctuations may have been much larger than changes in the evaporation-to-input ratio seen during the Holocene.

6. Conclusion

The stable isotopic composition of carbonates is useful in determining changes in the Lake Malawi basin between the late Pleistocene and early Holocene. A shift in the atmospheric composition of CO_2, changes in lake productivity, and possibly changes in the DIC entering the lake probably explain the 1.9‰ difference in $\delta^{13}C$ values in carbonates from the two time spans. Up to a 3.7 °C decrease in water temperature and variation in windspeed, the isotopic composition of inputs to the lake, and other climatic parameters may have created part of the change in $\delta^{18}O$ composition seen in the Pleistocene carbonate. High frequency shifts in the $\delta^{18}O$ composition of carbonates may be due to shifts in the evaporation-to-input ratio. If this is true, shifts in E/I during the late Pleistocene were much larger than during the early Holocene.

7. Acknowledgments

Thanks to T. C. Johnson for the use of cores from the core collection of the Large Lakes Observatory, University of Minnesota. Thanks also to the governments of Malawi and Tanzania for logistical assistance and for research permits issued to Project PROBE of Duke University. Funding for this research was provided by the Bradley Foundation to R.D. Ricketts and U.S. AID grant number HRN-5600-g-00-2056-00 and NSF grant ATM88-16615 to T. C. Johnson. Thanks to T. C. Johnson and two anonymous reviewers for helpful reviews.

8. References

Beadle, L. C. 1981. The inland waters of tropical Africa, 2nd edition. Longman.

Benson, L. V., and J. W. C. White. 1994. Stable isotopes of oxygen and hydrogen in the Truckee River-Pyramid Lake surface-water system. 3. Source of water vapor overlying Pyramid Lake. Limnology and Oceanography **39**: 1945-1958.

Billups, K, and H. J. Spero. 1996. Reconstructing the stable isotope geochemistry and paleotemperatures of the equatorial Atlantic during the last 150,000 years: results from individual foraminifera. Paleoceanography **11**: 217-238.

Broecker, W. S. 1993. The glacial world according to Wally. Columbia University.

Cerling, T. E. 1996. Pore water chemistry of an alkaline lake: Lake Turkana, Kenya, p. 225-240. *In* T. C. Johnson and E. O. Odata [eds.], The limnology, climatology and paleoclimatology of the East African Lakes. Gordon and Breach.

Chao, G. Y. 1969. 2-Theta (Cu) table for common minerals. Geological Paper 69-2, Carleton University Department of Geology.

CLIMAP Project Members. 1976. The surface of the ice-age Earth. Science **191**: 1131-1137.

Craig, H. 1957. Isotopic standards for carbon and oxygen and correction factors for mass spectrometric analysis of carbon dioxide. Geochimica et Cosmochimica Acta **12**: 133-149.

Craig, H. 1975. Lake Tanganyika geochemical and hydrographic study: 1973 expedition. Scripps Institution of Oceanography 75-5.

Crossley, R. 1984. Controls of sedimentation in the Malawi Rift Valley, Central Africa. Sedimentary Geology **40**: 33-50.

De Angelis, M., N. I. Barkov, and V. N. Petrov. 1987. Aerosol concentrations over the last climatic cycle (160 kyr) from an Antarctic ice core. Nature **325**: 318-321.

Debusk, G. H. 1994. Transport and stratigraphy of pollen in Lake Malawi, Africa. Ph.D. dissertation, Duke University.

Eccles, D. H. 1974. An outline of the physical limnology of Lake Malawi (Lake Nyasa). Limnology and Oceanography **19**: 730-742.

Finney, B. P., and T. C. Johnson. 1991. Sedimentation in Lake Malawi (East Africa) during the past 10,000 years: a continuous paleoclimatic record from the southern tropics. Palaeogeography, Palaeoclimatology, Palaeoecology **85**: 351-366.

Finney, B. P., C. A. Scholz, T. C. Johnson, and S. Trumbore. 1996. Late Quaternary lake-level changes of Lake Malawi, p. 495-508. *In* T. C. Johnson and E. O. Odata [eds.], The limnology, climatology and paleoclimatology of the East African Lakes. Gordon and Breach.

Flannery, J. W. 1988. The acoustic stratigraphy of Lake Malawi, East Africa. M.S. thesis, Duke University.

Friedman, I., and J. R. O'Neil. 1977. Compilation of stable isotope fractionation factors of geochemical interest, p. 1-12. *In* M. Fleischer [ed.], Data of geochemistry. U.S. Geological Survey Professional Paper 440-KK, 6th edition, USGS.

Fronval, T., N. Jensen, and B. Buchardt. 1995. Oxygen isotope disequilibrium precipitation of calcite in Lake Arresø, Denmark. Geology **23**: 463-466.

Goldsmith, J. R., and D. L. Graf. 1958. Relations between lattice constants and composition of the Ca-Mg carbonates. American Mineralogist **43**: 84-101.

Gonfiantini, R., G. M. Zuppi, D. H. Eccles, and W. Ferro. 1979. Isotope investigation of Lake Malawi, p. 195-207. *In* Isotopes in lake studies. Panel Proceedings Series. International Atomic Energy Agency.

Guilderson, T. P., R. G. Fairbanks, and J. L. Rubenstone. 1994. Tropical temperature variations since 20,000 years ago: modulating interhemispheric climate change. Science **263**: 663-665.

Halfman, J. D. 1993. Water column characteristics from modern CTD data, Lake Malawi, Africa. Journal of Great Lakes Research **19**: 512-520.

Halfman, J. D. 1996. CTD-transmissometer profiles from Lakes Malawi and Turkana, p. 169-182. *In* T. C. Johnson and E. O. Odata [eds.], The limnology, climatology and paleoclimatology of the East African Lakes. Gordon and Breach.

Herczeg, A. L., and R. G. Fairbanks. 1987. Anomalous carbon isotope fractionation between atmospheric CO_2 and dissolved inorganic carbon induced by intense photosynthesis. Geochimica et Cosmochimica Acta **51**: 895-899.

Johnson, T. C. 1996. Sedimentary processes and signals of past climatic change in the large lakes of the East African Rift Valley, p. 367-412. *In* T. C. Johnson and E. O. Odata [eds.], The limnology, climatology and paleoclimatology of the East African Lakes. Gordon and Breach.

Johnson, T. C., T. W. Davis, B. M. Halfman, and N. D. Vaughan. 1988. Sediment Core Descriptions: Malawi '86, Lake Malawi, East Africa. Duke University Marine Laboratory Technical Report.

Johnson, T. C., J. D. Halfman, and W. J. Showers. 1991. Paleoclimate of the past 4000 years at Lake Turkana, Kenya, based on the isotopic composition of authigenic calcite. Palaeogeography, Palaeoclimatology, Palaeoecology **85**: 189-198.

Johnson, T. C., and P. Ng'ang'a. 1990. Reflections on a rift lake, p. 113-135. *In* B. J. Katz [ed.], Lacustrine basin exploration: case studies and modern analogs. American Association of Petroleum Geologists Memoir 50.

Jones, G. A., and P. Kaiteris. 1983. A vacuum-gasometric technique for rapid and precise analysis of calcium carbonate in sediments and soils. Journal of Sedimentary Petrology **53**: 655-660.

Kelts, K., U. Briegel, K. Ghilardi, and K. Hsü. 1986. The limnogeology-ETH coring system. Schweizerische Zeitschrift für Hydrologie **48**: 104-115.

Kemp, A. L. W., R. L. Thomas, H. K. T. Wong, and L. M. Johnston. 1977. Nitrogen and C/N ratios in the sediments of Lakes Superior, Huron, St. Clair, Erie, and Ontario. Canadian Journal of Earth Science **14**: 2402-2413.

Leuenberger, M., U. Siegenthaler, and C. C. Langway. 1992. Carbon isotope composition of atmospheric CO_2 during the last ice age from an Antarctic ice core. Nature **357**: 488-490.

Livingstone, D. A. 1975. Late Quaternary climatic change in Africa. Annual Review of Ecology and Systematics **6**: 249-280.

Merlivat, L., and Jouzel, J. 1979. Global climatic interpretation of the deuterium-oxygen 18 relationship for precipitation. Journal of Geophysical Research **84**: 5029-5033.

Mix, A. C., W. F. Ruddiman, and A. McIntyre. 1986. Late Quaternary paleoceanography of the tropical Atlantic, 1: spatial variability of mean sea-surface temperatures, 0-20,000 years B.P. Paleoceanography **1**: 43-66.

Owen, R. B., R. Crossley, T. C. Johnson, D. Tweddle, I. Kornfield, S. Davison, D. H. Eccles, and D. E. Engstrom. 1990. Major low levels of Lake Malawi and their implications for speciation rates in cichlid fishes. Proceedings of the Royal Society of London, **Series B 240**: 519-553.

Patterson, G., and O. Kachinjika. 1995. Limnology and phytoplankton ecology, p. 1-68. *In* A. Menz [ed.], The fishery potential and productivity of the pelagic zone of Lake Malawi/Niassa. Natural Resources Institute.

Petit, J. R., L. Mounier, J. Jouzel, Y. S. Korotkevich, V. I. Kotlyakov, and C. Lorius. 1990. Palaeoclimatological and chronological implications of the Vostok core dust record. Nature **343**: 56-58.

Ricketts, R. D., and R. Anderson. in press. A direct comparison between the historical record of lake level and the $\delta^{18}O$ signal in carbonate sediments from Lake Turkana, Kenya. Limnology and Oceanography.

Ricketts, R. D. and T. C. Johnson. 1996a. Early Holocene changes in lake level and productivity in Lake Malawi as interpreted from oxygen and carbon isotopic measurements of authigenic carbonates p. 475-494. *In* T. C. Johnson and E. O. Odata [eds.], The limnology, climatology and paleoclimatology of the East African Lakes. Gordon and Breach.

Ricketts, R. D., T. C. and Johnson. 1996b. Climate change in the Turkana basin as deduced from a 4000 year long $d^{18}O$ record. Earth and Planetary Science Letters **142**: 7-17.

Romanek, C. S., E. L. Grossman, and J. W. Morse. 1992. Carbon isotopic fractionation in synthetic aragonite and calcite: effects of temperature and precipitation rate. Geochimica et Cosmochimica Acta **56**: 419-430.

Rosendahl, B. R. 1987. Architecture of continental rifts with special reference to East Africa. Annual Review of Earth and Planetary Science **15**: 445-503.

Rozanski, K., L. Araguás-Araguás, and R. Gonfiantini. 1996. Isotope patterns of precipitation in the East African region, p. 79-94. *In* T. C. Johnson and E. O. Odata [eds.], The limnology, climatology and paleoclimatology of the East African Lakes. Gordon and Breach.

Sikes, E. L., and L. D. Keigwin. 1994. Equatorial Atlantic sea surface temperature for the last 30 kyr: a comparison of U^{k}_{37}, $\delta^{18}O$, and foraminiferal assemblage temperature estimates. Paleoceanography **9**: 31-45.

Talbot, M. R. 1990. A review of the palaeohydrological interpretation of carbon and oxygen isotopic ratios in primary lacustrine carbonates. Chemical Geology **80**: 261-279.

Turner, J. V. 1982. Kinetic fractionation of carbon-13 during calcium carbonate precipitation. Geochimica et Cosmochimica Acta **46**: 1183-1191.

Vincens, A., F. Chalié, R. Bonnefille, J. Guiot, and J. J. Tiercelin. 1993. Pollen-derived rainfall and temperature estimates from Lake Tanganyika and their implication for Late Pleistocene water levels. Quaternary Research **40**: 343-350.

Webb, R.S., D. H. Rind, S. J. Lehman, R. J. Healy, and D. Sigman. 1997. Influence of ocean heat transport on the climate of the Last Glacial Maximum. Nature **385**: 695-699.

FLUCTUATIONS OF RIFT VALLEY LAKES MALAWI AND CHILWA DURING HISTORICAL TIMES: A SYNTHESIS OF GEOLOGICAL, ARCHAEOLOGICAL AND HISTORICAL INFORMATION

SHARON E. NICHOLSON
Department of Meteorology, Florida State University
Tallahassee, FL 32308 USA

1. Abstract

Synthesis of geological, historical, archaeological and geographical information is used to reconstruct detailed chronologies of the fluctuations of Lakes Malawi and Chilwa during the past two centuries and to produce less detailed, longer-term chronologies of these lakes. Fluctuations are compared with famine and drought records during historical times and rainfall data during modern times in order to understand their links with local and large-scale climatic conditions. This analysis underscores complexities in interpreting the historical and geological records of the lakes in terms of climate. It also points out the need to interpret the African lake record collectively, rather than focusing on individual lakes.

2. Introduction

The climatic history of equatorial Africa is strongly reflected in the fluctuations of the region's great lakes. The geological record of Lakes Chad, Victoria, Nakuru and others in East Africa (Butzer et al. 1972) unequivocally demonstrated the late Pleistocene aridity in the region and the humid period of the early Holocene, dramatically refuting the prevailing notion of the correspondence between tropical "pluvials" and higher latitude "glacials". Historical witness of the lakes has provided a wealth of information on fluctuations on time scales of centuries (Nicholson 1981, 1995). This, together with early quantitative records of the lakes, demonstrated two major shifts of the prevailing climatic regime in the tropics towards the end of the 19th century and during the 1960s (Kraus 1955; Lamb 1966).

In this paper, the fluctuations of the southern Rift Valley Lakes Malawi and Chilwa are examined. A twelve-hundred year chronology of Lake Malawi is produced that represents a synthesis of published archaeological, geological and historical information and new historical climate chronologies for the region. The record provides good detail of fluctuations during the last six centuries. A shorter chronology is developed for Lake Chilwa. Available evidence demonstrates that within the last one or two millenia,

J.T. Lehman (ed.), Environmental Change and Response in East African Lakes, 207–231.

fluctuations of at least 14 m and 12 m occurred in Lakes Malawi and Chilwa, respectively.

This work builds upon the records of these lakes that were compiled by Owen et al. (1990), Crossley et al. (1984), Pike (1965), Drayton (1979), Lamb (1966), Sieger (1887) and others. This is one in a series of papers that enhances the record of African lakes over the past few centuries, compares historical evidence with high resolution lake core studies, and compares climatic fluctuations in equatorial and subtropical regions of the continent and in the two hemispheres.

This article includes a summary of general trends over the last two millenia. It also includes detailed chronologies of both lakes from around the 18th century, based primarily on historical references, and compares them with famine and drought chronologies from areas of Malawi, Zambia, Zimbabwe and Mozambique in relatively close proximity to the lake basins. At the outset it was assumed that the reports of famine and drought could enhance the lake chronology developed from references directly pertinent to the conditions of the lakes and rivers. A comparison with the record for Lake Malawi demonstrated that this was not true. Instead, comparison of modern levels with regional rainfall demonstrates that Malawi responds mainly to rainfall in the northern catchment and thus is a better indicator of rainfall in the equatorial region than in its own subtropical latitudes. Also, the highlands may be subjected to more localized rainfall conditions than would be reflected in the lake levels. Lake Chilwa also reflects local rainfall conditions, at least on short time-scales.

2. Geographic Framework

Lake Malawi (Fig. 1) stretches from approximately 9° S to 14° S in the southern Rift Valley. It lies within the countries of Malawi, Tanzania and Mozambique. It occupies a relatively steep basin and has a surface area of 22,490 km^2. Its mean depth and maximum depth are 273 m and 706 m respectively. Its water balance is dominated by rainfall over the lake and evaporation, with river inflow and discharge making only a small contribution (Spigel and Coulter 1996). The fluctuations of Lake Malawi are a reasonably sensitive indicator of regional climate. Mean rainfall over the lake is 1350 mm yr^{-1}, but there is a strong north–south gradient. To the north, Lake Malawi is bordered by highlands where mean annual rainfall exceeds 2400 mm in some places, compared to 1000 to 1200 mm yr^{-1} over most of the rest of the region.

Lake Chilwa, almost directly south of Malawi, occupies a much smaller basin. It is a closed basin lake with a surface area of approximately 1300 km^2 and a depth of about 2 to 3 m. Rainfall in the region is on the order of 1000 to 1200 mm yr^{-1}.

Rainfall is these regions exhibits considerable interannual variability, but the patterns are complex. Lake Malawi straddles the boundary between two large-scale rainfall regions that tend to vary inversely with each other. To the north is eastern equatorial Africa, a region of very diverse mean climates ranging from deserts to rain forests. Despite this diversity, interannual fluctuations are remarkably homogeneous throughout the region spanning latitudes from 5° N to 10° or 12° S (Nicholson 1996).

To the south, in the subtropical latitudes of the southern hemisphere, is another relatively homogeneous area extending to 30° S (Nicholson 1987).

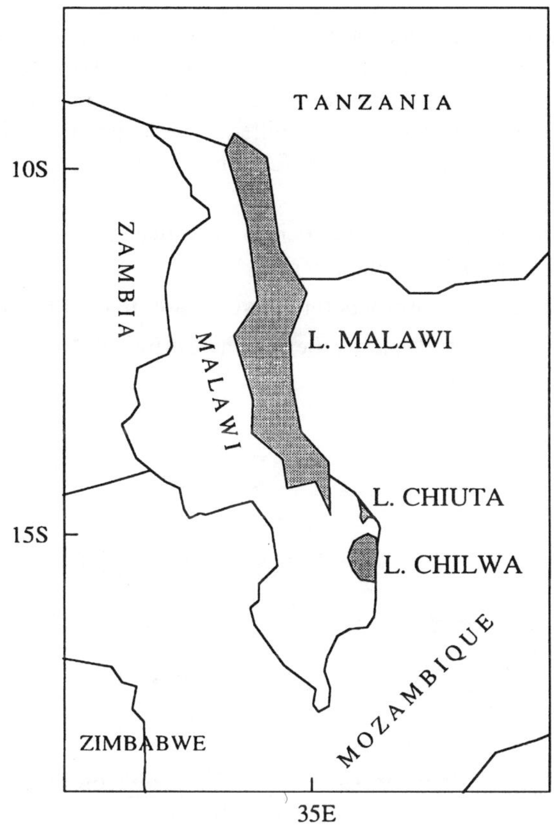

Figure 1. Map of the area of Lakes Malawi, Chilwa and Chiuta.

The most common spatial pattern of interannual variability (Fig. 2) is such that when rainfall is above average in one region, it is below average in the other (Nicholson 1986). This is true for both interannual and decadal or longer time scales. The sharp node separating the two regions generally runs just to the north of Lake Malawi in individual years and right through it on decadal time scales, but the lake can lie in either the "equatorial" or the "subtropical" zone. A secondary pattern that prevails is characterized by similar rainfall trends throughout both regions. This complexity of the interannual variability poses difficulties in interpreting fluctuations of the lake in climatic terms. In view of the much higher rainfall to the north, the lake may be more indicative of equatorial rainfall, despite its more southerly location.

Lake Chilwa lies well within the subtropical zone. For this reason, its fluctuations would not necessarily mirror those of Malawi, despite the proximity of the two lakes.

3. Lake Malawi

3.1 GENERAL TRENDS DURING THE LAST TWO MILLENIA

Over historical time scales, the fluctuations of Lake Malawi can be established in a very general sense from archaeological artifacts indicative of lakeshore settlement, from historical witness, and from three pre-modern beach units representing three periods of high stands of the lake within the last two millenia. A summary of the beach units, as dated from artifacts (Crossley and Davison-Hirschmann 1981), is given by Owen et al. (1990) as

1) a level of 476.1 m reached late in the period of Mawudzu pottery tradition (thought to span the mid-12th to mid-18th centuries),

2) a level of 475.6 m reached shortly before or very early in the Mawudzu period, and

3) a level of 476.8 m reached shortly before the period of Nkope pottery tradition, believed to span the 3rd to 10th centuries.

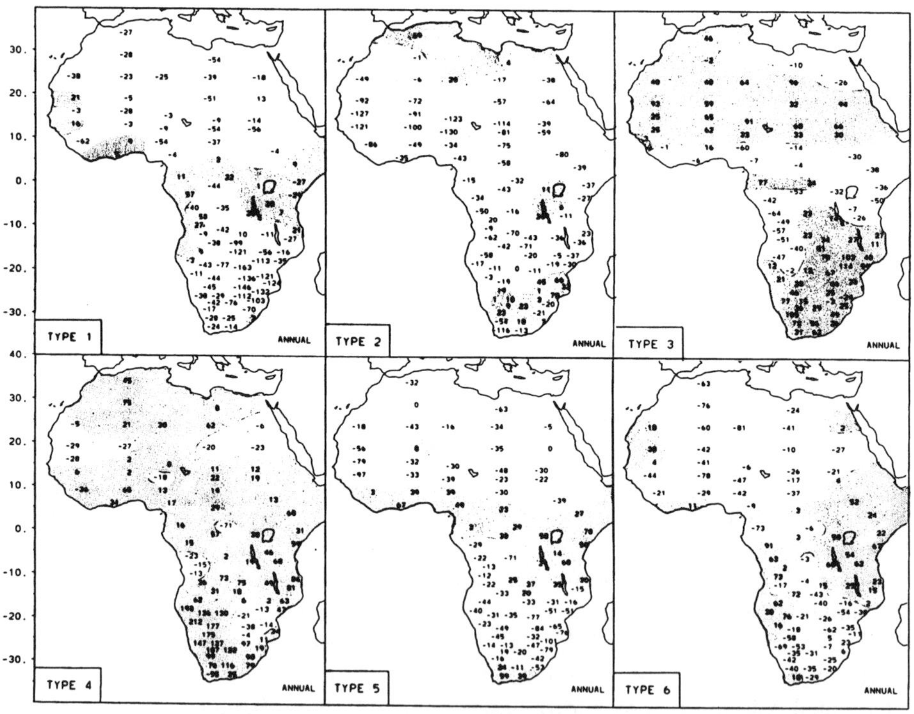

Figure 2. Dominant spatial modes of rainfall variability over Africa, as represented by the most frequently occurring anomaly types (from Nicholson 1986). Positive departures (shaded) indicate above-average rainfall, negative values indicate below-average rainfall. Numbers represent a normalized, regionally-averaged departure from the long-term mean.

Radiocarbon dates on charcoal samples help fix the timing of the first two of these levels (Crossley et al. 1984). A sample from the pre- or early- Mawudzu beach sand was dated as 990 ± 90 B.P., establishing a date of approximately the 10th century for a high stand of the lake. Owen et al. (1990) suggest a date of A.D. 950. Dates on samples post–dating the pre-Nkope beach deposits are 1340 ± 50 B.P. and 1720 ± 90 B.P. These suggest that this high stand would have ended by at least the 2nd century A.D.

The timing of the late Mawudzu high stand is more difficult to establish, in part because it appears to have been a double peaked event (Crossley et al. 1984). The maximum was probably attained sometime before the 15th century A.D., as charcoal within the late Mawudzu beach sand has a radiocarbon date of ca. 480 ± 80 B.P. This same date was obtained from charcoal associated with a stone structure on the Shire River. That structure, probably a wharf or a place for washing and water collecting, implies that at ca. 480 ± 80 B.P. the Shire River stood at a height of about 40 cm above the lake's 1980 level, but several meters below the previous high stand. This river level indicates that the overflow of the lake was not blocked during the late Mawudzu, so that the origin of the high levels was climatic (Crossley et al 1984).

Owen et al. (1990) assign ca. 1390 A.D. to this high stand of Malawi. Its end can be fixed to some extent by tree-growth around the lake. The age of the oldest baobabs, a tree particularly sensitive to rising water and flooding, does not exceed about 225 years. This is consistent with the age of baobabs around Lake Chilwa (see section 4) and indicates that the late Mawudzu high stand of Lake Malawi ended by 1750 A.D. One *Acacia albida* standing on the beach ridge has an estimated age of 360 years, suggesting the phase may have ended as early as 1620 A.D., or that lake levels were falling by then. Floods preventing the establishment of trees prior to these times could have represented a second peak within the late Mawudzu high stand. A more recent high stand, known from historical records, took place ca. 1980 A.D. (Owen et al. 1990; Eccles 1974) following a decade of abnormally high rainfall (Nicholson 1996).

In summary, three high stands of Lake Malawi occurred ca. A.D. 950, A.D. 1390 and A.D. 1980. Lake cores and other materials provide independent evidence for intervening low stands. A core taken near Nkhotakhota shows clear evidence of a recent low stand that historical information (see section 3.2.2) would date as the late 1700s and early 1800s. Plant materials and burrows indicate that the layer at depth 105 to 280 mm in the core was at the time above the level of the lake (Crossley et al. 1984). Three distinct sand layers in the core are probably remnants of three other low stands that occurred prior to the 18th century.

Radiocarbon dates of the archaeological record of shoreline settlements provide further evidence of low stands (Fig. 3). Gaps in the record indicate a settlement shift to follow the receding shore or to migrate to wetter highlands and presumably represent low stands of the lake. Gaps appear from late 15th century to mid-19th (except a single date ca. 1700) and in the 12th and 13th centuries (Owen et al. 1990). These dates are coincident with changes in ceramic traditions and occur mid-way between high stands dated by Crossley et al. (1984) to ca. 1980, 1390 and 950 A.D. The 1700 A.D. date may represent a second peak of the late Mawudzu high stand. Its decline would have been the recession that allowed the establishment of baobabs around the lake. Dating of material in the core by ^{210}Pb suggests that recovery from this dry phase had occurred by ca. 1860 A.D., in accord with historical information described in section 3.2.3.

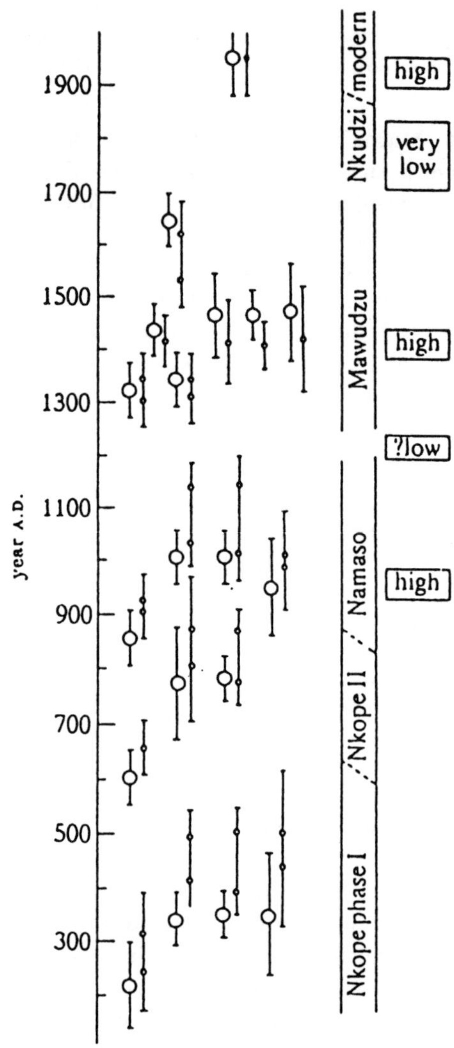

Figure 3. Long-term fluctuations of Lake Malawi, based on radiocarbon-dated archaeological data (from Owen et al. 1990). Large and small circles respectively indicate uncorrected and corrected ^{14}C dates; right axis indicates ceramic traditions. Also indicated on right are periods of high lake level (from Crossley et al. 1984) and inferred periods of low lake level.

Microfossil and lithological evidence of the second gap includes diatoms, dunes and erosional features in the sediment core. Diatoms sharply decline in abundance and diversity and there is a change in dominant species across this break, suggesting lower lake levels in the post-erosion period. During the low stage, exposed littoral sands reworked into aeolian dunes along windward shores; these evidence a major low stage within the last 600 years.

Sediment cores show an erosional hiatus stretching across the southern area of lake down to water depths of 121 m. This, coupled with desiccation cracks down to 121 m, is interpreted by Owen et al (1990) to mean a sharp 121 m or greater fall of the lake, and subsequent recovery, within the period between 1390 and 1860. The authors show that such a fall could occur within this interval if rainfall is reduced by only 30 to 50%, but historical evidence (section 3.2) suggests that the interval for this transgression/regression cycle would have been considerably shorter. Such rapidity of change in lake level and the desiccation of the Shire River, Lake Malombe and parts of Lake Malawi ought to have been so extreme as to warrant mention in the rich oral traditions of the region from this time, but such evidence is lacking. Moreover, T. Johnson (personal communication) finds no evidence in his Lake Malawi cores to indicate such a precipitous fall of the lake. Thus, such an extreme fluctuation is unlikely.

3.2. HISTORICAL INFORMATION AND LAKE LEVELS DURING THE LAST 600 YEARS

3.2.1. *Oral traditions of famines and droughts*

For the period post-dating the late Mawudzu high stand of Lake Malawi, several trends are apparent from sedimentary and archaeological records. These suggest a major decline of Lake Malawi beginning in the 15th century, possible recovery toward 1700, but otherwise persistence of the low levels and arid phase until the mid- or late 19th century. Further evidence of these fluctuations and their timing is presented in this section.

Information covering the 15th through 19th centuries comes from several sources, including oral traditions in the interlacustrine area to the north (Webster 1979a; Herring 1979), oral traditions of the Mutapa and other local peoples, and Portuguese documentary records (Webster 1979b). A fall in lake levels in the 15th century is supported by the historical traditions of drought. Although the genealogical dating of these episodes precludes precise knowledge of the times of occurrence, dates can usually be reliably established within two or three decades (Webster 1979b). Further back in time, there is less detail to the traditions and the droughts that do become part of oral history would have been particularly severe, creating much hardship among the peoples.

Much of the evidence derives from the East African "interlacustrine" region. That region extends from areas north of Lakes Albert and Kyoga to the southern end of Lake Victoria. Although it includes diverse conditions of climate, the region is homogeneous with respect to the interannual and interdecadal variability of rainfall. The Lake Malawi area is on the periphery of this climatic zone, but major droughts in that zone tend to extend as far south as Malawi (Nicholson 1987). Its levels correlate best with rainfall in this zone (see section 6). Webster (1979a) indicates that famines and droughts occurred in the interlacustrine region in the first decade of the fifteenth century, the 1430s and 1440s, and for a longer period beginning in the 1490s until about the 1530s. The area was relatively drought free between then and the late 16th century, when there appear reports of starvation, migration and the army crossing the Nile on bare ground during the generations 1580-1625. Arabic sources indicate that the Nile dried up in this region in 1621-1622. Reportedly half of the population of Egypt died then from starvation and disease. The lacustrine region was afterwards relatively drought-free until about the 1720s. Serious famines occurred then and in the 1750s, 1760s and 1780s (see Nicholson 1997). The last of these famines was particularly severe and was apparent throughout eastern Africa. It was followed by a decade of drought in the 1790s in southern Africa.

In areas closer to Lake Malawi there is specific evidence of drought during the 15th and 16th centuries. In the area northeast of the lake, the Mgulube experienced a three-year drought from 1561 to 1563. There was a great drought in the Mutapa empire within the period 1450 and 1480. This drought may have provided the impetus for Phiri migration to the Lake Malawi littoral (Webster 1979b), but the migration cannot be precisely dated and droughts that occurred in the 1400s and 1410s may have instead been a factor.

Webster's work suggests that the most important famines and droughts that affected the Lake Malawi region occurred in the 1580s, the 1620s and the 1720s. Famine and drought also because a common occurrence throughout the latter half of the 18th century. There was a notable absence of occurrences between the 1620s and 1720s. This chronology suggests that if a second peak in lake levels occurred during the late Mawudzu (Crossley et al. 1984), it would have been sometime between the 1630s and 1710s. This would place it as roughly coincident with the Lake Chilwa maximum (section 4) sometime within the period 1650 to 1750 A.D.

The record of the Nile flow (Toussoun 1925; Lamb 1966) measured at the Rodah gage in Cairo since 641 A.D. lends credence to the preceding scenario. The gage measurements provide an imperfect record of both the Nile flood and the summer minimum flow (Popper 1951), but the latter provides a reasonable indication of trends in equatorial rainfall. A more detailed analysis and a discussion of links to the African lakes appears in Nicholson (1997). The Nile record confirms the dry conditions in the early 15th century and then again beginning in the 1530s. This low stage is supported by a core from Lake Victoria (Stager et al. 1997) indicating a long arid interval commencing roughly about this time. The Nile flow also suggests relatively wet conditions from the 1630s to the 1710s, although the record is quite incomplete during this period.

The series of droughts that occurred late in the 18th century marked a trend towards increasing aridity that accelerated in the 1790s and resulted in a long arid interval throughout most of the early 19th century. Such a trend is evident throughout most of Africa (Nicholson 1995); its occurrence in southern Africa is detailed by Nicholson (1981). Evidence of its impact on Lake Malawi is described in 3.2.2; historical information related to rainfall, famines and droughts is examined in section 5.

3.2.2. *Period of low lake levels commencing toward 1800*

Historical references beginning around 1800 have been summarized into a lake chronology in Appendix A. The records show that Lake Malawi was unquestionably low throughout most of the first few decades of the 19th century and probably towards the end of the previous century as well An interesting reference indicative of a prolonged period of lower lake levels, in agreement with the sediment record, comes from Swann (1894). In 1892 and 1893 he noted "gigantic" trees standing some 3 feet deep in lake water. Known growth rates of the trees suggest a period of low levels lasting at least 50 to 100 years. This span probably included the late 1700s and early 1800s. Traditions of severe droughts and dried rivers also date to about that time. The location of villages, later flooded, suggests the lake stood at least 9 m below modern levels (Owen et al. 1990).

Other information suggests the persistence of this low stage during the 1820s, 1830s and 1840s. There are numerous reports of people traversing on foot formerly dry sections of the current lake, and of drying of the Rukuru River, as well as references to a much smaller lake and drought–induced tribal fighting and migrations. The lake gradually rose from around the late 1840s to the end of the next decade, at which time reports from European explorers become available.

There is corroboration from other lakes of the pronounced early 19th century arid interval. Lithological evidence in Lakes Chilwa (see next section) and Chiuta, about 100 km to the southeast, indicates that both lakes dried out during the period before 1850 A.D. (Owen and Crossley 1990). Although both lakes are shallow, considerable aridity would have been required for complete desiccation.

3.2.3. *Higher lake levels of the late 19th century and subsequent recession*

Lake Malawi maintained a relatively high stand from the late 1850s until the 1890s, with brief falls in some years. However, it had begun to recede in the mid-1870s. A dramatic decline occurred after ca. 1895 and the lake again reached and maintained very low levels throughout the early 20th century (Appendix A).

The stand of the lake in the 1860s is well established from reports of European travelers such as Livingstone and Kirk. Lake levels of the early 1860s can even be quantified on the basis of soundings made in Lake Malombe, which is hydrologically continuous with Malawi at moderate and high levels. A subsequent rise is attested by villages subjected to regular flooding during the 1870s. A subsequent decline was documented through measurements by the missionaries at Livingstonia and verified by numerous reports of visitors. The lake probably reached a maximum around 1875 or 1876.

Figure 4 shows the lake chronology beginning in 1800. The range of fluctuations during the last century has been about 5 meters, with conditions since the 1930s having been quite similar to those in the mid-19th century. Although during the 20th century the development of a sand bar has elevated Malawi's levels, its effect is small in comparison to the range of fluctuations exhibited by the lake.

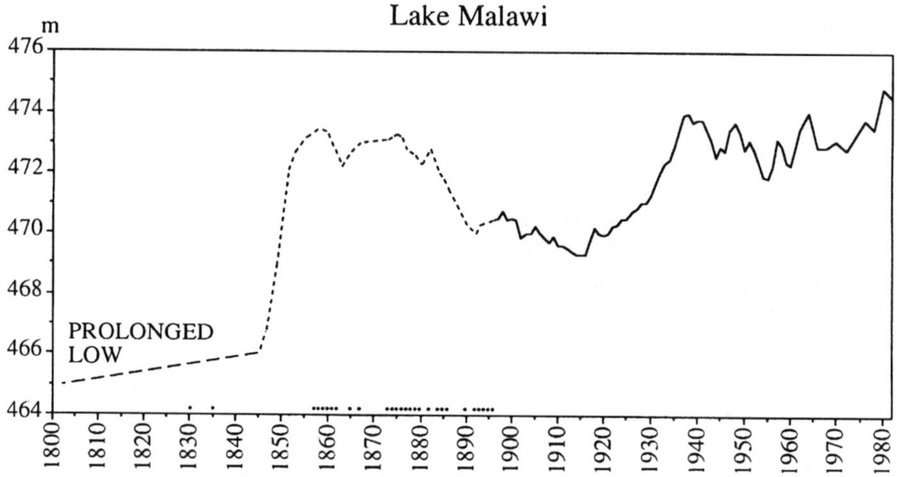

Figure 4. Fluctuations of Lake Malawi since 1800 based on historical and geographical information and, beginning in 1896, modern records. Long-dashed lines indicate general periods of low levels; short-dashed lines indicate trends based on proxy data; solid lines indicate the modern lake record. Modern data are corrected for build-up of a sand bar in the lake. Years for which specific references are available are indicated at the bottom with dots.

4. Lake Chilwa

Lancaster (1979) and Crossley et al. (1984) present historical evidence of levels of Lake Chilwa back to the early 1800s. Theirs and other information on the lake levels is summarized in the chronology in Appendix B. Missionaries indicated that the lake was dry in the early 1900s, as it seems to have been also in the 1940s. Chilwa was also dry more recently in 1968 and 1973. Around 1870 there was probably a brief low phase, with a prolonged low phase prior to ca. 1850. Lake Chilwa was high at the time of Livingstone's visit (1859); Lancaster suggests its level exceeded that of its modern 1978 peak, but Crossley et al. suggest that it did not.

Lake Chilwa probably maintained a prolonged low-level in the early 1800s. Evidence of that low stand includes a well-mixed, sandy clay, hard pan layer in sediment cores from the lake and an account that around 1900 one could walk across the dried lake floor to Nchisi Island, occasionally breaking through crust to ankle deep mud and a hard surface below. That hard surface would correspond to the hard layer in the cores and would represent a long arid interval. Based on the depth of the mud, the probable thickness of the 1903 crust, and estimated sedimentation rates, Crossley et al. (1984) suggest that it would have taken some 50 to 100 years to accumulate these sediments. This would suggest that the dry phase represented by the hard layer in the cores probably ended between 1800 and 1850. The degree of mixing in the hard layer indicates that the arid interval it represents was far more prolonged than any that occurred within the last 100 years.

There is a Yao tradition, reported by the headman Chendombo, of a very great drought and famine when the Mang'anja fled to the river (presumably Shire River) to escape the invading Yao (Stegman 1953). This would have occurred during the raids of the 1860s (Ajayi 1989) and fits with documentary reports of droughts in the 1860s (Webster 1979b). Another Yao tradition of this period states that Lake Chiuta nearly dried up during a great famine and drought. This would also imply at least a brief period of low levels of Lake Chilwa, something reported by Owen et al. (1990) as occurring around 1870. A brief recession of Lake Malawi also occurred in the 1860s, coincident with these droughts and the desiccation of Lake Chiuta.

Evidence of a recent high stand is beach gravel around the lake margin at a height of 631 m, 9 m above modern levels of the lake. This beach has a radiocarbon date of 160 B.P. (ca. 1750 A.D.). A lake of this depth would represent rainfall some 35 to 40% higher than at present. The recent date of this high stand is confirmed by archaeological evidence, including ceramic traditions and recent trade beads in sediment just below the gravel layer. It is also confirmed by the longevity of baobab trees near the lake, because the trees are killed rapidly by floods. Many trees have an age of 350 years, indicating growth extending back to 1650 A.D. or earlier in areas above 631 m, the old shoreline height, as shown by beach gravel. No tree in the lower areas has an age greater than 225 years, hence indicating no growth prior to ca. 1760 A.D. This suggests that the lower areas were flooded prior to 1760 and, thus, that the high stand at 631 m occurred some time between 1650 and 1750 A.D.

The beach gravel around the lake has been reworked so it is hard to get good evidence on the number and height of earlier high stands. Lake Chilwa probably had repeated pre-Iron Age high stands (pre-18th century). Evidence suggests a stand similar to today's (624 m) during part of the early Iron Age, perhaps around the 10th century (Shaw et al. 1984).

Figure 5 shows the chronology of Lake Chilwa since 1800. Although many similarities with the trends of Lake Malawi (Fig. 4) are apparent, there are some notable differences. Both lakes were extremely low or desiccated throughout most of the first half of the 18th century and rose to relatively high levels in mid-century. However, while Malawi retained these levels until the 1880s, a relatively major decline of Chilwa appeared to have occurred in the late 1860s, with low stands maintained until around 1870. The analogous fluctuation in Lake Malawi was of lower magnitude and shorter duration. Trends in the lakes remained similar until the 1930s: a rapid decline occurred in the 1880s, the lakes were relatively low during the 1890s and early 20th century, and the lakes began to rise in the 1920s, reaching maxima in the late 1930s. Both lakes experienced declines in the 1940s, but from the 1950s on the fluctuations of the lakes were in rough opposition. Chilwa was high in the 1950s, low in the 1960s and high in the 1970s, consistent with rainfall fluctuations occurring in the subtropics of the southern hemisphere at this time (Nicholson 1995). Levels of Lake Malawi follow the opposing pattern of rainfall in equatorial regions during these decades.

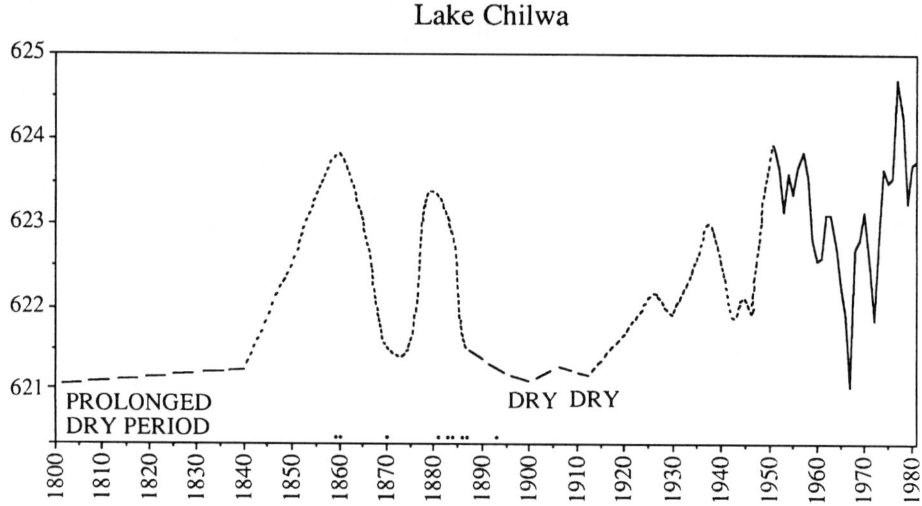

Figure 5. Fluctuations of Lake Chilwa since 1800, based on historical and geographical information and, beginning in 1949, modern records. Long-dashed lines indicate periods of low levels; short-dashed lines indicate trends based on proxy data; solid lines indicate the modern lake record. Years for which specific references are available are indicated at the bottom with dots.

5. Recent historical record of famines and droughts

The records of famines, droughts and other climate-related phenomena have been summarized in the form of chronologies for Malawi (Appendix C) and for areas of Mozambique (Appendix D) and Zambia/Zimbabwe (Appendix E) where the interannual variability of rainfall is well correlated with conditions in Malawi. While the chronology for Malawi is vague and generalized, numerous specific references are available for the other areas. Although many of the trends of the lakes parallel variations in rainfall conditions, as indicated by the famine and drought record, much of the time the lakes (particularly Malawi) show quite independent fluctuations.

Drought conditions commencing late in the 18th century are apparent, particularly in the 1790s. Likewise there is substantial evidence of a prolonged period of aridity, with frequent and severe drought, throughout at least the 1820s and 1830s, and possibly the 1840s. This period is evident in all three chronologies in Appendices C, D, and E, and it coincides with prolonged low phases of both Malawi and Chilwa. The period of aridity is evident in other Rift Valley lakes as well and throughout most of Africa (Nicholson 1995).

In the 1850s and 1860s, when lake levels were generally high, drought and famine occurred frequently in Malawi and northern Mozambique. The evidence for this includes the reliable witness of Livingstone (1875) and others. There is, however, a small recession of Lake Malawi during the 1860s, coincident with the most severe droughts, and Lake Chilwa began to recede, reaching a low level somewhere around the

1870s. Droughts also occurred during the lake maximum of the early and mid-1870s, but were infrequent during the time of the lake's recession from 1875 to at least 1886.

TABLE 1. Correlations between rainfall in select regions shown in Fig. 6 and the levels of Lakes Chilwa and Malawi.

	ST	ZM	NMZ	M	SM	WMZ	SMZ	ZM
Malawi	0.44	-0.05	0.10	0.20	-0.30	0.33	0.14	0.13
Chilwa	-0.10	0.18	-0.04	-0.04	0.25	0.00	-0.10	0.30

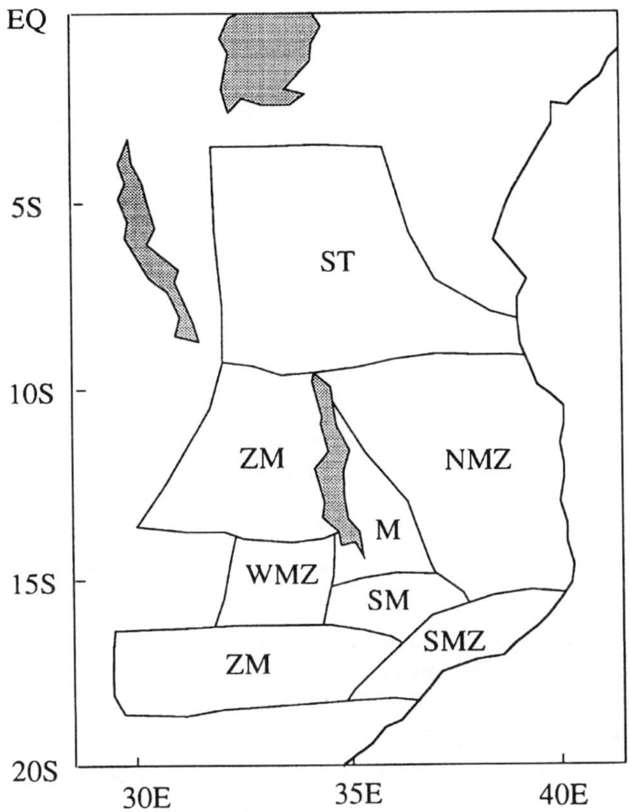

Figure 6. Map of rainfall regions correlated with lake levels. ST= southern Tanzania, ZM= Malawi/eastern Zambia, M= Malawi, SM= southern Malawi, SMZ= southern Mozambique, NMZ= northern Mozambique, WMZ= western Mozambique, ZM = Zimbabwe/Mozambique.

6. Relationship between lake-level fluctuations and rainfall

Lack of correspondence between records of famine and drought and trends in the levels of Lakes Malawi and Chilwa suggest the need for further assessment of the response of the lakes to rainfall fluctuations. To a first approximation, each lake is assumed to respond to the average rainfall over its catchment. It is questionable, however, to what extent these catchments reflect conditions on a broader scale, such as large regions within southern or eastern Africa. For Malawi, the question is complex because rainfall anomalies are frequently of opposite sign in the northern and southern portions of its catchment. Its levels likely reflect conditions to the north, over highlands where rainfall is substantially higher than in the southern lake area.

The relationship between rainfall variability and lake level is examined using rainfall from an archive described in Nicholson (1986). The modern lake records, commencing in 1896 for Malawi and 1949 for Chilwa are correlated with rainfall in the eight homogeneous regions shown in Figure 6. The results are shown in Table 1.

Lake Malawi is correlated with rainfall in only two regions: western Mozambique and southern Tanzania. The correlation coefficient with western Mozambique, 0.33, is significant at the 95% level. That for southern Tanzania, 0.44, is significant at the 99% level. The commonality of these two regions is that both include considerable areas of highlands with extreme rainfall conditions. Presumably the same factors that control rainfall in these highland regions control rainfall over the lake itself. This seems reasonable, because of the presence of local mountain-lake circulation systems that produce mesoscale patterns of rainfall over the lake (Hastenrath 1988). Also, stream flow from the higher regions affects lake levels.

Lake Chilwa is best correlated with rainfall in two regions: southern Malawi (0.25) and western Mozambique (0.30). However, neither correlation reaches the 95% significance level. The rainfall correlation is much lower than in the case of Lake Malawi. This probably reflects local influences on this small lake that are independent of those in larger-scale regions. Correlation between Lakes Chilwa and Malawi is 0.18 (n.s.) for the period 1949 to 1981. The relatively low degree of correlation reflects independent fluctuations, particularly high frequency fluctuations, of the lake during many intervals. Over the longer term the lake curves are considerably more similar than during the 1949 to 1981 period (see section 7, Fig. 7).

Despite relatively low correlations, it is apparent that the lake records mimic rainfall conditions when conditions prevail over very large areas. In particular, the correspondence is good when similar conditions arise in both eastern equatorial and southern Africa.

Owen et al. (1990) produced a simple water balance model for Lake Malawi and estimated the rainfall variations needed to account for a recession/transgression cycle with a 110 m change in lake level within historical times. They concluded that such a change could be accomplished with a change in rainfall on the order of 50%. Much smaller changes could account for the fluctuations shown in Figures 3 and 4.

7. Summary and Conclusions

Synthesis of sedimentary, geographical and historical information presents a fairly detailed picture of the fluctuations of Lakes Malawi and Chilwa since about 1800. Low lake levels prevailed during most of the first half of the 19th century while high levels generally prevailed during the mid- and late-19th century. Maxima were reached in the 1870s, but the lakes remained high in the early 1880s. A decline then occurred that culminated in low stands throughout the early twentieth century, with a return to higher lake-levels in mid-century. Highest recent levels occurred in the 1970s. Except for that maximum, these trends closely parallel those in other subtropical areas of Africa (Nicholson 1995), although the decline late in the 19th century may have commenced earlier here than elsewhere over Africa.

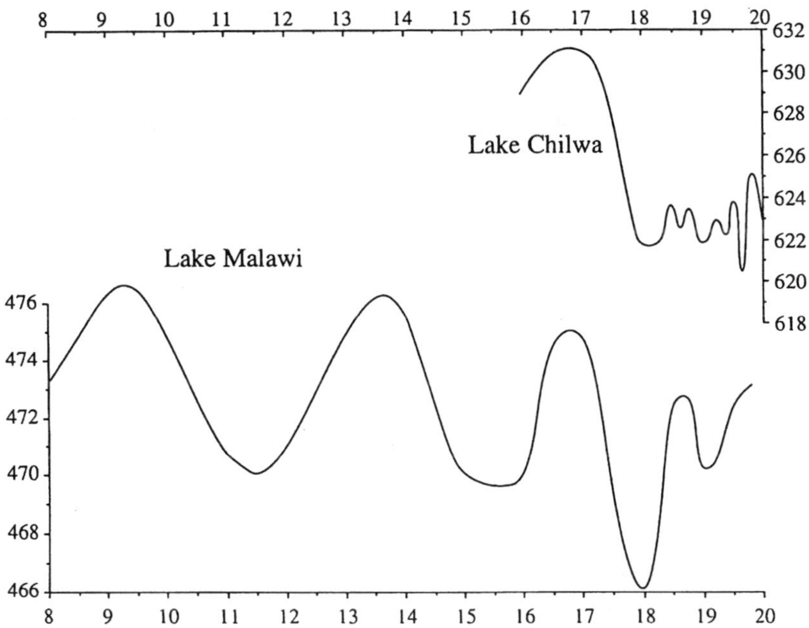

Figure 7. Long-term fluctuations of Lakes Malawi and Chilwa.

The long-term chronologies of Lakes Malawi and Chilwa are presented in Figure 7, based on information in the text and the lake chronologies in Appendices A and B. The long-term history of the lakes is reasonably similar, with high stands toward 1700, then a decline to comparatively low modern levels. The detailed historical information available to complement sedimentary and archaeological records of Malawi suggests a secondary maximum toward 1700 that is roughly coincident with the maximum of Lake Chilwa at that time. This is consistent with the suggestion of Owen et al. (1990) that the late Mawudzu high stand was a double peaked event.

It appears that lower frequency fluctuations of these lakes are roughly in phase, but that higher frequency fluctuations are often out of phase. Extreme dry periods, with major falls of the lakes, tend to occur synchronously in both lakes.

Independent higher frequency fluctuations reflect the geographical differences in the conditions of rainfall influencing lake levels. Lake Malawi is better correlated with rainfall to the north of the lake, in southern Tanzania, than in other areas surrounding the lake. However, local forcing is also apparent. Correlation analysis suggests than Lake Malawi is a better indicator of rainfall conditions in equatorial East Africa than in the outer-tropical latitudes in which it lies. Lake Chilwa instead appears to respond to relatively local conditions of rainfall. The complex climate dynamics governing the levels of these lakes underscores the difficulty in interpreting their histories in climatic terms. A more reliable interpretation can be gleaned from the collective histories of the East Africa Rift Valley lakes, rather than from any individual lake.

8. Acknowledgments

This work was supported by grants from NOAA (NA46GP0285) and the National Science Foundation (ATM9417063). A careful review by Curt Stager is greatly appreciated.

9. References

Ajayi, J. F. A., 1989. General History of Africa. University of California Press.

Brueckner, E. 1890. *Klimaschwankungen seit 1700*. E. Holzel.

Butzer, K. W., G. L. Isaac, J. L. Richardson, and C. Washbourne-Kamau. 1972. Radiocarbon dating of East African lake levels. Science **175**: 1069-1076.

Crossley, R., S. Davison-Hirschmann, R. B. Owen, and P. Shaw. 1984. Lake level fluctuations during the last 2,000 years in Malawi, p. 305-316. *In* J. C. Vogel [ed.], Late Cainozoic palaeoclimates of the southern hemisphere. Balkema.

Crossley, R., and S. Davison-Hirschmann. 1981. Hydrology and archaeology of Lake Malawi and its outlet during the Iron Age. Palaeoecology of Africa **13**: 123-126.

Curtin, P., S. Feierman, L. Thompson, and J. Vansina. 1978. African History. Little Brown and Company.

Dixey, F. 1924. Lake level in relationship to rainfall and sunspots. Nature **114**: 659-661.

Drayton, P. 1979. Study of the causes of the abnormally high levels of Lake Malawi in 1979. Technical Report No. 5, Malawi Government, Water Resources Division.

Eccles, D. H. 1974. An outline of the physical limnology of Lake Malawi (Lake Nyassa). Limnology and Oceanography **18**: 730-742.

Fisher, H. 1974. Presented at the African History Seminar, School of Oriental and African Studies. Unpublished manuscript, University of London.

Hastenrath, S. L. 1988. Climate and Circulation of the Tropics. Reidel.

Herring, R. S. 1979. Hydrology and chronology: The Rodah nilometer as an aid in dating interlacustrine history, p. 39-86. *In* J. B. Webster [ed.], Chronology, Migration and Drought in Interlacustrine Africa. Africana Publishing Company.

Johnston, H. H. 1897. British Central Africa. Metheun.

Kirk, J. 1865. Meteorological tables illustrating the climate of eastern tropical Africa. Proceedings of the British Meteorological Society **11**: 22.

Kraus, E. B. 1955. Secular changes of tropical rainfall regimes. Quarterly Journal of the Royal Meteorological Society **81**: 198-210.

Lamb, H. H. 1966. Climate in the 1960s. Geographical Journal **132**: 183-212.

Lancaster, N. 1979. The changes in the lake level, p. 27-34. *In* M. Kalk, A. J. McLachlan, and C. Howard-Williams [eds.], Lake Chilwa: Studies of Change in a Tropical Ecosystem. Monographae Biologicae 35, Junk.

Liesegang, G. 1978. Famines and Smallpox in South Eastern Africa 18th to 20th Centuries. Unpublished manuscript, Universida de Eduardo Mondlane, Maputo, 12 pp.

Livingstone, D. 1875. A Popular Account of Missionary Travels and Researches in South Africa. John Murray.

Livneh, A. 1974. Some notes on drought, famine and pestilence in Rhodesia. Unpublished manuscript, presented at the African History Seminar, School of Oriental and African Studies, University of London, 3 p.

Mohr, E. 1876. To the Victoria Falls of the Zambezi. Books of Rhodesia.

Nicholson, S. E. 1976. A climate chronology for Africa: synthesis of geological, historical and meteorological information and data. Ph.D. Thesis. Department of Meteorology, University of Wisconsin, Madison.

Nicholson, S. E. 1981. The historical climatology of Africa, p. 249-270. *In* T. M. L. Wigley, M. J. Ingram, and G. Farmer [eds.], Climate and History: Studies on Past Climates and Their Impact on Man. Cambridge University Press.

Nicholson, S. E. 1986. The spatial coherence of African rainfall anomalies - interhemispheric teleconnections. Journal of Climate and Applied Meteorology **25**: 1365-1381.

Nicholson, S.E. 1987. The nature of rainfall variability in Africa south of the equator. Journal of Climatology **6**: 515-530.

Nicholson, S. E. 1995. Environmental Change within the Historical Period, p. 60-75. *In* A. S. Goudie, W. M. Adams, and A. Orme [eds.], The Physical Geography of Africa. Oxford University Press.

Nicholson, S. E. 1996. A review of climate dynamics and climate variability in eastern Africa, p. 25-56. *In* T. C. Johnson and E. Odada [eds.], The limnology, climatology and paleoclimatology of the East African lakes. Gordon and Breach.

Nicholson, S. E. 1998. Historical fluctuations of Lake Victoria and other lakes in the northern Rift Valley of East Africa, p. 7-35. *In* J. T. Lehman [ed.], Environmental change and response in East African lakes. Kluwer.

Owen, R. B., and R. Crossley. 1990. Recent sedimentation in Lakes Chilwa and Chiuta, Malawi. Palaeoecology of Africa **20**: 109-117.

Owen, R. B., R. Crossley, T. C. Johnson, D. Tweddle, I. Kornfield, S. Davison, D. H. Eccles, and D. E. Engstrom. 1990. Major low levels of Lake Malawi and their implications for speciation rates in cichlid fishes. Proceedings of the Royal Society of London B **240**: 519-553.

Pike, J. G., 1965. The sunspot-lake level relationship and the control of Lake Nyasa. Journal of the Institute of Water Engineers **19**: 221-226.

Popper, W. 1951. The Cairo Nilometer. University of California Press.

Roberts, A.D., 1974. A note on drought, famine and pestilence in and around Zambia. Unpublished manuscript, presented at the African History Seminar, School of Oriental and African Studies, University of London.

Shaw, P., R. Crossley, and S. Davison-Hirschmann. 1984. A major fluctuation in the level of Lake Chilwa, Malawi, during the Iron Age. Palaeoecology of Africa **16**: 391-395.

Sieger, R. 1887. Schwankungen der innerafrikanischen Seen, p. 41-60. *In* Vereinigung der Geographie Jahresbericht. University of Vienna.

Spigel, R. H., and G. W. Coulter. 1996. Comparison of hydrology and physical limnology of the East African Great Lakes: Tanganyika, Malawi, Victoria, Kivu and Turkana, p. 103-140. *In* T. C. Johnson and E. Odada [eds.], The limnology, climatology and paleoclimatology of the East African lakes. Gordon and Breach.

Stager, J. C., B. Cumming, and L. Meeker. 1997. A high-resolution, 11,400-year diatom record from Lake Victoria, East Africa. Quaternary Research **47**: 81-89..

Stegman, J. J. 1953. Nyasaland droughts. Nyasaland Journal **4**: 67-69.

Swann, A. 1894. Two African Lakes. British Central Africa Gazette **1 October**: 2.

Toussoun, O. 1925. Memoire sur l'histoire du Nil. Memoire de l'Institut d'Egypt, vol. 9. Cairo: Français de l'Archaéologie Orientale.

Webster, J. B. 1979a. Noi! Noi! Famines as an aid to interlacustrine chronology, p. 1-38. *In* J. B. Webster [ed.], Chronology, Migration and Drought in Interlacustrine Africa. Africana Publishing Company.

Webster, J. B. 1979b. Drought and migration: the Lake Malawi Littoral as a region of refuge, p. 148-157. *In* M. T. Hinchey [ed.], Botswana Drought Symposium. The Botswana Society.

15th and 16th centuries

Frequent famines and droughts in the interlacustrine region, approximately dated to the 1430s and 1440s and from about the 1490s to 1530s (Webster 1979a, 1979b). The Nile was low in the early 15th century and again in the 1530s. The region was relatively drought free from about the 1540s to the 1570s. Reports of drought, starvation and exceedingly low Nile levels occurred again in generations from ca. 1580 to 1625.

In closer proximity to Lake Malawi, periods of severe droughts included the 1400s and 1410s, sometime within the period 1450 to 1480, and in the 1560s, 1580s and 1620s.

These reports collectively suggest relatively low lake levels, particularly in the early 1400s, the mid-15th century, the early 16th century, the late 16th century, and the very early 17th century.

17th and early 18th centuries

The interlacustrine region was relatively free of drought and famine free; Nile flow was relatively high from about the 1630s to the 1710s. The Lake Malawi region was also relatively drought free from then until the 1720s. This suggests relatively high lake levels within these intervals.

Late 1700s to early 1800s

Swann, between 1892 and 1893, noted "gigantic" trees standing in about 3 feet of water (Owen et al. 1990). The lake then stood at 470 m, lowest since Livinstone's visit of 1861. Trees suggest that some time before 1860, Lake Malawi was at a continuously lower level than 469 m for a long period. On basis of known growth rates of trees, the minimum period of continuous low levels must have lasted 50 to 100 years, which probably includes the late 1700s and early 1800s.

Traditions referring to severe droughts and dry rivers persist in the region from the early 1800s back to the late 18th century (Owen et al. 1990).

Early 19th century

Location of villages suggests lake levels at least 9 m below modern levels (Owen et al. 1990).

Chief Moses Ngosi, born around 1885, said that in much earlier times there was a village an hour's walk lakeward of the present Ruasho River delta. The site probably refers to a shoal area, which could only be reached on foot if the lake had been at least 9 m below modern levels (Owen et al. 1990).

ca. 1825 Johnston (1897) records a tradition that "some 70 year ago (i.e., about 1825) the north end of the lake became so shallow between Deep Bay and Amelia Bay that a chief and his men waded across it where it is now many fathoms deep" (Owen et al. 1990).

ca. 1820s Ngonde king Mwangonde, whose reign genealogically dated to about 1815 to 1835, walked across the northern part of Lake Malawi to marry a woman (Owen et al. 1990).

The North Rukuru River in northern Malawi dried up during the reign of Mwangonde (ca. 1815-1835) (Owen et al. 1990).

There was a general view that widespread aridity in the late 1820s was a cause of the Yao dispersal towards the southern lake area (Owen et al. 1990).

Stewart visited the lake in 1879 and spoke of flooding of a small land area that stood in the mouth of the Rukuru river some 60 yr previous. Although a change in the river could account for its flooding, this is consistent with other evidence of drier conditions, including the above reference to the Rukuru drying up (Sieger 1887).

1830 Low level (Nicholson 1976).

1835 The Ngoni Gama clan of northern Malawi, the followers of Zwangendaba (Curtin et al. 1978) have a tradition of crossing the Zambezi near Zumbo on a sandy causeway in the dry season of 1835 (date established via reference to a solar eclipse; Owen et al. 1990).

1830s Dixey reported very low levels for the 1830s, but gives no source of information (Owen et al. 1990).

Further south, tribal chief Amapunda walked across dry ground to Likoma Island at north end. Tribal chronology suggests his reign was approximately the 1830s (Owen et al. 1990).

1840s Informant at Chikowa's village near Mawudzu Hill on SE arm of lake, born ca. 1910, said her father's village had once been close to Boadzulu Island "when the lake was small" about the time of "fighting between the Yao and Ngoni". She may refer to period of alliance between the Msamala Yao and the Nyanja against the Ngoni around the 1840s. Since water near the island is now about 40 m deep, the report implies very low levels in 1840s (Owen et al. 1990).

References to tribal fighting ca. 1840s indicate very low lake levels (Owen et al. 1990).

late 1840s to late 1850s

Probably a gradual rise of Lake Malawi (Sieger 1887).

1857-63 Very high levels (Nicholson 1976)

1859 The lake had a relatively high stand when Livingstone visited because the missionaries found no evidence of formerly higher stands. Also the Shire was a broad and deep current (Sieger 1887). No evidence that the lake rose or fell much; Livingstone suggested the fluctuations did not exceed 3 or 4 feet. Stewart's report from 1879 suggested fluctuations of similar magnitude (Sieger 1887).

Evidence that the lake rose subsequently comes from descriptions of flooded villages in later years (see 1878 and 1879).

1860 Peak level is based on the coincidence between lake level deduced from Livingstone's soundings in Lake Malombe in 1861 and Dixey's (1924) report of marks, presumably aquatic algal lines, on lakeshore rocks up to that level (Owen et al. 1990).

1863 Level based on sounding by Kirk in Lake Malombe because that lake, when at moderate to high levels, is virtually identical to Lake Malawi (Owen et al. 1990).

1865 High stand according to Livingstone (Lamb 1966; Pike 1965). Remained relatively higher until about 1880.

1867 Drop in lake levels (Pike 1965; Owen et al. 1990)

1873 High lake level (Nicholson 1976)

1874 Young spoke of high water level of the Shire River (Sieger 1887).

1875-85 At the time of Laws in 1878 the lake was in recession; he traversed large stretches that had previously been underwater even during the dry season. Rivers along the plain he crossed dried up before reaching the lake. Recession of the lake was confirmed by measurements of missionaries at Livingstonia, showing a drop of 3 feet between December 1875 and December 1880. The level was so low as to pose a risk to ships on the southern end of the lake and on the Shire River. Although by about 1886 the lake appears to have lacked an outlet, both Livingstone in 1875 and Young in 1876 argued vigorously for the existence of an outlet, which was even mentioned by name by the natives (Sieger 1887). The outlet appears to have been lost in about 1886. Brueckner (1890) and Sieger (1887) suggest that the maximum stand of Lake Malawi occurred in 1875.

1875 Lake level falling (Nicholson 1976).

1876 Lake level falling (Nicholson 1976).

1876 Appears that a big fall occurred in 1876 (Sieger 1887), immediately after maximum stand. Later maps of Stewart, Kerr and Johnson showed shifts from islands to peninsulas, compared to earlier maps, indicating desiccation.

1877 Lake level very low and falling; apparently deficient rainfall in the region (Nicholson 1976).

1878 Laws reported in autumn of 1878 that the village of Matetes was so low-lying that the water from the lake covered the village nearly to a man's height during nearly every rainy season, driving away its inhabitants. This implies a much higher lake level than when the village was established.

Nevertheless, lake level was very low and falling compared to more recent years; apparently deficient rainfall in the region.

1879 Stewart reported the village of Karongas was covered by deep lake water during the rainy season, driving away its inhabitants (Sieger 1887).

1882 High lake level (Nicholson 1976).

End of 1870s to 1886

Recession of Lake Malawi (Sieger 1887).

1884 Kerr could see continued recession of the lake via drying of rivers that previously held much water and drying of land along the Shire and Pamalomba. The level of the lake was apparently already a few feet lower than during Livingstone's time (Sieger 1887).

1885 Moir confirmed the lowering of the lake also in 1885.

1886 Last mention of seeing sand and mud surfaces that had previously been covered by water of the lakeshore. This is the year that the lake apparently lost its second (northern) outlet.

1890 Very low level of the lake (Nicholson 1976).

1892-93 Swann, between 1892 and 1893, noted "gigantic" trees standing in about 3 feet of water (Webster 1979b). The lake stood at 470 m, lowest since Livingstone's visit of 1861.

1892-95 In these four years, the lake was rising rapidly (Nicholson 1976).

1896 to 1915

Steady decline of the lake (Nicholson 1976).

APPENDIX B. Long-term chronology of Lake Chilwa.

Early 1800s

Prolonged low level (Lancaster 1979; Crossley et al. 1984; Owen et al. 1990). Dry phase probably ended between 1800 and 1850.

Lake Chiuta almost dried up during reign of some king - probably indicates low levels of Chilwa as well (Webster 1979b).

Lithological evidence that both lakes dried out for long period prior to ca. 1850 (Owen and Crossley 1990).

pre-1850 Prolongued low phase prior to this (Owen et al. 1990)

1859 Lake was high at time of Livingstone's visit (Owen et al. 1990). However, he indicated the lake water tasted weakly bitter, indicating relatively high salt content and levels below maximum, when the basin contains fresh water (Sieger 1887).

Lake was apparently lower in 1859 than during the time of Burton's visit (1857/58), because he reported the water tasted fresh (Sieger 1887).

1860 Lake was high, according to Brueckner (1890).

ca. 1870 Brief low phase (Owen et al. 1990).

1881 Johnson assumed the lake basins to the north (apparently Lakes Chiuta and Amamba) were continuous with Chilwa. If so, this would indicate a high stand of the lake (Sieger 1887).

1883 O'Neill discovered that Lakes Chiuta and Amamba were separated from Chilwa by a 4.5 to 9 m wooded drainage divide, but indicated they may have been connected in some years, although natives said this had not been so in recent memory (Sieger 1887).

1884 Lake level was high (Brueckner 1890).

1886 Last found that water flowed between the lakes during the rainy season. He mentioned a decline of Chilwa levels since the beginning of the 1880s, presumably referring to differences in the observations of Johnson and O'Neill (Sieger 1887).

ca. 1886 The lake was in recession (Sieger 1887). He indicated it had been in recession since probably the end of the 1870s.

1887 Lake level was comparatively low, compared to 1884 and 1860 (Brueckner 1890).

Early 1900s

Missionaries indicated the lake was dry (Owen et al. 1990). In 1903 one could walk across the dry lake bed to Nchisi Island.

1968 and 1973

Low levels of the lake (Owen et al. 1990).

APPENDIX C. Historical chronology of famines, droughts and related phenomena in Malawi.

late 18th century/early 19th century

Rainmakers were called to Nkamanga (Rumphi area, northern Malawi) during a severe drought in the time of the first Mlowoka or his son (about 1780-1840). This may have been about the time Lake Rukwa dried up, ca. 1770 (Owen et al. 1990).

Kalinga records an increase in immigration from the highlands to the northeast of the lake into the Ngonde region of the lake shore, perhaps partly attributable to drier conditions, between 1740 and 1840 (Owen et al. 1990).

late 1790s

Raiding from Makua into Yao territory in the 1790s caused by drought (Webster 1979b). The Mahlatule famine of ca. 1799-1803 caused the Yao (NE Zambia, NW Mozambique) to disperse into the Malawi region.

1820s General view that widespread aridity in the late 1820s was a cause of the Yao dispersal towards the southern lake area (Owen et al. 1990).

early 1860s

Drought in Malawi produced chaos that was a major factor in the failure of the UMCA Magomero Mission (Webster 1979b).

1860s Chendombo, a Yao headman, indicated that a severe drought occurred when the Mang'anja fled to the river from invading Yao (Webster 1979b). Crops had been planted and were growing well when the drought occurred and killed them.

There is a Yao tradition from this period that Lake Chiuta almost dried up and that there was a very great drought and famine (Webster 1979b).

ca. 1714 Drought in Changamire's territory (Liesegang 1978).

1744-45 Drought and famine in the Zumbo area (region 57) (Liesegang 1978).

1758 In 1758 the gold washeries of Mixonga north of Tete had to be abandoned for a few years because the workers could not be fed (Liesegang 1978).

1759 Immediately before this year there was widespread drought in the Zambezi Valley and in neighboring areas; the usual Zambezi flood in the delta failed to appear (Liesegang 1978).

1794-98 Drought produced famine at Tete (16.5S 33.5E; Roberts 1974).

1822-30 Famine apparently induced by low rainfall near Sena (18S 36E) where agriculture was largely dependent on rainfall (Liesegang 1978). Drought in some years in Inhambane.

1828-29 Hunger induced most inhabitants of Sena (18S 36E) to abandon their town (Liesegang 1978). Groups of warriors from Barue and neighboring areas were looking for food. Problem probably extended to Sofala, as well, as grain was shipped there from Mozambique.

1844-45 Famine in Sofala (20S 35E) (no cause given) in 1844 and a drought in the beginning of 1845 (Liesegang 1978).

1854 Famine at Sena (18S 36E) in January and, according to Livingstone, at Tete (16.5S 33.5E; Liesegang 1978).

1855 Adequate rainfall at Tete (16.5S 33.5E), according to Livingstone (1875). This appears to relate to the 1855-56 rainy season.

1856 Famine in Quelimane (18S 36E) supposedly helped cause increased slave trade (Liesegang 1978).

1858-60 Drought and famine at Tete (16.5S 33.5E), according to Livingstone. At Inhambane (24S 35.5E) in November 1858 drought forced the population to rely on seafood and wild fruits, and in the Limpopo Valley (23S 33E) there was drought and famine in 1859-60 (Liesegang 1978).

1860-61 Rainfall at Tete totaled 852 mm in this year (Kirk 1865).

1861-63 Drought caused famine at Tete (16.5S 33.5E; region 56), according to Livingstone, and affected also the Mang'anja of the middle Shire Valley (ca. 16S 35E; region 55); the lower Zambezi (18S 35.5E; region 58) was unusually low. In mid-1863 crop failure was reported for Limpopo (ca. 23S 33E; region 67) and Nkomati Valleys, probably from irregular rainfall and wars (Liesegang 1978).

1864-70 No droughts or famines documented (Liesegang 1978).

ca. 1871-73
Famine along the Nkuna and further south (Liesegang 1978).

1875 Madimgele famine (Liesegang 1978).

1880s No droughts or famines reported in this area (Liesegang 1978)

1903 Drought for the lower Buzi Valley (ca. 20.5S 33E), the area near Inhambane (24.5S 35.5E), and the eastern Transvaal (ca. 24S 30; region 68; Liesegang 1978).

1910-16 Deficient rainfall widespread; famine near Inhambane (24.5S 35.5E) in 1812 (Liesegang 1978).

1917-19 Great floods in the Sena (18S 36E) - Tete (16.5S 33.5E) region of the lower Zambezi, which caused prolonged famine and probably resulted from early rather than high floods (Fisher 1974).

APPENDIX E. Chronology of famines, droughts and related phenomena: Zambia and Zimbabwe.

1790s NE Zambia, food shortages among the Bisa (ca. 12S 32E) caused migration to the Chewa country, but this may be due to intense showers (Roberts 1974).

1800s Frequent famines in early 19th century (probably ca. 16S 24E, assuming this refers to the Lozi people), supposedly caused by war with the Lunda and Yeke (Roberts 1974).

1827-29 For three years about this time there was drought and famine in Zambia (Bemba country; ca. 12S 32E; Roberts 1974). Drought in present southeastern Zambia too.

The drought that plagued Mozambique and Zambia in these years seems to have affected at least the northeastern parts of the Rhodesian plateau (ca. 17S 31E; Livneh 1974; Roberts 1974). The Duma area experienced drought and severe famine (Liesegang 1978).

1827-31 The first recorded locust invasion (1827), possibly associated with droughts and famines in the entire period (probably ca. 16S 24E). Locusts need a low lake level in order to multiply (Roberts 1974). Around 1827 there was supposedly a year without rain, followed by a locust invasion.

1833-36 Drought in Bemba country (Zambia), probably propelled the Ngoni in their trek northward (Webster 1979b; Roberts 1964).

1845 Drought in Bemba country (Zambia), probably propelled the Ngoni in their trek northward (Webster 1979b; Roberts 1964).

1869 Mohr (1876) experienced much rain on his travels through Zimbabwe.

ca. 1879 Drought in northeastern Zimbabwe (ca. 17S 30E; Liesegang 1978).

1880s No droughts reported except for a minor one among the Duma (Zimbabwe) between 1886 and 1890 (Liesegang 1978).

Post 1883 Famine and drought near Salisbury (i.e., Harare, 17 49S 31 06E; Livneh 1974).

1887 Mazoe Valley (ca. 17S 32E) famine (Livneh 1974).

1893-96 Partial drought and locusts in Matabeleland (ca. 17S 30E; Livneh 1974).

1899 Severe drought over the whole of Zimbabwe (Livneh 1974).

1913-18 Drought in Zimbabwe (Liesegang 1978).

INDEX